Lecture Notes in Computer Science　11807

More information about this series at http://www.springer.com/series/7409

Giuseppe Amato · Claudio Gennaro ·
Vincent Oria · Miloš Radovanović (Eds.)

Similarity Search
and Applications

12th International Conference, SISAP 2019
Newark, NJ, USA, October 2–4, 2019
Proceedings

 Springer

Editors
Giuseppe Amato 🆔
ISTI-CNR
Pisa, Italy

Claudio Gennaro 🆔
ISTI-CNR
Pisa, Italy

Vincent Oria
New Jersey Institute of Technology
Newark, NJ, USA

Miloš Radovanović 🆔
University of Novi Sad
Novi Sad, Serbia

ISSN 0302-9743   ISSN 1611-3349   (electronic)
Lecture Notes in Computer Science
ISBN 978-3-030-32046-1   ISBN 978-3-030-32047-8   (eBook)
https://doi.org/10.1007/978-3-030-32047-8

LNCS Sublibrary: SL3 – Information Systems and Applications, incl. Internet/Web, and HCI

Preface

This volume contains the papers presented at the 12th International Conference on Similarity Search and Applications (SISAP 2019) held in Newark, NJ, USA, during October 2–4, 2019.

SISAP is an annual forum for researchers and application developers in the area of similarity data management. It focuses on the technological problems shared by numerous application domains, such as data mining, information retrieval, multimedia, computer vision, pattern recognition, computational biology, geography, biometrics, machine learning, and many others that make use of similarity search as a necessary supporting service.

From its roots as a regional workshop in metric indexing, SISAP has expanded to become the only international conference entirely devoted to the issues surrounding the theory, design, analysis, practice, and application of content-based and feature-based similarity search. The SISAP initiative has also created a repository (http://www.sisap.org/) serving the similarity search community, for the exchange of examples of real-world applications, source code for similarity indexes, and experimental test beds and benchmark data sets.

The call for papers welcomed full papers, short papers, as well as demonstration papers, with all manuscripts presenting previously unpublished research contributions. All contributions were presented both orally and in a poster session, which facilitated fruitful exchanges between the participants. In addition, SISAP 2019 featured a doctoral consortium, accepting papers describing doctoral research and work in progress, providing students valuable feedback from experienced researchers in similarity search and related fields.

We received 42 submissions from authors based in 17 different countries. The Program Committee (PC) was composed of 63 international members. The papers and reviews were thoroughly discussed by the chairs and PC members: Each submission received three reviews. Based on these reviews and discussions among PC members, the PC chairs accepted 12 full papers to be included in the conference program and proceedings, resulting in an acceptance rate of 28% for the full papers. In addition, 18 short papers were accepted, and after separate review by SISAP chairs, two doctoral consortium papers were included in the program and proceedings as well.

The proceedings of SISAP are published by Springer as a volume in the *Lecture Notes in Computer Science* (LNCS) series. For SISAP 2019, as in previous years, extended versions of selected excellent papers were invited for publication in a special issue of the journal *Information Systems*. The conference also conferred a Best Paper Award, as judged by the PC co-chairs and Steering Committee.

Besides the presentations of the accepted papers, the conference program featured three keynote talks by exceptional researchers: Fabrizio Silvestri from Facebook, UK, Divesh Srivastava from AT&T Labs-Research, USA, and Prof. Alexander Tuzhilin from New York University, USA.

We would like to thank all the authors who submitted papers to SISAP 2019, as well as all members of the PC and the external reviewers for their effort and contribution to the conference. We want to extend our gratitude to the members of the Organizing Committee for the enormous amount of work they invested in making the SISAP series of conferences possible.

We also thank our sponsors and supporters for their generosity. All the submission, reviewing, and proceedings generation processes were made much easier through the EasyChair platform.

August 2019

Giuseppe Amato
Claudio Gennaro
Vincent Oria
Miloš Radovanović

Organization

Program Committee Chairs

Vincent Oria	NJIT, USA
Giuseppe Amato	ISTI-CNR, Italy
Miloš Radovanović	University of Novi Sad, Serbia

Steering Committee

Laurent Amsaleg	CNRS-IRISA, France
Edgar Chavez	CICESE, Mexico
Michael E. Houle	National Institute of Informatics, Japan
Pavel Zezula	Masaryk University, Czech Republic

Program Committee

Giuseppe Amato	ISTI-CNR, Italy
Laurent Amsaleg	CNRS-IRISA, France
Martin Aumüller	IT University of Copenhagen, Denmark
Ilaria Bartolini	University of Bologna, Italy
Virendra Bhavsar	University of New Brunswick, Canada
Panagiotis Bouros	Johannes Gutenberg University Mainz, Germany
Benjamin Bustos	University of Chile, Chile
K. Selcuk Candan	Arizona State University, USA
Guang-Ho Cha	Seoul National University of Science and Technology, South Korea
Aniket Chakrabarti	Microsoft, USA
Edgar Chavez	CICESE, Mexico
Richard Chbeir	University of Pau and Pays Adour, UPPA/E2S, LIUPPA Anglet, France
Paolo Ciaccia	University of Bologna, Italy
Richard Connor	University of Stirling, UK
Robson Cordeiro	ICMC-USP, Brazil
Michel Crucianu	CNAM, France
Petros Daras	Information Technologies Institute, Greece
Alberto Del Bimbo	Università degli Studi di Firenze, Italy
Dong Deng	Rutgers University, USA
Vlad Estivill-Castro	Griffith University, Australia
Andrea Esuli	ISTI-CNR, Italy
Fabrizio Falchi	ISTI-CNR, Italy
Joao Eduardo Ferreira	University of São Paulo, Brazil
Karina Figueroa	Universidad Michoacana, Mexico

Marcel Worring   University of Amsterdam, The Netherlands
Kaoru Yoshida   Sony Computer Science Laboratories, Inc., Japan
Pavel Zezula   Masaryk University, Czech Republic
Arthur Zimek   University of Southern Denmark, Denmark

Additional Reviewers

Vladimir Mic
Joe Tekli

Keynote Abstracts

Applications of Similarity Search to Socially Relevant Problems

Fabrizio Silvestri

Facebook, UK

Abstract. The Facebook AI team in London deals with applying artificial intelligence techniques to address societal problems such as the spread of online misinformation, or the integrity of election processes around the world. To do so, we have developed throughout the last years a set of tools that exploit similarity search technologies to efficiently and effectively run a very high number of classification tasks on a massive set of data.

In this talk, we are going to review some of the problems we have studied in the last year and we are going to show some of the solutions we have adopted in order to make the system run efficiently. We are also going to showcase some details of an internal project that uses similarity search as a core operation to allow efficient and effective inference operations.

Repairing Noisy Graphs

Divesh Srivastava

AT&T Labs-Research, USA

Abstract. Graphs are a flexible way to represent data in a variety of applications, with nodes representing domain-specific entities (e.g., records in record linkage, products and types in an ontology) and edges capturing a variety of relationships between these entities (e.g., an equivalence relationship between records in record linkage, a type-subtype relationship between types in an ontology). Often, the edges in this graph are inferred based on similarities between nodes and are noisy, in that some edges are missing (i.e., real-world relationships that do not have corresponding edges in the graph) and some edges are spurious (i.e., edges in the graph that do not have corresponding real-world relationships). Directly analyzing such graphs can lead to undesirable outcomes, making it important to repair noisy graphs. In this talk, we describe an approach that takes advantage of properties of real-world relationships and their estimated probabilities to ask oracle queries (an abstraction of crowdsourcing) to efficiently repair the noisy graphs. We illustrate this approach for the case of graphs that are unions of cliques (which is the case for record linkage) and graphs that are trees (which is the case for ontologies), and present theoretical and empirical results for these cases.

On Similarity Measures
in Recommender Systems

Alexander Tuzhilin

New York University, USA

Abstract. Measures of similarity between users and between items to be recommended to the users lie at the core of many recommendation algorithms, and numerous metrics have been proposed in the recommender systems field since its inception. This talk will explore evolution of various similarity-based measures from the initial class of rating-based measures to the more recently proposed latent metrics and the metric learning methods. We will also explore possible future research directions and novel applications of similarity measures in recommender systems.

Contents

Clustering and Outlier Detection

Subspaces and Embeddings

Applications

Doctoral Symposium Papers

Similarity Search and Retrieval

Fast Locality-Sensitive Hashing Frameworks for Approximate Near Neighbor Search

Tobias Christiani[(✉)]

Maersk Line, Copenhagen, Denmark
tobiaschristiani@gmail.com

Abstract. The Indyk-Motwani Locality-Sensitive Hashing (LSH) framework (STOC 1998) is a general technique for constructing a data structure to answer approximate near neighbor queries by using a distribution $\mathcal{H}$ over locality-sensitive hash functions that partition space. For a collection of n points, after preprocessing, the query time is dominated by $O(n^\rho \log n)$ evaluations of hash functions from $\mathcal{H}$ and $O(n^\rho)$ hash table lookups and distance computations where $\rho \in (0, 1)$ is determined by the locality-sensitivity properties of $\mathcal{H}$. It follows from a recent result by Dahlgaard et al. (FOCS 2017) that the number of locality-sensitive hash functions can be reduced to $O(\log^2 n)$, leaving the query time to be dominated by $O(n^\rho)$ distance computations and $O(n^\rho \log n)$ additional word-RAM operations. We state this result as a general framework and provide a simpler analysis showing that the number of lookups and distance computations closely match the Indyk-Motwani framework. Using ideas from another locality-sensitive hashing framework by Andoni and Indyk (SODA 2006) we are able to reduce the number of additional word-RAM operations to $O(n^\rho)$.

1 Introduction

The approximate near neighbor problem is the problem of preprocessing a collection P of n points in a space (X, dist) into a data structure such that, for parameters $r_1 < r_2$ and given a query point $q \in X$, if there exists a point $x \in P$ with $\text{dist}(q, x) \leq r_1$, then the data structure is guaranteed to return a point $x' \in P$ such that $\text{dist}(q, x') < r_2$.

Indyk and Motwani [24] introduced a general framework for constructing solutions to the approximate near neighbor problem using a technique known as locality-sensitive hashing (LSH). The framework takes a distribution over hash functions $\mathcal{H}$ with the property that near points are more likely to collide under a random $h \sim \mathcal{H}$. During preprocessing a number of locality-sensitive hash functions are sampled from $\mathcal{H}$ and used to hash the points of P into buckets.

The research leading to these results has received funding from the European Research Council under the European Union's 7th Framework Programme (FP7/2007-2013)/ERC grant agreement no. 614331.

© Springer Nature Switzerland AG 2019
G. Amato et al. (Eds.): SISAP 2019, LNCS 11807, pp. 3–17, 2019.
https://doi.org/10.1007/978-3-030-32047-8_1

The query algorithm evaluates the same hash functions on the query point and looks into the associated buckets to find an approximate near neighbor.

The locality-sensitive hashing framework of Indyk and Motwani has had a large impact in both theory and practice (see surveys [3] and [32] for an introduction), and many of the best known solutions to the approximate near neighbor problem in high-dimensional spaces, such as Euclidean space [2], the unit sphere under inner product similarity [4], and sets under Jaccard similarity [7] come in the form of families of locality-sensitive hash functions that can be plugged into the Indyk-Motwani LSH framework.

Definition 1 (Locality-sensitive hashing [24]). *Let (X, dist) be a distance space and let $\mathcal{H}$ be a distribution over functions $h \colon X \to R$. We say that $\mathcal{H}$ is (r_1, r_2, p_1, p_2)-sensitive if for $x, y \in X$ and $h \sim \mathcal{H}$ we have that:*

- *If $\mathrm{dist}(x, y) \leq r_1$ then $\Pr[h(x) = h(y)] \geq p_1$.*
- *If $\mathrm{dist}(x, y) \geq r_2$ then $\Pr[h(x) = h(y)] \leq p_2$.*

The Indyk-Motwani framework takes a (r_1, r_2, p_1, p_2)-sensitive family $\mathcal{H}$ and constructs a data structure that solves the approximate near neighbor problem for parameters $r_1 < r_2$ with some positive constant probability of success. We will refer to this randomized approximate version of the near neighbor problem as the (r_1, r_2)-near neighbor problem, where we require queries to succeed with probability at least $1/2$ (see Definition 2). To simplify the exposition we will assume throughout the introduction, unless otherwise stated, that $0 < p_1 < p_2 < 1$ are constant, that a hash function $h \in \mathcal{H}$ can be stored in $n/\log n$ words of space, and for $\rho = \log(1/p_1)/\log(1/p_2) \in (0, 1)$ that a point $x \in X$ can be stored in $O(n^\rho)$ words of space. The assumption of a constant gap between p_1 and p_2 allows us to avoid performing distance computations by instead using the 1-bit sketching scheme of Li and König [26] together with the family $\mathcal{H}$ to approximate distances (see Sect. 4.1 for details). In the remaining part of the paper we will state our results without any such assumptions to ensure, for example, that our results hold in the important case where p_1, p_2 may depend on n or the dimensionality of the space [2,4].

Theorem 1 (Indyk-Motwani [23,24], simplified). *Let $\mathcal{H}$ be (r_1, r_2, p_1, p_2)-sensitive and let $\rho = \frac{\log(1/p_1)}{\log(1/p_2)}$, then there exists a solution to the (r_1, r_2)-near neighbor problem using $O(n^{1+\rho})$ words of space and with query time dominated by $O(n^\rho \log n)$ evaluations of functions from $\mathcal{H}$.*

The query time of the Indyk-Motwani framework is dominated by the number of evaluations of locality-sensitive hash functions. To make matters worse, almost all of the best known and most widely used locality-sensitive families have an evalution time that is at least linear in the dimensionality of the underlying space [2,4,7,11,17]. Significant effort has been devoted to the problem of reducing the evaluation complexity of locality-sensitive hash families [4,15,16,19,25,28,29,31], while the question of how many independent locality-sensitive hash functions are actually needed to solve the (r_1, r_2)-near neighbor problem has received relatively little attention [1,15].

This paper aims to bring attention to, strengthen, generalize, and simplify results that reduce the number of locality-sensitive hash functions used to solve the (r_1, r_2)-near neighbor problem. In particular, we will extract a general framework from a technique introduced by Dahlgaard et al. [15] in the context of set similarity search under Jaccard similarity, showing that the number of locality-sensitive hash functions can be reduced to $O(\log^2 n)$ in general. We further show how to reduce the word-RAM complexity of the general framework from $O(n^\rho \log n)$ to $O(n^\rho)$ by combining techniques from Dahlgaard et al. and Andoni and Indyk [1]. Reducing the number of locality-sensitive hash functions allows us to spend time $O(n^\rho / \log^2 n)$ per hash function evaluation without increasing the overall complexity of the query algorithm — something which is particularly useful in Euclidean space where the best known LSH upper bounds offer a trade-off between the ρ-value that can be achieved and the evaluation complexity of the locality-sensitive hash function [2,4,25].

1.1   Related Work

Indyk-Motwani. The Indyk-Motwani framework uses $L = O(n^\rho)$ independent partitions of space, each formed by overlaying $k = O(\log n)$ random partitions induced by k random hash functions from a locality-sensitive family $\mathcal{H}$. The parameter k is chosen such that a random partition has the property that a pair of points $x, y \in X$ with $\mathrm{dist}(x, y) \leq r_1$ has probability $n^{-\rho}$ of ending up in the same part of the partition, while a pair of points with $\mathrm{dist}(x, y) \geq r_2$ has probability n^{-1} of colliding. By randomly sampling $L = O(n^\rho)$ such partitions we are able to guarantee that a pair of near points will collide with constant probability in at least one of them. Applying these L partitions to our collection of data points P and storing the result of each partition of P in a hash table we obtain a data structure that solves the (r_1, r_2)-near neighbor problem as outlined in Theorem 1 above. Sections 3 and 3.1 contains a more complete description of LSH-based frameworks and the Indyk-Motwani framework.

Andoni-Indyk. Many locality-sensitive hash functions have a super-constant evaluation time. This motivated Andoni and Indyk to introduce a replacement to the Indyk-Motwani framework in a paper on substring near neighbor search [1]. The key idea is to re-use hash functions from a small collection of size $m \ll L$ by forming all combinations of $\binom{m}{t}$ hash functions. This technique is also known as tensoring and has seen some use in the work on alternative solutions to the approximate near neighbor problem, in particular the work on locality-sensitive filtering [6,12,18]. By applying the tensoring technique the Andoni-Indyk framework reduces the number of hash functions to $O(\exp(\sqrt{\rho \log n \log \log n})) = n^{o(1)}$ as stated in Theorem 2.

Theorem 2 (Andoni-Indyk [1], simplified). *Let $\mathcal{H}$ be (r_1, r_2, p_1, p_2)-sensitive and let $\rho = \frac{\log(1/p_1)}{\log(1/p_2)}$, then there exists a solution to the (r_1, r_2)-near* *neighbor problem using $O(n^{1+\rho})$ words of space and with query time dominated by $O(\exp(\sqrt{\rho \log n \log \log n}))$ evaluations of functions from $\mathcal{H}$ and $O(n^{\rho})$ other word-RAM operations.*

The paper by Andoni and Indyk did not state this result explicitly as a theorem in the same form as the Indyk-Motwani framework; the analysis made some implicit restrictive assumptions on p_1, p_2 and ignored integer constraints. Perhaps for these reasons the result does not appear to have received much attention, although it has seen some limited use in practice [30]. In Sect. 3.2 we present a slightly different version of the Andoni-Indyk framework together with an analysis that satisfies integer constraints, providing a more accurate assessment of the performance of the framework in the general, unrestricted case.

Dahlgaard-Knudsen-Thorup. The paper by Dahlgaard et al. [15] introduced a different technique for constructing the L hash functions/partitions from a smaller collection of m hash functions from $\mathcal{H}$. Instead of forming all combinations of subsets of size t as the Andoni-Indyk framework they instead sample k hash functions from the collection to form each of the L partitions. The paper focused on a particular application to set similarity search under Jaccard similarity, and stated the result in terms of a solution to this problem. In Sect. 3.3 we provide a simplified and tighter analysis to yield a general framework:

Theorem 3 (DKT [15], simplified). *Let $\mathcal{H}$ be (r_1, r_2, p_1, p_2)-sensitive and let $\rho = \frac{\log(1/p_1)}{\log(1/p_2)}$, then there exists a solution to the (r_1, r_2)-near neighbor problem using $O(n^{1+\rho})$ words of space and with query time dominated by $O(\log^2 n)$ evaluations of functions from $\mathcal{H}$ and $O(n^{\rho} \log n)$ other word-RAM operations.*

The analysis of [15] indicates that the Dahlgaard-Knudsen-Thorup framework, when compared to the Indyk-Motwani framework, would use at least 50 times as many partitions (and a corresponding increase in the number of hash table lookups and distance computations) to solve the (r_1, r_2)-near neighbor problem with success probability at least $1/2$. Using elementary tools, the analysis in this paper shows that we only have to use twice as many partitions as the Indyk-Motwani framework to obtain the same guarantee of success.

Number of Hash Functions in Practice. To provide some idea of what the number of hash functions H used by the different frameworks would be in practice, Fig. 1 shows the value of $\log_2 H$ that is obtained by actual implementations of the Indyk-Motwani (IM), Andoni-Indyk (AI), and Dahlgaard-Knudsen-Thorup (DKT) frameworks according to the analysis in Sect. 3 for $p_1 = 1/2$ and every value of $0 < p_2 < 1/2$ for a solution to the (r_1, r_2)-near neighbor problem on a collection of $n = 2^{30}$ points. Figure 1 reveals that the number of hash functions used by the Indyk-Motwani framework exceeds 2^{30}, the size of the collection of points P, as p_2 approaches p_1. In addition, locality-sensitive hash

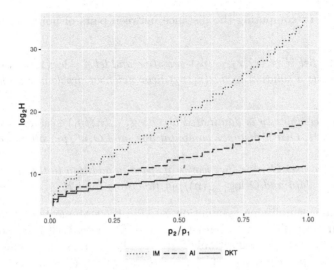

Fig. 1. The exact number of locality-sensitive hash functions from a $(r_1, r_2, 0.5, p_2)$-sensitive family used by different frameworks to solve the (r_1, r_2)-near neighbor problem on a collection of 2^{30} points according to the analysis in this paper.

functions used in practice such as Charikar's SimHash [11] and p-stable LSH [17] have evaluation time $O(d)$ for points in $\mathbb{R}^d$. These two factors might help explain why a linear scan over sketches of the entire collection of points is a popular approach to solve the approximate near neighbor problem in practice [21,33]. The Andoni-Indyk framework reduces the number of hash functions by several orders of magnitude, and the Dahlgaard-Knudsen-Thorup framework presents another improvement of several orders of magnitude. Since the word-RAM complexity of the DKT framework matches the number of hash functions used by the IM framework, the gap between the solid line (DKT) and the dotted line (IM) gives some indication of the time we can spend on evaluating a single hash function in the DKT framework without suffering a noticeable increase in the query time.

1.2   Contribution

Improved Word-RAM Complexity. In addition to our work on the Andoni-Indyk and Dahlgaard-Knudsen-Thorup frameworks as mentioned above, we show how the word-RAM complexity of the DKT framework can be reduced by a logarithmic factor. The solution is a simple combination of the DKT sampling technique and the AI tensoring technique: First we use the DKT sampling technique twice to construct two collections of $\sqrt{L}$ partitions. Then we use the AI tensoring technique to form $L = \sqrt{L} \times \sqrt{L}$ pairs of partitions from the two collections. Below we state our main Theorem 4 in its general form where we make no implicit assumptions about $\mathcal{H}$ (p_1 and p_2 are not assumed to be constant and can depend on for example n) or about the complexity of storing a point or a hash function, or computing the distance between pairs of points in the space (X, dist).

Theorem 4. *Let $\mathcal{H}$ be (r_1, r_2, p_1, p_2)-sensitive and let $\rho = \log(1/p_1)/\log(1/p_2)$, then there exists a solution to the (r_1, r_2)-near neighbor problem with the following properties:*

- *The query complexity is dominated by $O(\log^2_{1/p_2}(n)/p_1)$ evaluations of functions from $\mathcal{H}$, $O(n^\rho)$ distance computations, and $O(n^\rho/p_1)$ other word-RAM operations.*
- *The solution uses $O(n^{1+\rho}/p_1)$ words of space in addition to the space required to store the data and $O(\log^2_{1/p_2}(n)/p_1)$ functions from $\mathcal{H}$.*

Under the same simplifying assumptions used in the statements of Theorems 1, 2, and 3, our main Theorem 4 can be stated as Theorem 3 with the word-RAM complexity reduced by a logarithmic factor to $O(n^\rho)$. This improvement in the word-RAM complexity comes at the cost of a (rather small) constant factor increase in the number of hash functions, lookups, and distance computations compared to the DKT framework. By varying the size m of the collection of hash functions from $\mathcal{H}$ and performing independent repetitions we can obtain a tradeoff between the number of hash functions and the number of lookups.

Distance Sketching Using LSH. Finally, we combine Theorem 4 with the 1-bit sketching scheme of Li and König [26] where we use the locality-sensitive hash family to create sketches that allow us to leverage word-level parallelism and avoid direct distance computations. This sketching technique is well known and has been used before in combination with LSH-based approximate similarity search [13], but we believe there is some value in the simplicity of the analysis and in a clear statement of the combination of the two results as given in Theorem 5, for example in the important case where $0 < p_2 < p_1 < 1$ are constant.

Theorem 5. *Let $\mathcal{H}$ be (r_1, r_2, p_1, p_2)-sensitive and let $\rho = \log(1/p_1)/\log(1/p_2)$, then there exists a solution to the (r_1, r_2)-near neighbor problem with the following properties:*

- *The query complexity is dominated by $O(\log^2(n)/(p_1 - p_2)^2)$ evaluations of hash functions from $\mathcal{H}$ and $O(n^\rho/(p_1 - p_2)^2)$ other word-RAM operations.*
- *The solution uses $O(n^{1+\rho}/p_1 + n/(p_1 - p_2)^2)$ words of space in addition to the space required to store the data and $O(\log^2(n)/(p_1 - p_2)^2)$ hash functions from $\mathcal{H}$.*

2   Preliminaries

Problem and Dynamization. We begin by defining the version of the approximate near neighbor problem that the frameworks presented in this paper will be solving:

Definition 2. *Let $P \subseteq X$ be a collection of $|P| = n$ points in a distance space (X, dist). A solution to the (r_1, r_2)-near neighbor problem is a data structure that supports the following query operation: Given a query point $q \in X$, if there exists a point $x \in P$ with $\text{dist}(q, x) \leq r_1$, then, with probability at least $1/2$, return a point $x' \in P$ such that $\text{dist}(q, x') < r_2$.*

We aim for solutions with a failure probability that is upper bounded by $1/2$. The standard trick of using η independent repetitions of the data structure allows us to reduce the probability of failure to $1/2^\eta$. For the sake of simplicity we restrict our attention to static solutions, meaning that we do not concern ourselves with the complexity of updates to the underlying set P, although it is simple to modify the static solutions presented in this paper to dynamic solutions where the update complexity essentially matches the query complexity [23,27].

LSH Powering. The Indyk-Motwani framework and the Andoni-Indyk framework will make use of the following standard powering technique described in the introduction as "overlaying partitions". Let $k \geq 1$ be an integer and let $\mathcal{H}$ denote a locality-sensitive family of hash functions as in Definition 1. We will use the notation $\mathcal{H}^k$ to denote the distribution over functions $h \colon X \to R^k$ where

$$g(x) = (h_1(x), \ldots, h_k(x))$$

and $h_1, \ldots, h_k$ are sampled independently at random from $\mathcal{H}$. It is easy to see that $\mathcal{H}^k$ is (r_1, r_2, p_1^k, p_2^k)-sensitive. To deal with some special cases we define $\mathcal{H}^0$ to be the family consisting of a single constant function.

Model of Computation. We will work in the standard word-RAM model of computation [22] with a word length of $\Theta(\log n)$ bits where n denotes the size of the collection P to be searched in the (r_1, r_2)-near neighbor problem. During the preprocessing stage of our solutions we will assume access to a source of randomness that allows us to sample independently from a family $\mathcal{H}$ and to seed pairwise independent hash functions [9,10]. The latter can easily be accomplished by augmenting the model with an instruction that generates a uniformly random word in constant time and using that to seed the tables of a Zobrist hash function [34].

3 Frameworks

Overview. We will describe frameworks that take as input a (r_1, r_2, p_1, p_2)-sensitive family $\mathcal{H}$ and a collection P of n points and constructs a data structure that solves the (r_1, r_2)-near neighbor problem. The frameworks described in this paper all use the same high-level technique of constructing L hash functions $g_1, \ldots, g_L$ that are used to partition space such that a pair of points x, y with $\text{dist}(x, y) \leq r_1$ will end up in the same part of one of the L partitions with probability at least $1/2$. That is, for x, y with $\text{dist}(x, y) \leq r_1$ we have that

$\Pr[\exists l \in [L] \colon g_l(x) = g_l(y)] \geq 1/2$ where $[L]$ is used to denote the set $\{1, 2, \ldots, L\}$. At the same time we ensure that the expected number of collisions between pairs of points x, y with $\mathrm{dist}(x, y) \geq r_2$ is at most one in each partition.

Preprocessing and Queries. During the preprocessing phase, for each of the L hash functions $g_1, \ldots, g_L$ we compute the partition of the collection of points P induced by g_l and store it in a hash table in the form of key-value pairs $(z, \{x \in P \mid g_l(x) = z\})$. To reduce space usage we store only a single copy of the collection P and store references to P in our L hash tables. To guarantee lookups in constant time we can use the perfect hashing scheme by Fredman et al. [20] to construct our hash tables. We will assume that hash values $z = g_l(x)$ fit into $O(1)$ words. If this is not the case we can use universal hashing [8] to operate on fingerprints of the hash values.

We perform a query for a point q as follows: for $l = 1, \ldots, L$ we compute $g_l(q)$, retrieve the set of points $\{x \in P \mid g_l(x) = g_l(q)\}$, and compute the distance between q and each point in the set. If we encounter a point x' with $\mathrm{dist}(q, x') < r_2$ then we return x' and terminate. If after querying the L sets no such point is encountered we return a special symbol $\varnothing$ and terminate.

We will proceed by describing and analyzing the solutions to the (r_1, r_2)-near neighbor problem for different approaches to sampling, storing, and computing the L hash functions $g_1, \ldots, g_L$, resulting in the different frameworks as mentioned in the introduction.

3.1   Indyk-Motwani

To solve the (r_1, r_2)-near neighbor problem using the Indyk-Motwani framework we sample L hash functions $g_1, \ldots, g_L$ independently at random from the family $\mathcal{H}^k$ where we set $k = \lceil \log(n)/\log(1/p_2) \rceil$ and $L = \lceil (\ln 2)/p_1^k \rceil$. Correctness of the data structure follows from the observation that the probability that a pair of points x, y with $\mathrm{dist}(x, y) \leq r_1$ does not collide under a randomly sampled $g_l \sim \mathcal{H}^k$ is at most $1 - p_1^k$. We can therefore upper bound the probability that a near pair of points does not collide under any of the hash functions by $(1 - p_1^k)^L \leq \exp(-p_1^k L) \leq 1/2$.

In the worst case, the query operation computes L hash functions from $\mathcal{H}^k$ corresponding to Lk hash functions from $\mathcal{H}$. For a query point q the expected number of points $x' \in P$ with $\mathrm{dist}(q, x') \geq r_2$ that collide with q under a randomly sampled $g_l \sim \mathcal{H}^k$ is at most $np_2^k \leq np_2^{\log(n)/\log(1/p_2)} = 1$. It follows from linearity of expectation that the total expected number of distance computations during a query is at most L. The result is summarized in Theorem 6 from which the simplified Theorem 1 follows.

Theorem 6 (Indyk-Motwani [23,24]). *Given a (r_1, r_2, p_1, p_2)-sensitive family $\mathcal{H}$ we can construct a data structure that solves the (r_1, r_2)-near neighbor problem such that for $k = \lceil \log(n)/\log(1/p_2) \rceil$ and $L = \lceil (\ln 2)/p_1^k \rceil$ the data structure has the following properties:*

- *The query operation uses at most Lk evaluations of hash functions from $\mathcal{H}$, expected L distance computations, and $O(Lk)$ other word-RAM operations.*
- *The data structure uses $O(nL)$ words of space in addition to the space required to store the data and Lk hash functions from $\mathcal{H}$.*

Theorem 6 gives a bound on the expected number of distance computations while the simplified version stated in Theorem 1 uses Markov's inequality and independent repetitions to remove the expectation from the bound by treating an excessive number of distance computations as a failure.

3.2 Andoni-Indyk

In 2006 Andoni and Indyk, as part of a paper on the substring near neighbor problem, introduced an improvement to the Indyk-Motwani framework that reduces the number of locality-sensitive hash functions [1]. Their improvement comes from the use of a technique that we will refer to as tensoring: setting the hash functions $g_1, \ldots, g_L$ to be all t-tuples from a collection of m functions sampled from $\mathcal{H}^{k/t}$ where $m \ll L$. The analysis in [1] shows that by setting $m = n^{\rho/t}$ and repeating the entire scheme $t!$ times, the total number of hash functions can be reduced to $O(\exp(\sqrt{\rho \log n \log \log n}))$ when setting $t = \sqrt{\frac{\rho \log n}{\log \log n}}$. This analysis ignores integer constraints on t, k, and m, and implicitly place restrictions on p_1 and p_2 in relation to n (e.g. $0 < p_2 < p_1 < 1$ are constant). We will introduce a slightly different scheme that takes into account integer constraints and analyze it without restrictions on the properties of $\mathcal{H}$.

Assume that we are given a (r_1, r_2, p_1, p_2)-sensitive family $\mathcal{H}$. Let the parameters $\eta, t, k_1, k_2, m_1, m_2$ be non-negative integers. Each of the L hash functions $g_1, \ldots, g_L$ will be formed by concatenating one hash function from each of t collections of m_1 hash functions from $\mathcal{H}^{k_1}$ and concatenating a last hash function from a collection of m_2 hash functions from $\mathcal{H}^{k_2}$. We take all $m_1^t m_2$ hash functions of the above form and repeat η times for a total of $L = \eta m_1^t m_2$ hash functions constructed from a total of $H = \eta(m_1 k_1 t + m_2 k_2)$ hash functions from $\mathcal{H}$. See the Appendix of the online version of this paper for a more complete analysis [14].

Setting t. It remains to show how to set t to obtain a good bound on the number of hash functions H. Note that in practice we can simply set $t = \arg \min_t H$ by trying $t = 1, \ldots, k$. If we ignore integer constraints and place certain restrictions of $\mathcal{H}$ as in the original tensoring scheme by Andoni and Indyk we want to set t to minimize the expression $t^t n^{\rho/t}$. This minimum is obtained when setting t such that $t^2 \log t = \rho \log n$. We therefore cannot do much better than setting $t = \sqrt{\rho \log(n)/\log \log n}$ which gives the bound $H = O(\exp(\sqrt{\rho \log(n) \log \log n}))$ as shown in [1]. To allow for easy comparison with the Indyk-Motwani framework without placing restrictions on $\mathcal{H}$ we set $t = \lceil \sqrt{k} \rceil$, resulting in Theorem 7.

Theorem 7. *Given a* (r_1, r_2, p_1, p_2)-*sensitive family* $\mathcal{H}$ *there exists a solution to the* (r_1, r_2)-*near neighbor problem such that for* $k = \lceil \log(n)/\log(1/p_2) \rceil$, $H = k(\sqrt{k}/p_1)^{\sqrt{k}}$, *and* $L = \lceil 1/p_1^k \rceil$ *the data structure has the following properties:*

- *The query operation uses* $O(H)$ *evaluations of functions from* $\mathcal{H}$, $O(L)$ *distance computations, and* $O(L + H)$ *other word-RAM operations.*
- *The data structure uses* $O(nL)$ *words of space in addition to the space required to store the data and* $O(H)$ *hash functions from* $\mathcal{H}$.

Thus, compared to the Indyk-Motwani framework we have gone from using $O(k(1/p_1)^k)$ locality-sensitive hash functions to $O(k(\sqrt{k}/p_1)^{\sqrt{k}})$ locality-sensitive hash functions. Figure 1 shows the actual number of hash functions of the revised version of the Andoni-Indyk scheme when t is set to minimize H.

3.3   Dahlgaard-Knudsen-Thorup

In a recent paper Dahlgaard et al. [15] introduce a different technique for reducing the number of locality-sensitive hash functions. The idea is to construct each hash value $g_l(x)$ by sampling and concatenating k hash values from a collection of km pre-computed hash functions from $\mathcal{H}$. Dahlgaard et al. applied this technique to provide a fast solution for the approximate near neighbor problem for sets under Jaccard similarity. In this paper we use the same technique to derive a general framework solution that works with every family of locality-sensitive hash functions, reducing the number of locality-sensitive hash functions compard to the Indyk-Motwani and Andoni-Indyk frameworks.

Let $[n]$ denote the set of integers $\{1, 2, \ldots, n\}$. For $i \in [k]$ and $j \in [m]$ let $h_{i,j} \sim \mathcal{H}$ denote a hash function in our collection. To sample from the collection we use k mutually independent and pairwise independent hash functions [10] of the form $f_i \colon [L] \to [m]$ and set

$$g_l(x) = (h_{1, f_1(l)}(x), \ldots, h_{k, f_k(l)}(x)).$$

To show correctness of this scheme we will make use of an elementary one-sided version of Chebyshev's inequality stating that for a random variable Z with mean $\mu > 0$ and variance $\sigma^2 < \infty$ we have that $\Pr[Z \leq 0] \leq \sigma^2/(\mu^2 + \sigma^2)$. We will apply this inequality to lower bound the probability that there are no collisions between close pairs of points. For two points x and y let $Z_l = \mathbb{1}\{g_l(x) = g_l(y)\}$ so that $Z = \sum_{l=1}^{L} Z_l$ denotes the sum of collisions under the L hash functions. To apply the inequality we need to derive an expression for the expectation and the variance of the random variable Z. Let $p = \Pr_{h \sim \mathcal{H}}[h(x) = h(y)]$ then by linearity of expectation we have that $\mu = \mathrm{E}[Z] = Lp^k$. To bound $\sigma^2 = \mathrm{E}[Z^2] - \mu^2$ we proceed by bounding $\mathrm{E}[Z^2]$ where we note that $Z_l = \Pi_{i=1}^{k} Y_{l,i}$ for $Y_{l,i} = \mathbb{1}\{h_{i, f_i(l)}(x) = h_{i, f_i(l)}(x)\}$ and make use of the independence between $Y_{l,i}$ and $Y_{l',i'}$ for $i \neq i'$.

$$E[Z^2] = \sum_{\substack{l,l' \in [L] \\ l \neq l'}} E[Z_l Z_{l'}] + \sum_{l=1}^{L} E[Z_l] = (L^2 - L)E[Z_l Z_{l'}] + \mu$$

$$\leq L^2 E\left[\Pi_{i=1}^{k} Y_{l,i} Y_{l',i}\right] + \mu = L^2 \left(E[Y_{l,i} Y_{l',i}]\right)^k + \mu.$$

We have that $E[Y_{l,i} Y_{l',i}] = \Pr[f_i(l) = f_i(l')]p + \Pr[f_i(l) \neq f_i(l')]p^2 = (1/m)p + (1 - 1/m)p^2$, i.e., with probability $1/m$ we have that our pairwise independent hash functions choose the same underlying locality sensitive hash function from our pool of m functions and the probability of collision ($Y_{l,i} = 1$ and $Y_{l',i} = 1$) is given by the collision probability of a single locality-sensitive hash function p. With probability $1 - 1/m$ we sample two independent locality-sensitive hash function and the probability that they both collide is then given by p^2.

Let $\varepsilon > 0$ and set $m = \lceil \frac{1-p_1}{p_1} \frac{k}{\ln(1+\varepsilon)} \rceil$ then for $p \geq p_1$ we have $(E[Y_{l,i} Y_{l',i}])^k \leq (1+\varepsilon)p^{2k}$. This allows us to bound the variance of Z by $\sigma^2 \leq \varepsilon\mu^2 + \mu$ resulting in the following lower bound on the probability of collision between similar points.

Lemma 1. *For $\varepsilon > 0$ let $m \geq \lceil \frac{1-p_1}{p_1} \frac{k}{\ln(1+\varepsilon)} \rceil$, then for every pair of points x, y with $\mathrm{dist}(x, y) \leq r_1$ we have that*

$$\Pr[\exists l \in [L] : g_l(x) = g_l(y)] \geq \frac{1 + \varepsilon\mu}{1 + (1+\varepsilon)\mu}.$$

By setting $\varepsilon = 1/4$ and $L = \lceil (2\ln(2))/p_1^k \rceil$ we obtain an upper bound on the failure probability of $1/2$. Setting the size of each of the k collections of precomputed hash values to $m = \lceil 5k/p_1 \rceil$ is sufficient to yield the following solution to the (r_1, r_2)-near neighbor problem where provide exact bounds on the number of lookups L and hash functions H:

Theorem 8 (Dahlgaard-Knudsen-Thorup [15]). *Given a family $\mathcal{H}$ that is (r_1, r_2, p_1, p_2)-sensitive we can construct a data structure that solves the (r_1, r_2)-near neighbor problem such that for $k = \lceil \log(n)/\log(1/p_2) \rceil$, $H = k\lceil 5k/p_1 \rceil$, and $L = \lceil (2\ln(2))/p_1^k \rceil$ the data structure has the following properties:*

- *The query operation uses at most H evaluations of hash functions from $\mathcal{H}$, expected L distance computations, and $O(Lk)$ other word-RAM operations.*
- *The data structure uses $O(nL)$ words of space in addition to the space required to store the data and H hash functions from $\mathcal{H}$.*

Compared to the Indyk-Motwani framework we have reduced the number of locality-sensitive hash functions H from $O(k(1/p_1)^k)$ to $O(k^2/p_1)$ at the cost of using twice as many lookups. To reduce the number of lookups further we can decrease ε and perform several independent repetitions. This comes at the cost of an increase in the number of hash functions H.

4   Reducing the Word-RAM Complexity

One drawback of the DKT framework is that each hash value $g_l(x)$ still takes $O(k)$ word-RAM operations to compute, even after the underlying locality-sensitive hash functions are known. This results in a bound on the total number of additional word-RAM operations of $O(Lk)$. We show how to combine the DKT universal hashing technique with the AI tensoring technique to ensure that the running time is dominated by $O(L)$ distance computations and $O(H)$ hash function evaluations. The idea is to use the DKT scheme to construct two collections of respectively L_1 and L_2 hash functions, and then to use the AI tensoring approach to form $g_1, \ldots, g_L$ as the $L = L_1 \times L_2$ combinations of functions from the two collections. The number of lookups can be reduced by applying tensoring several times in independent repetitions, but for the sake of simplicity we use a single repetition. For the usual setting of $k = \lceil \log(n)/\log(1/p_2) \rceil$ let $k_1 = \lceil k/2 \rceil$ and $k_2 = \lfloor k/2 \rfloor$. Set $L_1 = \lceil 6(1/p_1)^{k_1} \rceil$ and $L_2 = \lceil 6(1/p_1)^{k_2} \rceil$. According to Lemma 1 if we set $\varepsilon = 1/6$ the success probability of each collection is at least $3/4$ and by a union bound the probability that either collection fails to contain a colliding hash function is at most $1/2$. This concludes the proof of our main Theorem 4.

4.1   Sketching

The theorems of the previous section made no assumptions on the word-RAM complexity of distance computations and instead stated the number of distance computations as part of the query complexity. We can use a (r_1, r_2, p_1, p_2)-sensitive family $\mathcal{H}$ to create sketches that allows us to efficiently approximate the distance between pairs of points, provided that the gap between p_1 and p_2 is sufficiently large. In this section we will re-state the results of Theorem 4 when applying the family $\mathcal{H}$ to create sketches using the 1-bit sketching scheme of Li and König [26]. Let b be a positive integer denoting the length of the sketches in bits. The advantage of this scheme is that we can use word level parallelism to evaluate a sketch of b bits in time $O(b/\log n)$ in our word-RAM model with word length $\Theta(\log n)$.

For $i = 1, \ldots, b$ let $h_i \colon X \to R$ denote a randomly sampled locality-sensitive hash function from $\mathcal{H}$ and let $f_i \colon R \to \{0, 1\}$ denote a randomly sampled universal hash function. We let $s(x) \in \{0, 1\}^b$ denote the sketch of a point $x \in X$ where we set the ith bit of the sketch $s(x)_i = f_i(h_i(x))$. For two points $x, y \in X$ the probability that they agree on the ith bit is 1 if the points collide under h_i and $1/2$ otherwise. The probability that two sketch bits collide is therefore given by $\Pr[s(x)_i = s(y)_i] = \Pr[h_i(x) = h_i(y)] + (1 - \Pr[h_i(x) = h_i(y)])/2 = (1 + \Pr[h_i(x) = h_i(y)])/2$. We will apply these sketches during our query procedure instead of direct distance computations when searching through the points in the L buckets, comparing them to our query point q. Let $\lambda \in (0, 1)$ be a parameter that will determine whether we report a point or not. For sketches of length b we will return a point x if $\|s(q) - s(x)\|_1 > \lambda b$. An application of Hoeffding's inequality gives us the following properties of the sketch:

Lemma 2. *Let $\mathcal{H}$ be a (r_1, r_2, p_1, p_2)-sensitive family and let $\lambda = (1 + p_2)/2 + (p_1 - p_2)/4$, then for sketches of length $b \geq 1$ and for every pair points $x, y \in X$:*

- *If $\mathrm{dist}(x, y) \leq r_1$ then $\Pr[\|s(x) - s(y)\|_1 \leq \lambda b] \leq e^{-b(p_1 - p_2)^2/8}$.*
- *If $\mathrm{dist}(x, y) \geq r_2$ then $\Pr[\|s(x) - s(y)\|_1 > \lambda b] \leq e^{-b(p_1 - p_2)^2/8}$.*

If we replace the exact distance computations with sketches we want to avoid two events: Failing to report a point with $\mathrm{dist}(q, x) \leq r_1$ and reporting a point x with $\mathrm{dist}(q, x) \geq r_2$. By setting $b = O(\ln(n)/(p_1 - p_2)^2)$ and applying a union bound over the n events that the sketch fails for a point in our collection P we obtain Theorem 5.

5   Conclusion and Open Problems

We have shown that there exists a simple and general framework for solving the (r_1, r_2)-near neighbor problem using only few locality-sensitive hash functions and with a reduced word-RAM complexity matching the number of lookups. The analysis in this paper indicates that the performance of the Dahlgaard-Knudsen-Thorup framework is highly competitive compared to the Indyk-Motwani framework in practice, especially when locality-sensitive hash functions are expensive to evaluate, as is often the case.

An obvious open problem is the question of whether the number of locality-sensitive hash functions can be reduced even below $O(k^2/p_1)$. Another possible direction for future research would be to obtain similar framework results in the context of solutions to the (r_1, r_2)-near neighbor problem that allow for space-time tradeoffs [5, 12].

References

1. Andoni, A., Indyk, P.: Efficient algorithms for substring near neighbor problem. In: Proceedings of the SODA 2006, pp. 1203–1212 (2006)
2. Andoni, A., Indyk, P.: Near-optimal hashing algorithms for approximate nearest neighbor in high dimensions. In: Proceedings of the FOCS 2006, pp. 459–468 (2006)
3. Andoni, A., Indyk, P.: Near-optimal hashing algorithms for approximate nearest neighbor in high dimensions. Commun. ACM **51**(1), 117–122 (2008)
4. Andoni, A., Indyk, P., Laarhoven, T., Razenshteyn, I., Schmidt, L.: Practical and optimal LSH for angular distance. In: Proceedings of the NIPS 2015, pp. 1225–1233 (2015)
5. Andoni, A., Laarhoven, T., Razenshteyn, I.P., Waingarten, E.: Optimal hashing-based time-space trade-offs for approximate near neighbors. In: Proceedings of the SODA 2017, pp. 47–66 (2017)
6. Becker, A., Ducas, L., Gama, N., Laarhoven, T.: New directions in nearest neighbor searching with applications to lattice sieving. In: Proceedings of the SODA 2016, pp. 10–24 (2016)
7. Broder, A.Z., Charikar, M., Frieze, A.M., Mitzenmacher, M.: Min-wise independent permutations. J. Comput. Syst. Sci. **60**(3), 630–659 (2000)

8. Carter, J.L., Wegman, M.N.: Universal classes of hash functions. In: Proceedings of the STOC 1977, pp. 106–112 (1977)
9. Carter, J.L., Wegman, M.N.: Universal classes of hash functions. J. Comput. Syst. Sci. **18**(2), 143–154 (1979)
10. Carter, J.L., Wegman, M.N.: New hash functions and their use in authentication and set equality. J. Comput. System Sci. **22**(3), 265–279 (1981)
11. Charikar, M.: Similarity estimation techniques from rounding algorithms. In: Proceedings of the STOC 2002, pp. 380–388 (2002)
12. Christiani, T.: A framework for similarity search with space-time tradeoffs using locality-sensitive filtering. In: Proceedings of the SODA 2017, pp. 31–46 (2017)
13. Christiani, T., Pagh, R., Sivertsen, J.: Scalable and robust set similarity join. CoRR abs/1707.06814 (2017)
14. Christiani, T.: Fast locality-sensitive hashing for approximate near neighbor search. CoRR abs/1708.07586 (2017). http://arxiv.org/abs/1708.07586
15. Dahlgaard, S., Knudsen, M.B.T., Thorup, M.: Fast similarity sketching. CoRR abs/1704.04370 (2017). http://arxiv.org/abs/1704.04370
16. Dasgupta, A., Kumar, R., Sarlós, T.: Fast locality-sensitive hashing. In: Proceedings of the SIGKDD 2011, pp. 1073–1081 (2011)
17. Datar, M., Immorlica, N., Indyk, P., Mirrokni, V.S.: Locality-sensitive hashing scheme based on p-stable distributions. In: Proceedings of the SOCG 2004, pp. 253–262 (2004)
18. Dubiner, M.: Bucketing coding and information theory for the statistical high-dimensional nearest-neighbor problem. IEEE Trans. Inf. Theory **56**(8), 4166–4179 (2010)
19. Eshghi, K., Rajaram, S.: Locality sensitive hash functions based on concomitant rank order statistics. In: Proceedings of the KDD 2008, pp. 221–229 (2008)
20. Fredman, M.L., Komlós, J., Szemerédi, E.: Storing a sparse table with 0(1) worst case access time. J. ACM **31**(3), 538–544 (1984)
21. Gong, Y., Kumar, S., Verma, V., Lazebnik, S.: Angular quantization-based binary codes for fast similarity search. In: NIPS, pp. 1205–1213 (2012)
22. Hagerup, T.: Sorting and searching on the word RAM. In: Morvan, M., Meinel, C., Krob, D. (eds.) STACS 1998. LNCS, vol. 1373, pp. 366–398. Springer, Heidelberg (1998). https://doi.org/10.1007/BFb0028575
23. Har-Peled, S., Indyk, P., Motwani, R.: Approximate nearest neighbor: towards removing the curse of dimensionality. Theory Comput. **8**(1), 321–350 (2012)
24. Indyk, P., Motwani, R.: Approximate nearest neighbors: towards removing the curse of dimensionality. In: Proceedings of the STOC 1998, pp. 604–613 (1998)
25. Kennedy, C., Ward, R.: Fast cross-polytope locality-sensitive hashing. In: Proceedings of the ITCS 2017, pp. 53:1–53:16 (2017)
26. Li, P., König, A.C.: Theory and applications of b-bit minwise hashing. Commun. ACM **54**(8), 101–109 (2011)
27. Overmars, M.H., van Leeuwen, J.: Worst-case optimal insertion and deletion methods for decomposable searching problems. Inf. Process. Lett. **12**(4), 168–173 (1981)
28. Shrivastava, A.: Simple and efficient weighted minwise hashing. In: NIPS, pp. 1498–1506 (2016)
29. Shrivastava, A.: Optimal densification for fast and accurate minwise hashing. In: ICML. Proceedings of Machine Learning Research, vol. 70, pp. 3154–3163. PMLR (2017)
30. Sundaram, N., et al.: Streaming similarity search over one billion tweets using parallel locality-sensitive hashing. PVLDB **6**(14), 1930–1941 (2013)

31. Terasawa, K., Tanaka, Y.: Spherical LSH for approximate nearest neighbor search on unit hypersphere. In: Dehne, F., Sack, J.-R., Zeh, N. (eds.) WADS 2007. LNCS, vol. 4619, pp. 27–38. Springer, Heidelberg (2007). https://doi.org/10.1007/978-3-540-73951-7_4

32. Wang, J., Shen, H.T., Song, J., Ji, J.: Hashing for similarity search: A survey. CoRR abs/1408.2927 (2014). http://arxiv.org/abs/1408.2927

33. Weiss, Y., Torralba, A., Fergus, R.: Spectral hashing. In: NIPS, pp. 1753–1760. Curran Associates, Inc. (2008)

34. Zobrist, A.L.: A new hashing method with application for game playing. ICCA J. **13**(2), 69–73 (1970)

Storing Data Once in M-tree and PM-tree

Humberto Razente$^{(\boxtimes)}$ and Maria Camila Nardini Barioni

Faculdade de Computação (FACOM),
Universidade Federal de Uberlândia (UFU), Uberlândia, Brazil
{humberto.razente,camila.barioni}@ufu.br

Abstract. Since the publication of M-tree, several enhancements were proposed to its structure. One of the most exciting is the use of additional global pivots that resulted in the PM-tree. In this paper, we revisit both M-tree and PM-tree to propose a new construction algorithm that stores data elements once in their trees hierarchies. The main challenge is to select data elements when an inner node split is needed. The idea is that as a data element is evaluated for pruning during traversal, it can become part of the result set, allowing faster convergence of nearest neighbor algorithms. The new insert and query algorithms enable faster retrieval, the decrease in node occupation of trees built with the same parameters, and also a reduction in the overlap among nodes, as shown in the experimental evaluation.

Keywords: Metric access method · Ball partitioning indexing · M-tree · PM-tree · k-nearest neighbor query

1 Introduction

Since the development of B+trees focused on secondary storage, even though some keys are used as routing information in the inner nodes, all keys are stored in the leaf nodes of tree-based methods. The motivation for this tree organization is that the space in an inner node is so valuable that it is better to use it to partition the data than to store the location of the data represented by that key [4]. Moreover, the leaves are connected and form a sequential set, which is of great interest when searching for a range of keys based on the total ordering relation.

M-tree stores all data elements in the leaf nodes, although a few are also stored in the inner nodes for routing purposes. The leaves are not interconnected. For indexes built for similarity search, rather than having numeric or small text keys, metric data elements may occupy up to a few kilobytes. Although the purpose of an inner node is to allow data partitioning, storing an 8 bytes numeric

This work has been supported by CNPq (Brazilian National Council for Scientific and Technological Development), by CAPES (Brazilian Coordination for Improvement of Higher Level Personnel), by FAPEMIG (Minas Gerais State Research Foundation) and by PROPP/UFU.

G. Amato et al. (Eds.): SISAP 2019, LNCS 11807, pp. 18–31, 2019.
https://doi.org/10.1007/978-3-030-32047-8_2

identifier with each entry may result in a minimal disturbance in the indexing structure and allows retrieving the full tuples after retrieving the metric instances in a range or k-nearest neighbor query.

In this paper, we propose not to duplicate elements promoted during node splits. Instead, the pair of elements promoted to the upper level during a split is removed from their respective nodes. This algorithm is easily defined for leaf nodes by removing the elements selected for promotion. When splitting an inner node, it is not possible to remove a local pivot that needs to be promoted, as it represents a branch. Instead, we have the opportunity to select a better pivot to be promoted. We propose the use of an aggregate nearest query to solve this issue, aiming to find an element that better minimizes the covering radius considering the set of ball entries (each composed of an element and a radius) that form an inner node.

The contributions of the work described in this paper can be summarized as:

- a new indexing algorithm for M-tree and PM-tree that allows building more efficient indexes for k-nearest neighbor querying operations;
- a refined aggregate nearest neighbor algorithm that allows finding better elements to be promoted during inner node splits;
- an extensive experimentation and discussion that evaluates diverse aspects of the use of the new indexing algorithms in the state-of-the-art methods.

The remaining of the paper is organized as follows. In Sect. 2 we describe the fundamental concepts. Section 3 details the new construction algorithms. Section 4 discusses the experimental results. Section 5 presents the final considerations.

2 Fundamental Concepts

A metric space is a pair $<\mathbb{S}, \delta()>$, where $\mathbb{S}$ is a data domain, and $\delta()$ is a distance function that satisfies the following axioms for any element of x, y, z in $\mathbb{S}$: $\delta(x,x) = 0$ (*identity*); $\delta(x,y) = \delta(y,x)$ (*symmetry*); $0 \leq \delta(x,y) < \infty$ (*non-negativity*); and $\delta(x,y) \leq \delta(x,z) + \delta(z,y)$ (*triangle inequality*). The triangle inequality is used to determine if a ball defined by a data element and a radius covers another element or intersects another ball. It is employed to avoid reading data and computing distances from data elements that certainly the user is not looking for.

Existing metric access methods are divided into compact partitioning techniques and pivot-based techniques [2]. M-tree [3] is the landmark of the compact partitioning techniques. It is a ball-partitioning method that results in a hierarchy composed of inner and leaf nodes, built in a bottom-up fashion, such as the B+tree. An inner node contains a set of entries of the form $<pivot, radius>$, where *pivot* is a data element and *radius* is the branch covering radius. Each entry defines a ball that covers all the data elements in the tree branch it represents. The leaf nodes contain all data elements.

The tree is created with an empty leaf node initially defined as the root. In the case of leaf overflow, a split algorithm is used to create a new node and to distribute the elements between them. Each node promotes one element to the upper level, that stores it and the coverage radius. The upper levels may be updated recursively, if necessary. This process guarantees the structure is always balanced. After the first split, the insertion process starts finding out a path from the root to a leaf. Space is not exclusively partitioned as node coverage may intersect other nodes.

Insertion on leaf nodes may result in an overflow. In case of overflow, a split algorithm is employed to distribute the elements between the node and a new node. In the same way, promotion may result in an inner node overflow and thus, an inner node split. The work [3] proposes the use of m_RAD, an algorithm that finds a pair of pivots that splits a node by minimizing the sum of the covering radii. Its time complexity is $O(n^2)$, where n is the number of elements of a node. Another interesting split strategy [20] computes a minimum spanning tree and removes its longest edge to split a node with time complexity of $O(n \cdot log\ n)$.

Several related works have been proposed to enhance the M-tree performance. Among them, there are works that explored the reorganization of the trees [20], the reinsertion of elements [8] and the use of short-term memories during the construction [13]. There are also other works that aimed to explore the metric properties to propose interesting new data structures, such as the Dynamic Spatial Approximation Trees [10], iDistance [6], GroupSim [14], Omni-R [19], M-Index [11], and PM-tree [7,17]. All of them presented great performance considering different scenarios. A comprehensive review of pivot-based methods can be found at [2] and an extensive review of the area can be found at [15].

Pivot-based techniques consider a static dataset to find a constant set of pivots. A naive pivot selection algorithm is to randomly select n elements as pivots. Finding the optimal pivot set takes polynomial time, thus unpractical. Several heuristics with linear time complexity were proposed recently, such as Maximum of Minimum Distances (MMD) and Maximum of Sum of Distances (MSD) [18]. These heuristics start by randomly selecting the first pivot and then they select pivots incrementally, maximizing $p_i = argmax_{s \in T-\{p_1,...,p_{i-1}\}} \min_{j=1}^{i-1} \delta(s,p_j)$ (MMD) or $p_i = argmax_{s \in T-\{p_1,...,p_{i-1}\}} \sum_{j=1}^{i-1} \delta(s,p_j)$ (MSD) regarding the previously selected pivots.

In addition to creating hierarchies based on local minimum bound rectangles (spatial) or pivot (metric) representations, Omni-R [19] and PM-tree [17] proposed the use of a set of static global pivots to dynamically store cut-region information applied to R-tree and M-tree, respectively. The cut-region concepts and algorithms were later formalized in [7]. PM-tree is the state-of-the-art of dynamic ball-partitioning metric access methods.

3 New M-tree and PM-tree Algorithms

A metric index is an instance of a metric access method that organizes the elements s_i of a dataset $S \in \mathbb{S}$ regarding the metric $\delta()$. Both M-tree and PM-tree allow dynamic insertion of elements and allow the optimization of similarity range and k-nearest neighbor queries from a query element $s_q \in \mathbb{S}$ based on the limit ℓ that is either the radius τ or the number of neighbors (k). The construction of a PM-tree also considers a set of static global pivots $P \in \mathbb{S}$.

In order to store data elements once in the hierarchy of a M-tree or a PM-tree, we must propose a new insert algorithm. When inserting a new element, the algorithm starts from the root node and searches for a leaf to hold the element, employing a heuristic to choose the suitable branch to follow. If the node has enough space, the element is inserted and the coverage radius may be updated in the upper levels, if necessary. So far, the same behavior of the original M-tree and PM-tree insert algorithms. However, in case of a leaf overflow, the element promoted to the upper level is removed from the leaf node, as it will be stored in the upper level.

When inserting the first elements in a new tree, the leaf node is also the root. In case of overflow, a new leaf is created and data is distributed between them. A new inner node is also created and will receive the pivots promoted from the pair of leaves. From this moment on, for every leaf split, there will be a prior pivot that used to represent that leaf. Instead of maintaining the original pivot, we propose to select a new one, allowing to find an element that better represents that portion of the data that remained in the node, considering that part of the entries was distributed to the new node. For the new node, its pivot is promoted, while for the original node, we must replace the promoted pivot. In this case, as there is no other copy of the original pivot in any leaf node, we propose to reinsert it, allowing it to find a suitable leaf node. In the recursive version of the insert algorithm, this is performed when the function returns from every recursion, as a split may promote an element that may cause another split in the upper level until we get to the root.

The main challenge is to select a pair of pivots when an inner node split is needed. When splitting an inner node, selecting an element to be promoted and remove it from the node is not possible, as each element is a pivot that represents a branch. The algorithm employs the aggregate nearest neighbor query to solve this issue, in order to find an element that minimizes the covering radius considering the set of ball entries (each composed of an element and a radius) that form an inner node. The algorithm searches the branch for this element and removes it from its leaf. The aggregate nearest neighbor query allows finding, for instance, the element in the branch that minimizes the sum of distances to the set of ball pivots, among other aggregation functions. Following, this element is promoted and stored in the upper level. This strategy can be applied to both M-tree and PM-tree insert algorithms.

Figure 1 illustrates the selection of promoted pivots when splitting an inner node. It represents the set of balls stored in the node. In (a) the node entries $\{<s_1, r_1>, <s_2, r_2>, <s_3, r_3>\}$ represent the pivots and the covering radii, each one covering a branch of the tree. In the standard M-tree (b), s_1 is also promoted to the upper level with radius r_p, as among the options $\{s_1, s_2, s_3\}$, the element s_1 results in the minimization of r_p that covers all entries in this node. The new strategy (c) searches downward to find the aggregate first-nearest neighbor p

regarding the query $Q = \{s_1, s_2, s_3\}$ by minimizing an aggregation of distances, such as the sum or the mean square distance.

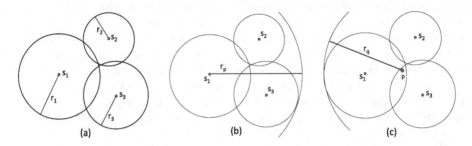

Fig. 1. M-tree inner node representation. (a) Node entries $\{<s_1, r_1>, <s_2, r_2>, <s_3, r_3>\}$. (b) in the standard M-tree, s_1 is also promoted to the upper level with radius r_p: among the options $\{s_1, s_2, s_3\}$, promoting s_1 results in the minimization of r_p. (c) New strategy: search downward to find the aggregate first-nearest neighbor p regarding the query $Q = \{s_1, s_2, s_3\}$, remove it from its leaf and promote with radius r_q.

Notice the promoted radius r_p from (b) is greater than the promoted radius r_q from (c), thus (c) results in less dead space. The optimization of the aggregate nearest neighbor query employs the triangle inequality to prune branches that certainly do not overlap with the search space, thus it is straightforward to implement a general best-first aggregate k-nearest neighbor query algorithm, as presented in Sect. 3.1.

3.1 The Aggregate k-Nearest Neighbor Query

An aggregate similarity query [12] is a relational selection operation that retrieves the most similar elements of a dataset $S \in \mathbb{S}$ to a query composed of the constant values Q (called the query set) taken from the domain $\mathbb{S}$. The ranking requires the definition of a similarity aggregation function $d_g()$, which evaluates the aggregate similarity of each element $s_i \in S$ regarding its similarity measured by the metric $\delta()$ to every element $s_q \in Q$. Limits can be expressed as a similarity threshold ξ (aggregated radius) or based on a number k of elements. In this paper, we present a refined version of the general algorithm.

In this generalization, the well-known similarity range and k-nearest neighbor queries turn into special cases of the aggregate queries, where the set of query centers has only one element $Q = \{s_q\}$ and the limit ℓ is either the range or the number of neighbors. As the set of query centers Q may have more than one element, the distances $\delta(s_i, s_q)$ from each query center $s_q \in Q$ to the element $s_i \in S$ must be aggregated. Consider $\delta()$ is a distance function, Q is the set of query centers, s_i is a dataset element, and the power $g \in \mathbb{R}^*$ is a non-zero real value, the similarity aggregation function $d_g()$ is defined in Eq. 1. The aggregation provides interesting functions, such as: $g = 1$ defines the minimization of the sum of the distances; $g = 2$ defines the minimization of the mean square distance; $g = \infty$ defines the minimization of the maximum distance; and $g = -\infty$ defines the minimization of the minimum distance.

$$d_g(Q, s_i) = \sqrt[g]{\sum_{s_q \in Q} \delta(s_q, s_i)^g} \tag{1}$$

The time complexity of a sequential scan to solve the aggregate range and aggregate k-NN is $O(n * |Q|)$ distance calculations, where n is the number of elements in the dataset and $|Q|$ is the cardinality of Q. As a generalization of range queries, the triangle inequality property can be employed to discard branches of ball-partitioning based metric access methods, and so the composite triangle inequalities related to the set Q. Consider Fig. 2 as an example of an aggregate range query in a 2-dimensional Euclidean space, the query centered at $\{q_1, q_2\}$ and a branch centered at s_t with covering radius r_t. Also, let h_1 be an unknown element that minimizes $d_g()$ with respect to q_1 and q_2. The challenge is to compute the lower bound aggregate similarity from centers q_1 and q_2 to h_1 to decide if the aggregate range overlaps the region covered by the ball centered at s_t.

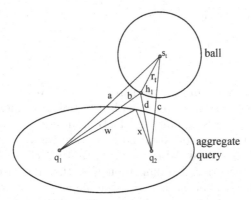

Fig. 2. Ball region and aggregate range, $Q = \{q_1, q_2\}$ (query), s_t is a branch representative and radius r_t. Adapted from [12].

From Fig. 2, r_t is known and $\{a, c\}$ can be computed, but $\{b, d\}$ cannot. To assure that a branch centered at s_t with covering radius r_t can be pruned, we need to determine if $d_g(Q, h_1) = \sqrt[g]{b^g + d^g}$ is less than or equal to the limiting aggregated radius $\xi = \sqrt[g]{w^g + x^g}$ that generates the query region, i.e., if the two regions overlap. If they do not overlap, the branch centered at s_t can be pruned. From the definition of distance functions on a metric space, the following triangle inequality property always holds. For $g = 1$, $d_g = b + d$:

$$b \geq |a - r_t| \tag{2}$$

$$d \geq |c - r_t| \tag{3}$$

$$b + d \geq |a - r_t| + |c - r_t| \tag{4}$$

Generalizing, it can be stated that, for $g \neq 0$:

$$b^g \geq |a - r_t|^g \tag{5}$$

$$d^g \geq |c - r_t|^g \tag{6}$$

$$b^g + d^g \geq |a - r_t|^g + |c - r_t|^g \tag{7}$$

From Eq. 7 we built Eq. 8, where Q is the set of query centers, ξ is aggregate query radius, s_t is a branch representative, and r_t is a branch covering radius. It allows verifying if an aggregated range overlaps a ball with no false dismissals, providing exact answers. Its use is straightforward to replace the single center triangle inequality comparisons to discard branches during a depth-first traversal in a range query or in a best-first approach employed for nearest-neighbor algorithms.

$$\sqrt[g]{\sum_{s_q \in Q} | \, \delta(s_q, s_t) - r_t \, |^g} \leq \xi \tag{8}$$

4   Experiments

The access methods were implemented in C++ [1]. We run the experiments on a Linux 64 bits personal computer with 8 GB of main memory, Intel Core $i7-4770@3.40\,GHz$ processing unit, and 1 TB hard disk.

Table 1 presents the datasets. Nasa and Colors are available at [5] while Cloud is a synthetic 20-dimensional dataset built with 1000 clusters of randomized points. We present the embedded dimension E as the number of dimensions of their spaces. The existing correlations between dimensions decrease the intrinsic dimensionality, that is, the intrinsic characteristics of the data, regardless of the space where it is embedded. The approximation of the intrinsic Hausdorff dimensionality D can be computed by the Distance Exponent (DEX) algorithm [21]. We employed DEX to determine the number of global pivots ($\lceil D \rceil + 1$), as stated in [9,21]. DEX is based on the box-counting method, which divides each dimension several times, creating n-dimensional boxes in a hierarchical structure. At each level, it counts the number of boxes that contain data instances. As the size of the boxes becomes very small, $logN(r)/log(1/r)$ converges to a finite value, i.e., the Hausdorff dimension D [16].

Considering it is not possible to compute the volume of the intersections of generic metric spaces, we computed the relative fat-factor [20], a measurement based on counting the elements in the intersections of overlapped leaf nodes. The

Table 1. Datasets.

Dataset	Cardinality	Embedded dimensionality E	Distance exponent D	Number of pivots ($\lceil D \rceil + 1$)
Nasa	40,150	20	1.7	3
Colors	112,682	112	4.8	6
Clouds	1,000,000	20	5.2	7

relative fat-factor of index T takes into consideration the minimum theoretical number of nodes ($Mmin$), the minimum theoretical height ($Hmin$), the number of nodes read to answer a point query Ic, the height H, the number of nodes M, and the total number of elements N: $fat_{rel}(T) = \frac{Ic-Hmin.N}{N} \frac{1}{Mmin-Hmin}$.

All experiments were run with the Euclidean distance. We employed the MaxSum algorithm [18] to select $\lceil D \rceil + 1$ global pivots for the experiments presented in Sects. 4.1 and 4.2. In the experiments of Sect. 4.3 we built indexes varying the number of pivots from 3 up to 10.

In the next sections, we refer to the standard algorithms as M-tree and PM-tree, and the newly proposed algorithms as M#tree and PM#tree. All indexes were built considering the optimistic forced reinsertions strategy with parameters $maxRemoved = 5$ and $recursionDepth = 10$ [8]. PM-tree and PM#tree also employed SingleWayForCutRegions strategy [7].

4.1   Effect of the Size of the Nodes

In this experiment, we evaluate the methods regarding the size of the nodes during index construction and searches. Each node is stored in a disk page and as the size grows there may be more elements per node, potentially resulting in hierarchies with lower heights. On the other hand, node split becomes expensive, due to the time complexity of the split strategies. Figure 3 presents the heights of the indexes. In these experiments, the proposed strategy resulted in indexes with the same height of the standard algorithms, allowing a fair comparison of the query features. Although they presented the same height, the indexes overlap decreased. Figure 4 presents their relative fat-factors.

Figure 5 presents the indexes file sizes. Although PM-tree and PM#tree indexes store more information on inner nodes (the cut-regions) than M-tree and M#tree, they resulted in more compact indexes. As it was expected, M#tree and PM#tree spent less space than the standard algorithms.

Figure 6 presents the time spent to build. Although the new indexes are more compact, the new algorithms did not reduce the time spent to build them.

Figures 7 and 8 present the average number of distance calculations and the time to run 100 k-nearest neighbor queries with k = 10. We randomly selected 100 elements of each dataset as the query elements. In Fig. 7, the average number of distances computed by PM#tree reduced up to 31.6% for Nasa, up to 40.6% for Colors, and up to 60.7% for Cloud datasets, when compared to PM-tree.

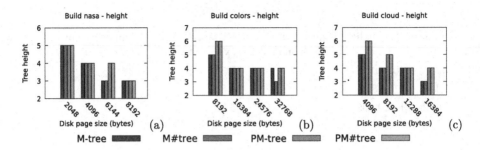

Fig. 3. Heights regarding different disk page sizes. (a) Nasa. (b) Colors. (c) Cloud.

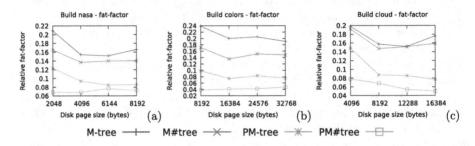

Fig. 4. Fat-factors regarding different disk page sizes. (a) Nasa. (b) Colors. (c) Cloud.

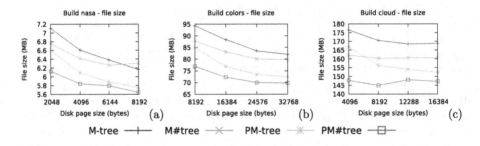

Fig. 5. File sizes regarding different disk page sizes. (a) Nasa. (b) Colors. (c) Cloud.

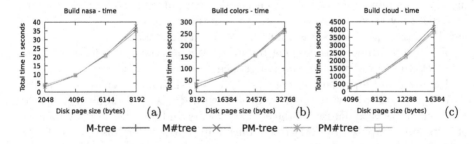

Fig. 6. Time to build the indexes. (a) Nasa. (b) Colors. (c) Cloud.

In Fig. 8, the total time to run 100 queries in PM#tree reduced up to 30.3% for Nasa, up to 40.8% for Colors, and up to 78.6% for Cloud datasets, when compared to PM-tree.

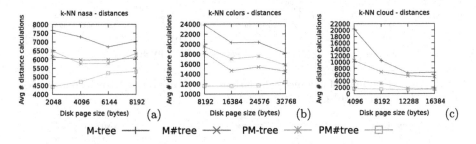

Fig. 7. Average number of distance calculations to run 100 k-nearest neighbor queries with k = 10 regarding different disk page sizes. (a) Nasa. (b) Colors. (c) Cloud.

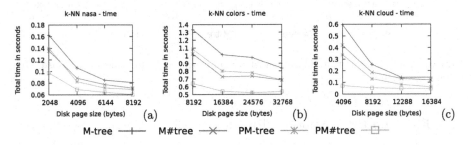

Fig. 8. Query total time to run 100 k-nearest neighbor queries with k=10 regarding different disk page sizes. (a) Nasa. (b) Colors. (c) Cloud.

4.2   Effect of k

In this set of experiments, we evaluate the k-nearest neighbor queries with respect to k. The indexes were created with page sizes of 4 KB for Nasa, 16 KB for Colors, and 8 KB for Cloud, i.e., the second page size variation of the experiments described in Sect. 4.1 (see the abscissa of Figs. 3, 4, 5, 6, 7 and 8). In Fig. 9, the average number of distance calculations computed by PM#tree when compared to PM-tree decreased 18.0%, 32.1% and 55.4% for $k = 10$, as well as 13.6%, 27.1%, and 84.8% for $k = 100$, for Nasa, Colors and Cloud, respectively. Accordingly, in Fig. 10, the time spent to run these queries by PM#tree when compared to PM-tree decreased 16.6%, 31.4%, and 59.6% for $k = 10$, as well as 12.5%, 26.9%, and 73.9% for $k = 100$, for Nasa, Colors and Cloud, respectively.

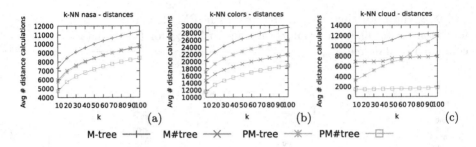

Fig. 9. Average number of distance calculations to run 100 k-nearest neighbor queries regarding k. (a) Nasa. (b) Colors. (c) Cloud.

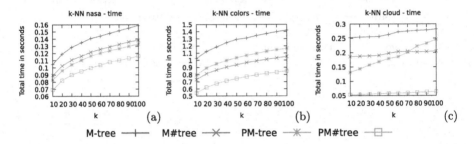

Fig. 10. Total time to run 100 k-nearest neighbor queries regarding k. (a) Nasa. (b) Colors. (c) Cloud.

4.3   Effect on the Number of Pivots

In this set of experiments, we evaluate the behavior of PM-tree and PM#tree while increasing the number of pivots. We also compare them with the embedding methods Omni-B [19] and iDistance [6]. Omni-B and iDistance employ $B+trees$ to optimize similarity queries based only on a static set of global pivots. In these methods, the data elements offsets (element position or address in a random access file) are indexed by the distances of the elements to the set of global pivots. The triangle inequality applied to the intersection of multiple embeddings allows the retrieval of the offsets, such as in the computation of the cut-regions. While Omni-B index each embedding in a $B+tree$, iDistance partitions space and employs a single $B+tree$ for all partitions. The selected offsets that were not discarded by triangle inequality are employed to retrieve the real data elements from a random access file.

We compute a distance as double precision floating point, (8 bytes). The space needed to store a cut-region on index nodes is $2 * |P| * 8$ *bytes* for each routing entry, where $|P|$ is the cardinality of the set of global pivots. Thus, a PM-tree built with 10 global pivots stores an extra 160 bytes of data for each entry of an inner node. As the number of global pivots that form a cut-region increases, the pruning ability increases but the routing capacity (the number of entries) of an inner node decreases. We aim at finding a trade-off between $|P|$ and construction/query performance.

In this experiment, we built PM-tree and PM#tree using 8 KB pages for Nasa and Cloud datasets. For Colors dataset, we employed 16 KB pages. Figures 11 and 12 presents the total time for construction and for querying the indexes. Notice the ordinate axes are presented in log scale. If we consider construction, we can notice in Fig. 11 that indexing $<distance, offset>$ in $B+trees$ and appending the real data element to a file is faster than maintaining a hierarchical ball-partitioning index. On the other hand, Fig. 12 shows that the higher build cost of PM-tree and PM#tree compared to Omni-B and iDistance allows faster retrieval of k-nearest neighbor queries.

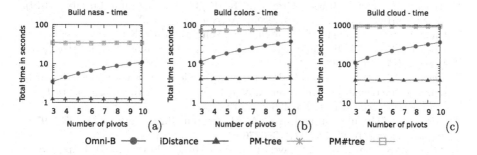

Fig. 11. Time (seconds) to build the indexes regarding different number of global pivots. (a) Nasa. (b) Colors. (c) Cloud.

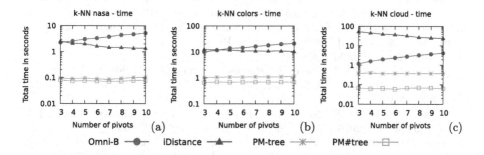

Fig. 12. Time (seconds) to run 100 k-nearest neighbor queries with $k = 10$ regarding different number of global pivots. (a) Nasa. (b) Colors. (c) Cloud.

5   Conclusion

The design of efficient dynamic metric access methods is fundamental for many search and analysis processes based on similarity comparison operations. We propose a new construction strategy for M-tree and PM-tree that does not duplicate elements during the split of nodes. To achieve this goal, we employed an aggregate k-nearest neighbor query to select the elements to be promoted during an inner node split. We also present an optimized algorithm to solve this query, based on the aggregation of the triangle inequality relations.

In our experiments, we compared the standard M-tree and PM-tree against these indexing methods built with our new strategy. We empirically show that our strategy allows building compact indexes that increase the performance of k-nearest neighbors. This is achieved due to the faster convergence of the query algorithms.

References

1. Arboretum: The Database Group at ICMC/USP Arboretum Library (2019). https://bitbucket.org/gbdi/arboretum. Accessed May 2019
2. Chen, L., Gao, Y., Zheng, B., Jensen, C.S., Yang, H., Yang, K.: Pivot-based metric indexing. Proc. VLDB Endow. (PVLDB) **10**(10), 1058–1069 (2017). https://doi.org/10.14778/3115404.3115411
3. Ciaccia, P., Patella, M., Zezula, P.: M-tree: an efficient access method for similarity search in metric spaces. In: International Conference on Very Large Data Bases (VLDB), Greece, Athens, pp. 426–435 (1997)
4. Comer, D.: The ubiquitous b-tree. ACM Comput. Surv. **11**(2), 121–137 (1979). https://doi.org/10.1145/356770.356776
5. Figueroa, K., Navarro, G., Chaves, E.: Metric spaces library (2007). http://www.sisap.org/metricspaceslibrary.html
6. Jagadish, H.V., Ooi, B.C., Tan, K.L., Yu, C., Zhang, R.: iDistance: an adaptive B+-tree based indexing method for nearest neighbor search. ACM Trans. Database Syst. (TODS) **30**(2), 364–397 (2005). https://doi.org/10.1145/1071610.1071612
7. Lokoc, J., Mosko, J., Cech, P., Skopal, T.: On indexing metric spaces using cut-regions. Inf. Syst. **43**, 1–19 (2014). https://doi.org/10.1016/j.is.2014.01.007
8. Lokoc, J., Skopal, T.: On reinsertions in m-tree. In: International Workshop on Similarity Search and Applications (SISAP), pp. 121–128. IEEE (2008). https://doi.org/10.1109/SISAP.2008.10
9. Navarro, G., Paredes, R., Reyes, N., Bustos, C.: An empirical evaluation of intrinsic dimension estimators. Inf. Syst. **64**, 206–218 (2017). https://doi.org/10.1016/j.is.2016.06.004
10. Navarro, G., Reyes, N.: New dynamic metric indices for secondary memory. Inf. Syst. **59**, 48–78 (2016). https://doi.org/10.1016/j.is.2016.03.009
11. Novak, D., Batko, M., Zezula, P.: Metric index: an efficient and scalable solution for precise and approximate similarity search. Inf. Syst. **36**(4), 721–733 (2011). https://doi.org/10.1016/j.is.2010.10.002
12. Razente, H.L., Barioni, M.C.N., Traina, A.J.M., Faloutsos, C., Traina-Jr, C.: A novel optimization approach to efficiently process aggregate similarity queries in metric access methods. In: International Conference on Information and Knowledge Management (CIKM), Napa Valley, California, pp. 193–202. ACM (2008). https://doi.org/10.1145/1458082.1458110
13. Razente, H., Sousa, R.M.S., Barioni, M.C.N.: Metric indexing assisted by short-term memories. In: Marchand-Maillet, S., Silva, Y.N., Chávez, E. (eds.) SISAP 2018. LNCS, vol. 11223, pp. 107–121. Springer, Cham (2018). https://doi.org/10.1007/978-3-030-02224-2_9

14. Razente, H.L., Lima, R.L.B., Barioni, M.C.N.: Similarity search through one-dimensional embeddings. In: ACM Symposium on Applied Computing (SAC), Marrakech, Morocco, pp. 874–879. ACM (2017). https://doi.org/10.1145/3019612.3019674
15. Samet, H.: Foundations of Multidimensional and Metric Data Structures. Morgan Kaufmann, San Francisco (2006)
16. Schroeder, M.: Fractals, Chaos, Power Laws: Minutes from an Infinite Paradise. W. H. Freeman and Company, New York (1990)
17. Skopal, T., Pokorný, J., Snášel, V.: Nearest neighbours search using the PM-tree. In: Zhou, L., Ooi, B.C., Meng, X. (eds.) DASFAA 2005. LNCS, vol. 3453, pp. 803–815. Springer, Heidelberg (2005). https://doi.org/10.1007/11408079_73
18. Socorro, R., Mico, L., Oncina, J.: A fast pivot-based indexing algorithm for metric spaces. Pattern Recogn. Lett. **32**(11), 1511–1516 (2011). https://doi.org/10.1016/j.patrec.2011.04.016
19. Traina-Jr, C., Filho, R.F.S., Traina, A.J.M., Vieira, M.R., Faloutsos, C.: The omni-family of all-purpose access methods: a simple and effective way to make similarity search more efficient. VLDB J. **16**(4), 483–505 (2007). https://doi.org/10.1007/s00778-005-0178-0
20. Traina-Jr, C., Traina, A., Faloutsos, C., Seeger, B.: Fast indexing and visualization of metric data sets using slim-trees. IEEE Trans. Knowl. Data Eng. (TKDE) **14**(2), 244–260 (2002). https://doi.org/10.1109/69.991715
21. Traina-Jr, C., Traina, A., Wu, L., Faloutsos, C.: Fast feature selection using fractal dimension. J. Inf. Data Manag. (JIDM) **1**(1), 3–16 (2010)

Index Maintenance Strategy and Cost Model for Extended Cluster Pruning

Anders Munck Højsgaard, Björn Þór Jónsson[✉], and Philippe Bonnet

IT University of Copenhagen, Copenhagen, Denmark
{amuh,bjth,phbo}@itu.dk

Abstract. With today's dynamic multimedia collections, maintenance of high-dimensional indexes is an important, yet understudied topic. Extended Cluster Pruning (eCP) is a highly-scalable approximate indexing approach based on clustering, that is targeted at stable performance in a disk-based scenario. In this work, we propose an index maintenance strategy for the eCP index, which utilizes the tree structure of the index and its approximate nature. We then develop a cost model for the strategy and evaluate its cost using a simulation model.

Keywords: High-dimensional indexing · Index maintenance · eCP

1 Introduction

In recent years, the scale and availability of multimedia collections has grown rapidly, spurring interest in scalable high-dimensional indexing methods. In some cases, including copyright protection and multimedia analytics applications, these media collections can be quite dynamic, with the most recently added material of particular interest. It is thus of interest to propose and study methods for dynamic index maintenance [7].

Overall, high-dimensional indexing structures fall into one of three main categories: tree-based, quantization-based, and hashing-based [7]. As many tree-based indexes inherit properties of their lower-dimensional counterparts, index maintenance has been discussed in some early works based on the seminal R-trees and kd-trees, as well as some more recent works [3,11]. In particular, dynamic index maintenance with transactional properties has been proposed for the NV-tree, the most scalable tree-based indexing structure [6,8]. Dynamic maintenance of index structures in the other categories, however, remains understudied.

Quantization-based methods have shown significant promise for scalability [1, 5,10]; exploring strategies for dynamic maintenance is therefore of interest. In this paper, we consider dynamic maintenance of the Extended Cluster Pruning (eCP) indexing strategy [2,4]. Unlike many other quantization-based methods, eCP focuses on disk-based scalability scenarios by targeting a balanced cluster size and using an approximate hierarchical index to facilitate access to clusters.

© Springer Nature Switzerland AG 2019
G. Amato et al. (Eds.): SISAP 2019, LNCS 11807, pp. 32–39, 2019.
https://doi.org/10.1007/978-3-030-32047-8_3

In this paper, we propose a new strategy for maintaining eCP clusters that relies on the approximate nature of the hierarchical index. The strategy maintains the balanced distribution of cluster sizes while also allowing gradual reorganization of the high-dimensional space. We describe this new strategy and show, using a cost model-based simulator, that the strategy is efficient.

The remainder of the paper is organized as follows. In Sect. 2 we review the eCP high-dimensional indexing strategy. In Sect. 3 we outline the index maintenance strategy and its cost model. We briefly outline and analyze the performance of the strategy in Sect. 4, and give concluding remarks in Sect. 5.

2 Extended Cluster Pruning

Extended Cluster Pruning (eCP) is an approximate clustering-based high-dimensional search index. This index takes a dataset D of n vectors, where each vector consists of S_v bytes (including an identifier), and forms a set of clusters by randomly selecting a set of l cluster leaders. All vectors in D are then assigned to the closest leader; this process is essentially a single round of the k-means algorithm. When the collection is queried with a query vector q, the query vector is compared to the l cluster leaders to find the nearest leader l'. Then q is compared to all the feature vectors in the cluster of l' to find the (approximate) k nearest neighbors. Additionally, eCP has a parameter b used to expand the cluster search process, such that a search of the index returns the b clusters nearest to q, from which the k nearest neighbors then are found.

The motivation of eCP is to perform well in disk-based scenarios, where data only partially fits in memory and must thus often be read from disk. The overall goal of eCP is that each cluster read should typically result only in a single disk read, and three main techniques are used to achieve this goal.

First, the eCP targets a well balanced distribution of data across clusters by only performing a single round of assignments to cluster leaders. Experimental results with real datasets have shown that using a full k-means algorithm results in a highly skewed cluster size distribution, which in turn results in sub-optimal performance [4,9].

Secondly, eCP tries to have most clusters fit within the size of a single disk block read S_{io}, which has a default value of 128 KB for Unix systems.[1] In order to find this "optimal cluster size" in which we can fit the most possible feature vectors within the size of one S_{io}, one must calculate the following:

$$S_c = \lfloor S_{io}/S_v \rfloor \tag{1}$$

This will then give the maximum number of feature vectors that can fit within a singular S_{io}. Then, to find the necessary amount of cluster leaders l, which would fit all n vectors, one must determine this by utilizing the S_c value. This gives the equation:

$$l = \left\lceil \frac{n}{S_c} \right\rceil \tag{2}$$

---

[1] E.g., see: https://git.savannah.gnu.org/cgit/coreutils.git/tree/src/ioblksize.h.

The third method to ensure efficiency and accuracy is the creation of an index tree of cluster leaders in the eCP. When the l cluster leaders have been determined, the actual structure of the index tree can be chosen and created. The eCP has a height parameter L, which is set at indexing time, that determines how the index should be structured internally.

Each level in the tree is based on a similar clustering approach as with the feature vectors. Each leader in the index structure represents on average

$$S_n = \sqrt[L]{l} \qquad (3)$$

leaders below itself, which are either actual cluster leaders or leaders of internal nodes. The process works by first selecting l/S_n group leaders from the collection of cluster leaders, and assigning each cluster leader to the nearest group leader, thus grouping the l cluster leaders into groups of (on average) S_n cluster leaders. These internal nodes are then also grouped into groups of S_n leaders; this process continues recursively until the top internal group has fewer than S_n leaders and becomes the root of the index tree.

The L parameter indirectly controls how large the internal nodes should be in the structure. A too-large L value would lead to many small internal nodes in the structure, which would incur a large overhead cost and a loss of result quality [2], while a too-small L would create large internal nodes leading to worse performance, as this would cause more comparisons performed within each internal node during query processing.

3   Index Maintenance Strategy

We now consider the scenario where additional vectors are inserted into the eCP index after the initial index construction has completed. The goal of any index maintenance strategy should be that query processing is affected as little as possible both in terms of efficiency and accuracy. Consider a strategy where these vectors are simply inserted into the appropriate clusters without updating the local structure in the index. The clusters would eventually become overly large, leading to an increase in query processing time and loss of accuracy. Therefore it is paramount that the clustering is dynamically maintained along with the index of cluster leaders in order to keep the cluster sizes and the distribution of data as stable as possible.

A second goal of an index maintenance strategy is that the index maintenance itself should be efficient. This goal in turn implies two things: (a) that the insertion operations should be aggregated in order to avoid the costs associated with repeatedly updating the same clusters and (b) that the structure of the index itself should be maintained, rather than rebuilt. Therefore it is necessary to perform local maintenance of the index in order to avoid having to make global changes to the clustering and the index structure.

The main topic of this section is a proposal for such a maintenance strategy. In Sect. 3.1 the procedure of aggregating insertions is detailed and in Sect. 3.2 the maintenance strategy for the cluster index is described. A cost model for this maintenance strategy is then developed in Sect. 3.3 and then the potential impacts on the effectiveness of the eCP is briefly outlined in Sect. 3.4.

3.1   Insertions

When a cluster is chosen for insertions in eCP, there are two different scenarios that can happen in terms of IO operations. The first scenario is that the insertion is performed directly to disk, which means that the insertion operation will require a disk read and a disk write operation in order to update the chosen cluster with the inserted data. Since this cluster may be anywhere on the disk, this will incur a random read operation and a random write operation.

Alternatively, insertions could be buffered, where a given number of inserted feature vectors is buffered in RAM. When the buffer is then filled, some (or all) of the buffered data must be written to disk. In this case, multiple insertions to the same cluster would only require one disk read and one disk write, thus saving some IOs compared to direct insertions.

3.2   Index Maintenance

In order to maintain the index, the index structure must grow. This implies firstly that L cannot be a static parameter. Instead, the average size of the internal nodes S_n is given as a parameter and used to determine a suitable L, by reversing Eq. 3:

$$L = \lceil \log_{S_n}(l) \rceil \tag{4}$$

As the size n of the data collection grows, then so does l and therefore L will also eventually grow, resulting in a deeper index. Based on previous results with eCP, a suitable value for S_n is about 100 [4].

A dynamically growing index implies secondly that we should over-allocate the number of cluster leaders, to leave free space in the clusters for insertions. We propose using a new percentage parametrer lo, such that clusters (and internal nodes) are only filled to lo capacity; based on industry experience with index maintenance, a decent lo value is about 70% of full capacity. Likewise, to avoid aggressively reorganizing clusters, we propose using a corresponding hi parameter, which is used to determine when to re-cluster a part of the index tree. Together, the two parameters help to avoid frequent local re-clustering.

The index maintenance strategy we propose is indeed based on local reorganization of a sub-tree in the index. Consider a set of cluster leaders, which are grouped together under the same internal node. If an insertion into any of these clusters causes the average cluster size in the group to grow beyond $hi \times S_c$, then all of the clusters in the internal node are re-clustered together.

This re-clustering process works by determining the number of feature vectors represented in that internal node, by summing the number of feature vectors in each cluster in the group; this sum is called n^*. Then the number of clusters l^*, which should be represented in the internal node, is determined using a variant of Eq. 2, modified to use n^*:

$$l^* = \left\lceil \frac{n^*}{\lfloor S_c \rfloor} \right\rceil \tag{5}$$

The l^* cluster leaders are selected randomly and the feature vectors are assigned to them, as in the original clustering process. These new l^* cluster leader vectors now form the updated internal node.

It is possible that the internal node grows so much that the parent node now has more than $hi \times S_n$ nodes on average. Then the parent node can also be reorganized in the exact same manner, by determining how many internal nodes should be used and grouping the representatives into these new internal nodes. This process can propagate all the way to the top of the index structure; if the root node grows to have more than $hi \times S_n$ children, then it must be split using the same reorganization process with a new root node created at the top of the tree; when this happens, the height of the tree L grows.

There are some important properties of this process worth noting. First, every re-clustering is local, as only one internal node and its children are reorganized at a time. The index maintenance therefore only flows upwards in the index tree and never downwards. As a result, the only clusters of feature vectors, that are affected, are the ones that originally caused the re-clustering that then propagated up the tree. Second, the reorganization of internal nodes can cause clusters to move within the tree to a closer parent node, which then will make the cluster more likely to be correctly found during the search process. This strategy should therefore maintain overall result quality over time.

3.3   Cost Model

We now describe a simple cost model for this strategy, that we will use to evaluate the efficiency of the strategy. As previously discussed, we assume each disk operation reads or writes $S_{io} = 128$ KB. The IO operations may be either sequential or random in nature, and we denote the costs of the operations C_{SR} and C_{SW} for sequential reads and writes, respectively, and C_{RR} and C_{RW} for random reads and writes.

Direct insertions of a feature vector into a cluster requires reading of that cluster from disk, adding the new feature vector, and then writing the cluster back to disk. Assuming that the cluster contains n feature vectors, the cost of an insertion is:

$$C_I = \lceil n/S_c \rceil * C_{RR} + \lceil (n+1)/S_c \rceil * C_{RW} \tag{6}$$

When a feature vector is inserted into the aggregation buffer, no cost is assigned. Once the buffer is full, in our cost model a full flush of the insertion buffer is forced. We model two different types of flushes, where (a) individual clusters are read as needed and updated, with the same cost as above, or (b) where all the clusters are read and written sequentially, resulting in a cost of $l \times (C_{SR} + C_{SW})$.

If a cluster forces a re-clustering, the leader will take all of its l_n clusters and their n^* descriptors and re-cluster these into l_n^* new clusters. The cost associated

Table 1. IO latencies for the two devices modelled in our experiment.

Device	C_{RR}	C_{RW}	C_{SR}	C_{SW}
Toshiba 7200RPM HDD	4.930 ms	2.110 ms	0.642 ms	0.645 ms
Intel P3700 PCI-E SSD	0.055 ms	0.122 ms	0.056 ms	0.121 ms

with such a re-clustering is modelled as:

$$C_R = l_n \times C_{RR} + l_n^* \times C_{RW} \tag{7}$$

In the case where a internal node forces a re-clustering of its parent, the parent node will then take its l_n representatives and reorganize them into l_n^* new leader groups. Assuming that each node fits within one disk IO, the cost of such reorganization is also given by Eq. 7. The cost of reorganizing the root is similar.

3.4 Discussion

We have proposed a new insertion strategy for the eCP high-dimensional index. As the analysis of the next section shows, the strategy achieves our two efficiency goals: the strategy itself can be implemented efficiently and it leads to a balanced cluster size distribution, which is key to the retrieval efficiency. We have also argued that because both feature vectors and clusters can dynamically move within the index tree, the result quality is likely to be maintained. Testing this latter hypothesis is part of our future work.

4 Experiments

In order to explore the performance of the index maintenance strategy, we have implemented a simulation model, which is a modified edition of the one used in previous experiments with the NV-tree [6,8].

The simulator starts by instantiating an index in an initial state, with a given number of feature vectors. Then the simulator inserts vectors into the index, using the index maintenance strategy, as long as desired. During this process the simulator uses the cost model above to keep track of the total IO cost.

For the experiment reported below, an IO size of $S_{io} = 128$ KB is used. The initial index contains 50 million feature vectors, with 1.5 billion subsequent insertions performed to measure efficiency. We consider disk IO latencies for the two devices presented in Table 1, which represent a competitive HDD and SSD, respectively [6]. We model SIFT feature vectors ($s_v = 132$ bytes) and set the internal node parameter $S_n = 100$, which has been shown to give good results [4]. The simulator can model both types of insertion buffer flushes, but for eCP a full scan of the index was more efficient.

Table 2. Simulation results: total time to insert 1.5 billion feature vectors.

Index	HDD	SSD
NV-tree	895.3 h	2.8 h
eCP	396.4 h	20.6 h

4.1 NV-tree

As a baseline, we compare to the NV-tree, which was already implemented in the simulator [6]. The NV-tree is a tree-based high-dimensional index, which utilizes a combination of projection of data points along random lines, and partitioning of the projected space, to separate the dataset into partitions which are designed to fit within a single IO. In order to maintain the index while insertions are performed, the NV-tree must select new random lines and re-project the data contained within its partitions, in order to maintain the small partitions.

The NV-tree only stores the feature vector identifier in the leaves of the tree. To re-project feature vectors, they must therefore be retrieved from disk separately. For an SSD, the most efficient way to do this is simply to issue many small reads for the feature vectors, as small random reads are efficient with SSDs. For an HDD, however, the NV-tree must maintain an auxiliary data-structure, called partition files, which contain the feature vectors for each partition. This leads to significant additional cost of maintaining this auxiliary structure.

4.2 Results

Table 2 shows the estimated time for inserting 1.5 billion feature vectors into both the eCP index and the NV-tree index. Overall, the results show that the newly proposed strategy for eCP index maintenance is competitive; with a HDD, eCP outperforms the NV-tree, while the NV-tree performs better on an SSD.

The reason why eCP performs better for the HDD is due to the overhead for the NV-tree, which has to maintain both the index and the partition files separately during re-projection maintenance, which is very costly in terms of IOs. Meanwhile, the eCP is a simpler index which stores the features within the index, which is acceptable on an HDD. With eCP, nearly 1 billion disk operations are issued (20% are random reads, 40% are sequential reads, and 40% are sequential writes) while for the NV-tree, about 5 billion operations are issued (evenly split between sequential reads and writes). The reason why the NV-tree is more efficient than eCP on an SSD, is more complex. Even though the NV-tree still requires more disk operations (about 3.2 billion), almost all these operations are small reads, which are very efficient on an SSD.

We also considered the cluster size distribution for eCP. After inserting 1.5 billion vectors, the smallest cluster contained 688 vectors and the largest 1083 vectors, with an average of 894 vectors. As the intended average cluster size is $S_c = 992$ vectors, these results indicate that the strategy will maintain a good cluster size distribution.

5 Conclusion

In this paper we argued for the importance of dynamic maintenance of high-dimensional index structures. We presented a novel index maintenance strategy for the scalable eCP index structure. This strategy aims at maintaining the balanced cluster sizes, which are crucial for the eCP to maintain its disk performance, while also implementing local reorganizations of clusters to reduce maintenance cost and preserve the accuracy of the index. We have implemented this strategy in a simulation model and compared it to the very efficient NV-tree structure, showing that while index maintenance of the NV-tree is more efficient with SSDs, the new index maintenance strategy for eCP is nevertheless quite competitive, with only ≈ 21 h required to insert 1.5 billion feature vectors.

References

1. Babenko, A., Lempitsky, V.: The inverted multi-index. In: Proceedings of the Conference on Computer Vision and Pattern Recognition (CVPR), Providence, RI, USA, pp. 3069–3076 (2012)
2. Chierichetti, F., Panconesi, A., Raghavan, P., Sozio, M., Tiberi, A., Upfal, E.: Finding near neighbors through cluster pruning. In: Proceedings of the Symposium on Principles of Database Systems (PODS), Beijing, China, pp. 103–112 (2007)
3. Fagin, R., Kumar, R., Sivakumar, D.: Efficient similarity search and classification via rank aggregation. In: Proceedings of the ACM SIGMOD International Conference on Management of Data, San Diego, CA, USA, pp. 301–312 (2003)
4. Guðmundsson, G.Þ., Jónsson, B.Þ., Amsaleg, L.: A large-scale performance study of cluster-based high-dimensional indexing. In: Proceedings of the Workshop on Very-Large-Scale Multimedia Corpus, Mining and Retrieval (co-located with ACM Multimedia), Firenze, Italy (2010)
5. Jegou, H., Douze, M., Schmid, C.: Product quantization for nearest neighbor search. IEEE Trans. Pattern Anal. Mach. Intell. **33**(1), 117–128 (2011)
6. Jónsson, B.Þ., Amsaleg, L., Lejsek, H.: SSD technology enables dynamic maintenance of persistent high-dimensional indexes. In: Proceedings of the ACM International Conference on Multimedia Retrieval (ICMR), New York, NY, USA, pp. 347–350 (2016)
7. Lejsek, H., Ásmundsson, F.H., Jónsson, B.Þ., Amsaleg, L.: Transactional support for visual instance search. In: Marchand-Maillet, S., Silva, Y.N., Chávez, E. (eds.) SISAP 2018. LNCS, vol. 11223, pp. 73–86. Springer, Cham (2018). https://doi.org/10.1007/978-3-030-02224-2_6
8. Ólafsson, A., Jónsson, B.Þ., Amsaleg, L., Lejsek, H.: Dynamic behavior of balanced NV-trees. Multimedia Syst. **17**, 83–100 (2011)
9. Sigurðardóttir, R., Hauksson, H., Jónsson, B.Þ., Amsaleg, L.: Quality vs. time tradeoff for approximate image descriptor search. In: Proceedings of the IEEE EMMA Workshop (co-located with ICDE), Tokyo, Japan (2005)
10. Sivic, J., Zisserman, A.: Video Google: a text retrieval approach to object matching in videos. In: Proceedings of the IEEE International Conference on Computer Vision (ICCV), Nice, France, pp. 1470–1477 (2003)
11. Tao, Y., Yi, K., Sheng, C., Kalnis, P.: Quality and efficiency in high dimensional nearest neighbor search. In: Proceedings of the ACM SIGMOD International Conference on Management of Data, Boston, MA, USA, pp. 563–576 (2009)

SPLX-Perm: A Novel Permutation-Based Representation for Approximate Metric Search

Lucia Vadicamo[1(✉)], Richard Connor[2], Fabrizio Falchi[1], Claudio Gennaro[1], and Fausto Rabitti[1]

[1] Institute of Information Science and Technologies (ISTI), Pisa, Italy
{lucia.vadicamo,fabrizio.falchi,claudio.gennaro,
fausto.rabitti}@isti.cnr.it
[2] Division of Mathematics and Computing Science, University of Stirling, Stirling, Scotland
richard.connor@stir.ac.uk

Abstract. Many approaches for approximate metric search rely on a permutation-based representation of the original data objects. The main advantage of transforming metric objects into permutations is that the latter can be efficiently indexed and searched using data structures such as inverted-files and prefix trees. Typically, the permutation is obtained by ordering the identifiers of a set of pivots according to their distances to the object to be represented. In this paper, we present a novel approach to transform metric objects into permutations. It uses the object-pivot distances in combination with a metric transformation, called n-Simplex projection. The resulting permutation-based representation, named *SPLX-Perm*, is suitable only for the large class of metric space satisfying the n-point property. We tested the proposed approach on two benchmarks for similarity search. Our preliminary results are encouraging and open new perspectives for further investigations on the use of the n-Simplex projection for supporting permutation-based indexing.

Keywords: Approximate metric search ·
Permutation-based indexing · Metric embedding · n-point property ·
n-Simplex projection

1 Introduction

Searching a data set for the most similar objects to a given query is a fundamental task in computer science. Over the years several methods for *exact similarity search* were proposed in the literature. These approaches guarantee to find the true result set. However, they scale poorly with the dimensionality of the data (a phenomenon known as *"curse of dimensionality"*) and mostly they are not convenient to deal with very large data sets. To overcome these

G. Amato et al. (Eds.): SISAP 2019, LNCS 11807, pp. 40–48, 2019.
https://doi.org/10.1007/978-3-030-32047-8_4

issues, the research community has developed a wide spectrum of techniques for *approximate similarity search*, which have higher efficiency though at the price of some imprecision in the results (e.g. some relevant results might be missing or some ranking errors might occur). Among them, we can distinguish between (1) approaches specialised for a particular kind of data (e.g. Euclidean vectors), and (2) techniques applicable to generic metric data objects. assessing the dissimilarity of any two objects. The advantage of the former class of approaches, like the Product Quantization [13] and the Inverted Multi-Index [5], is that they have very high efficiency and effectiveness. However, the engineering effort to design a method specialised for any particular data or application is typically too high. The metric approaches, instead, overcome this issue since they are applicable to generic metric objects without assuming a prior knowledge of the nature of the data. Successful examples of metric approximate indexing and searching techniques are the *Permutation-based Indexing* (PBI) ones, such as [4,7,14].

PBI techniques leverage the idea of transforming each metric object into a permutation of a finite set of integers in such a way that similar objects have similar permutations. The main advantage is that the permutations can be efficiently indexed and searched, e.g., using inverted files. The similarity queries are then performed in the permutation space by selecting objects whose permutations are the most similar to the query permutation. The common approach to generate a permutation-based representation of a data object is based on selecting a finite set of *pivots* (reference objects) and measuring the distances of each pivot to the object to be represented: the permutation is obtained as the list of the pivot identifiers ordered according to their distance to the object.

The main contribution of this paper is describing a novel approach to generate permutations associated with metric data objects. The proposed technique is applicable only to the large class of metric spaces satisfying the so-called *n-point property* [6,8]. This class encompasses many commonly used metric spaces, such as Cartesian spaces of any dimensionality regarded with the Euclidean, Cosine, Jensen-Shannon or Quadratic Form distances, and more generally any Hilbert-embeddable space [6]. Our technique exploits the *n-Simplex projection* [10], which is a metric transformation that allows projecting the data objects into a finite-dimensional Euclidean space. Starting from the idea that this space transformation maps similar objects into similar Euclidean vectors, we propose to process each projected vector to further generate a permutation-based representation. We show that, in most of the tested cases, our permutations are more effective than traditional permutations. Therefore, we believe that our technique may be relevant for many permutation-based indexing and searching techniques, even though we are aware that it may require more work to mature.

2   Background

We are interested in searching a (large) finite subset of a metric space (D, d), where D is a domain of objects and $d : D \times D \to \mathbb{R}^+$ is a metric function [15]. Many methods for approximate metric search rely on transforming the original data objects into a more tractable space, e.g. by exploiting the distances to a set of pivots. In the following, we summarise key concepts of two pivot-based approaches that transform metric objects into permutations and Euclidean vectors, respectively.

Permutation-Based Representation. For a given metric space (D, d) and a set of pivots $\{p_1, \ldots, p_n\} \subset D$, the traditional permutation-based representation Π_o (briefly *permutation*) of an object $o \in \mathcal{D}$ is the sequence of the pivots identifiers $\{1, \ldots, n\}$ ordered by their distance to o. Formally, the permutation $\Pi_o = [\Pi_o(1), \Pi_o(2), ..., \Pi_o(n)]$ lists the pivot identifiers in an order such that $\forall i \in \{1, \ldots, n-1\}$, $d(o, p_{\Pi_o(i)}) \leq d(o, p_{\Pi_o(i+1)})$. An equivalent representation is the *inverted permutation* Π_o^{-1} whose i-th element denotes the position of the pivot p_i in the permutation Π_o.

Most of the PBI methods, e.g. [4,11,14], use only a fixed-length prefix of the permutations to represent and compare objects. It means that only the positions of the nearest l out of n pivots are used for the data encoding. In this work, we do the same since often the prefix-permutations have better or similar effectiveness than the full-length permutations [4], resulting also in a more compact data encoding. The prefix permutations are compared using *top-l distances* [12]. We use the Spearman Rho with location parameter l, defined as $S_{\rho,l}(\Pi_{o_1}, \Pi_{o_2}) = \ell_2(\Pi_{o_1,l}^{-1}, \Pi_{o_2,l}^{-1})$, where $\Pi_{o,l}^{-1}$ is the *inverted prefix permutation*:

$$\Pi_{o,l}^{-1}(i) = \begin{cases} \Pi_o^{-1}(i) & \text{if } \Pi_o^{-1}(i) \leq l \\ l+1 & \text{otherwise} \end{cases}. \tag{1}$$

n-Simplex Projection. Recently, Connor et al. [8–10] investigated how to enhance the metric search on a class of spaces meeting the so-called n-point property, which is a geometrical property stronger than the triangle inequality. A metric space has the n-point property if for any finite set of n objects there exists an *isometric* embedding of those objects into a $(n-1)$-dimensional Euclidean space. They exploited this property to define a space transformation, called *n-Simplex projection*, that allows a metric space to be trasformed into a finite-dimensional Euclidean space. It uses the distances to a set of pivots $\mathcal{P}_n = \{p_1, \ldots, p_n\}$ for mapping metric objects to Euclidean vectors. Formally, the n-Simplex projection associated with the pivot set $\mathcal{P}_n$ is the transformation

$$\phi_{\mathcal{P}_n} : (D, d) \to (\mathbb{R}^n, \ell_2)$$
$$o \mapsto v_o$$

where v_o is the only vector with a positive last component that preserves the distances of the data object to the pivots, i.e. $\ell_2(v_o, v_{p_i}) = d(o, p)$, $\forall i \in \{1, \ldots, n\}$. The algorithm to compute the n-Simplex projected vectors is described in [10].

We recall that one interesting outcome of this space transformation is that the Euclidean distance between any two projected vectors is a lower-bound of the actual distance, and that the lower-bound converges to the actual distance for increasing number of pivots n. Thus, the larger the n the better the preservation of the similarities between the data objects.

Algorithm 1. SPLX-Perm computation

Input : $\mathcal{P}_n = \{p_1, \ldots, p_n\} \subset D, o \in D$
Output: The SPLX-Perm Π_o associated to the the object o
1 $v_o \leftarrow \phi_{\mathcal{P}_n}(o)$; // n-Simplex projection into $\mathbb{R}^n$
2 $v_o \leftarrow R\,v_o$; // Rotate the vector using a random rotation matrix R
3 $[v_{sorted}, v_{index}] = \mathrm{sort}(v_o, \text{ascending})$;   // sorts the vector elements of v_o
   in ascending order; v_{sorted} is the sorted array, v_{index} is the sort
   index vector describing the rearrangement of each element of v_o
4 $\Pi_o \leftarrow v_{index}$

---

3   SPLX-Perm Representation

As recalled above, the traditional approach to associate a permutation to a data object is sorting a set of pivot identifiers in ascending order with respect to the distances of those pivots to the object to be represented. This approach is justified by the observation that objects very close to each other should have similar relative distances to the pivots, and thus, similar permutations.

The main goal of such kind of metric transformation is that the similarity between the permutations reflects as much as possible the similarity of the original data objects. Starting from this concept we observe that, on one hand, the traditional permutation representation takes in consideration only the relative distances to the pivots, i.e. which is the closest pivot, the second closest pivot, etc. On the other hand, the recently proposed n-Simplex projection maps the data objects to Euclidean vectors by taking into consideration both object-pivot and pivot-pivot distances. Moreover, the Euclidean distance between those projected vectors well approximates the actual distance, especially when using a large number of pivots. Therefore, our idea is to start from these good approximations of the data objects and further transform them into permutations. Since we are now working in a Euclidean space, and the Euclidean distance does not mix the contribution of values in different dimensions of the vectors, it is reasonable to think that two vectors are very close to each other if they have similar components in each dimension. By exploiting this idea, we propose to generate the permutations by ordering the dimensional indexes of the n-Simplex projected vectors in ascending order with respect to their corresponding values. For example, the Euclidean vector $[0.4, 1.6, 0.3, 0.5]$ is transformed into the permutation $[3, 1, 4, 2]$, since the third element of the vector is the smallest one, the first element is the second smallest one, and so on.

The idea of generating a permutation from a Euclidean vector by ordering its dimensional indexes was investigated also in [2], where only the case of features extracted from images using a deep Convolutional Neural Network was analysed. Moreover, in [2] the intuition was that individual dimensions of the deep feature vectors represent some sort of visual concepts and that the value of each dimension specifies the importance of that visual concept in the image. Here we observe that a similar approach can be applied to Euclidean data in general, and thanks to the use of the n-Simplex projection it can be extended to a large class of metric objects as well. The only problem on applying this approach on general Euclidean vectors is that the variance of the values in a given dimensional position might be very different when varying the considered position. This happens, for example, in the case of vectors obtained using the Principal Component Analysis where elements in the first dimensional positions have higher variance than elements in the other dimensions. Other examples are the vectors obtained with the n-Simplex projection that, by construction, have higher values in top position and values that decrease to zero in the last components. To overcome this issue we propose to randomly rotate the vectors before transforming them into permutations. In facts, the random rotation distributes the information equally along all the dimensions of the vectors while preserving the Euclidean distance.

In summary, given a set of pivots $\mathcal{P}_n$, the proposed approach to associate a permutation to an object $o \in D$ is (1) compute the n-Simplex projected vector $\phi_{\mathcal{P}_n}(o)$; (2) randomly rotate the obtained vector (the same rotation matrix is used for all the data objects); (3) generate the permutation by ordering the values of the rotated vectors. We use the term *SPLX-Perms* for referring to the so obtained permutations (a pseudo-code is reported in Algorithm 1).

4   Experiments

We compared our permutation representations (SPLX-Perms) with the traditional permutation-based representations (Perms) in an approximate similarity search scenario. The experiments were conducted on two publicly available data sets:

SISAP colors is a benchmark for metric indexing. It contains about 113 K color histograms of medical images, each represented as 112-dimensional vector.

YFCC100M is a collection of almost 100M images from Flickr. We used a subset of 1M deep Convolutional Neural Network features extracted by Amato et al. [1] and available at http://www.deepfeatures.org/. Specifically, we used the activations of the *fc6* layer of the HybridNet [16] after ReLu and ℓ_2 normalization. The resulting features are 4,096-dimensional vectors.

The metrics used in the experiments are the *Jensen-Shannon distance* for the SISAP colors data, and the *Euclidean distance* for the YFCC100M deep features. For each data set, we considered 1,000 randomly selected queries and we built the ground-truth for the exact k-NN query search. The approximate results set for a given query is selected by performing the k-NN search in the permutation space. The quality of the approximate results was evaluated using the *recall@k*, that is $|\mathcal{R} \cap \mathcal{R}^A|/k$ where $\mathcal{R}$ is the result set of the exact k-NN search, and $\mathcal{R}^A$ is the set of the k approximate results.

To have a better overview of the tested approaches, we also consider the case in which the permutations are used to select a *candidate result set* to be re-ranked using the original distance d. In such cases, a k'-NN search (with $k' > k$) is performed in the permutation space in order to select the candidate result set.

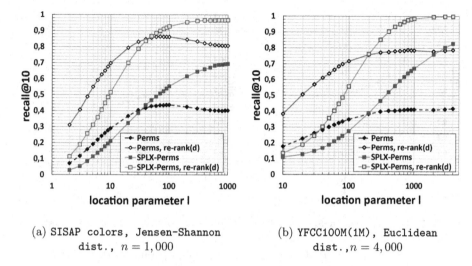

(a) SISAP colors, Jensen-Shannon
dist., $n = 1,000$

(b) YFCC100M(1M), Euclidean
dist.,$n = 4,000$

Fig. 1. *Recall*@10 varying the location parameter l (i.e. the prefix length).

Then the candidate results are re-ranked according to the actual distance d, and
the top-k objects are selected to form the final approximate result set $\mathcal{R}^A$. In
the experiments, we used $k' = 100$ and $k = 10$, if not specified otherwise.

To generate the permutation-based representations we used $n = 1,000$ pivots
for the SISAP Colors data set, and $n = 4,000$ pivots for the YFCC100M data
set. We tested the quality of the results obtained using either the full-length
permutations or a fixed-length prefix of the permutations. The metric used in
the permutation space is the *Spearman's rho with location parameter l*, where
the location parameter l is the length of the prefix permutation.

4.1   Results

Figure 1a and b show the *recall*@10 for the SISAP Colors and YFCC100M
data sets, respectively. Lines "`Perms`" and "`SPLX-Perms`" refer to the cases in
which the permutations are used to select the approximate result set by performing a 10-NN search in the permutation space. Lines "`Perms, re-rank(d)`"
and "`SPLX-Perms, re-rank(d)`" refer to the cases in which the permutation-
representation are used to select a candidate result set (obtained by performing
a 100-NN search) that is then re-ranked using the actual distance d. It is interesting to note that on YFCC100M data, our *full-length* SPLX-Perms representation allowed us to achieve a recall that not only is better than that achieved
using the traditional full-length permutation, but it is even better than that
obtained by the re-ranked approach. However, we also observe that for very short
prefix-lengths the traditional permutations shown better performance than our
technique. Another interesting aspect is that when considering the traditional
permutation-based representation there is usually an optimal prefix length $l < n$

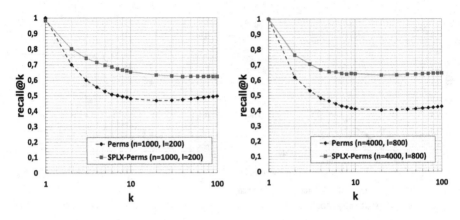

(a) SISAP Colors, Jensen-Shannon dist.   (b) YFCC100M(1M), Euclidean dist.

Fig. 2. *Recall@k* varying k (fixed location parameter l)

for which the best recall is achieved or for which the recall curve shows a plateau. This is evident in Fig. 1, where the recall lightly decrease as the location parameter l grows. This is a phenomenon experimentally observed also in other data sets as shown in several works (see e.g., [3,4]). Our SPLX-Perms seems to be not affected by this phenomenon since its recall increases when considering larger l. Moreover, the re-ranking of candidate results selected using our permutations achieved a recall very close to one for large prefix-lengths.

In Fig. 2, we also report the *recall@k* with k ranging from 1 to 100 for the baselines approaches (i.e. without considering the re-ranking phase) using a fixed prefix-length l. We can see that the improvement of the proposed approach over the traditional permutation-based representation holds for all ks. Our SPLX-Perm representation seems also to be more stable and provides recall values that are up to 1.6 times higher than that obtained using traditional permutations.

5   Conclusions

In this paper, we presented a novel permutation-based representation for metric objects, called SPLX-Perm. It exploits the n-Simplex projection to map the data object to Euclidean vectors, which are in turn transformed into permutations. The approach used to transform the Euclidean vectors into permutations has some analogies with the Deep Permutation approach that was proposed in [2] for associating permutations to visual deep features. To some extent, our work can be viewed as a generalisation of this technique to the large class of metric space meeting the n-point property. Our preliminary results show that our SPLX-Perms are more effective than the traditional permutations, even if there are some drawbacks with respect to the traditional permutations: (1) worse performance for very small prefix permutation; (2) higher cost for generating the

SPLX-Perm since for each object we need to compute both the object-pivot distances and the n-Simplex projection. Nevertheless, we believe that our technique as a lot of potentialities and deserves further investigations. In this perspective, we plan to extend our experimental evaluation on more data sets and metrics, using a different prefix length for the query object (to reduce the search cost) and using a pivot selection specifically designed for the n-simplex projection.

Acknowledgements. This work was partially supported by VISECH ARCO-CNR, CUP B56J17001330004, the AI4EU project, funded by the EC (H2020 - Contract n. 825619), and the Short-Term-Mobility (STM) program of the CNR.

References

1. Amato, G., Falchi, F., Gennaro, C., Rabitti, F.: YFCC100M-HNfc6: a large-scale deep features benchmark for similarity search. In: Amsaleg, L., Houle, M.E., Schubert, E. (eds.) SISAP 2016. LNCS, vol. 9939, pp. 196–209. Springer, Cham (2016). https://doi.org/10.1007/978-3-319-46759-7_15
2. Amato, G., Falchi, F., Gennaro, C., Vadicamo, L.: Deep permutations: deep convolutional neural networks and permutation-based indexing. In: Amsaleg, L., Houle, M.E., Schubert, E. (eds.) SISAP 2016. LNCS, vol. 9939, pp. 93–106. Springer, Cham (2016). https://doi.org/10.1007/978-3-319-46759-7_7
3. Amato, G., Falchi, F., Rabitti, F., Vadicamo, L.: Some theoretical and experimental observations on permutation spaces and similarity search. In: Traina, A.J.M., Traina, C., Cordeiro, R.L.F. (eds.) SISAP 2014. LNCS, vol. 8821, pp. 37–49. Springer, Cham (2014). https://doi.org/10.1007/978-3-319-11988-5_4
4. Amato, G., Gennaro, C., Savino, P.: MI-file: using inverted files for scalable approximate similarity search. Multimed. Tools Appl. **71**(3), 1333–1362 (2014)
5. Babenko, A., Lempitsky, V.: The inverted multi-index. IEEE Trans. Pattern Anal. Mach. Intell. **37**(6), 1247–1260 (2015)
6. Blumenthal, L.M.: Theory and Applications of Distance Geometry. Clarendon Press, Oxford (1953)
7. Chavez, E., Figueroa, K., Navarro, G.: Effective proximity retrieval by ordering permutations. IEEE Trans. Pattern Anal. Mach. Intell. **30**(9), 1647–1658 (2008)
8. Connor, R., Cardillo, F.A., Vadicamo, L., Rabitti, F.: Hilbert exclusion: improved metric search through finite isometric embeddings. ACM Trans. Inf. Syst. **35**(3), 17:1–17:27 (2016)
9. Connor, R., Vadicamo, L., Cardillo, F.A., Rabitti, F.: Supermetric search. Inf. Syst. **80**, 108–123 (2018)
10. Connor, R., Vadicamo, L., Rabitti, F.: High-dimensional simplexes for supermetric search. In: Beecks, C., Borutta, F., Kröger, P., Seidl, T. (eds.) SISAP 2017. LNCS, vol. 10609, pp. 96–109. Springer, Cham (2017). https://doi.org/10.1007/978-3-319-68474-1_7
11. Esuli, A.: Use of permutation prefixes for efficient and scalable approximate similarity search. Inf. Process. Manag. **48**(5), 889–902 (2012)
12. Fagin, R., Kumar, R., Sivakumar, D.: Comparing top k lists. In: Proceedings of SODA 2003, pp. 28–36. Society for Industrial and Applied Mathematics (2003)
13. Jégou, H., Douze, M., Schmid, C.: Product quantization for nearest neighbor search. IEEE Trans. Pattern Anal. Mach. Intell. **33**(1), 117–128 (2011)

48   L. Vadicamo et al.

14. Novak, D., Zezula, P.: PPP-codes for large-scale similarity searching. In: Hameurlain, A., Küng, J., Wagner, R., Decker, H., Lhotska, L., Link, S. (eds.) Transactions on Large-Scale Data- and Knowledge-Centered Systems XXIV. LNCS, vol. 9510, pp. 61–87. Springer, Heidelberg (2016). https://doi.org/10.1007/978-3-662-49214-7_2
15. Zezula, P., Amato, G., Dohnal, V., Batko, M.: Similarity Search: The Metric Space Approach, vol. 32. Springer, Boston (2006). https://doi.org/10.1007/0-387-29151-2
16. Zhou, B., Lapedriza, A., Xiao, J., Torralba, A., Oliva, A.: Learning deep features for scene recognition using places database. In: Proceedings of NIPS 2014, pp. 487–495. Curran Associates, Inc. (2014)

Fast and Exact Nearest Neighbor Search in Hamming Space on Full-Text Search Engines

Cun (Matthew) Mu[1]($\boxtimes$), Jun (Raymond) Zhao[1], Guang Yang[1], Binwei Yang[2], and Zheng (John) Yan[1]

[1] Jet.com/Walmart Labs, Hoboken, NJ, USA
{matthew.mu,raymond,guang,john}@jet.com
[2] Walmart Labs, Sunnyvale, CA, USA
BYang@walmartlabs.com

Abstract. A growing interest has been witnessed recently from both academia and industry in building nearest neighbor search (NNS) solutions on top of full-text search engines. Compared with other NNS systems, such solutions are capable of effectively reducing main memory consumption, coherently supporting multi-model search and being immediately ready for production deployment. In this paper, we continue the journey to explore specifically how to empower full-text search engines with fast and exact NNS in Hamming space (i.e., the set of binary codes). By revisiting three techniques (bit operation, subs-code filtering and data preprocessing with permutation) in information retrieval literature, we develop a novel engineering solution for full-text search engines to efficiently accomplish this special but important NNS task. In the experiment, we show that our proposed approach enables full-text search engines to achieve significant speed-ups over its state-of-the-art term match approach for NNS within binary codes.

Keywords: Full-text search engine · Nearest neighbor search · Hamming space · Semantic binary embedding · Elasticsearch · Lucene

1 Introduction

Full-text search engines, based on first-order document-term statistics such as TF-IDF and Okapi BM25, have been deployed ubiquitously in nowadays web applications to help customers find textual documents that match their specified keywords.

Recently, active efforts from both academia and industry [1,7,9,13,14] have been witnessed to empower full-text search engines with the capability of nearest neighbor search (NNS). Compared with other NNS solutions (e.g., Annoy [2], FLANN [10] and FAISS [5]), such full-text search engine based ones have a number of clear advantages.

© Springer Nature Switzerland AG 2019
G. Amato et al. (Eds.): SISAP 2019, LNCS 11807, pp. 49–56, 2019.
https://doi.org/10.1007/978-3-030-32047-8_5

Implemented in Secondary Memory. As demonstrated by Amato et al. [1], unlike other NNS solutions implemented in main memory, due to the highly optimized disk-based index mechanics behind full-text search engines, NNS systems established on full-text search engines substantially reduce main-memory consumption. This makes such systems more cost-effective and thus more suitable to big-data applications.

Flexible in Multi-model Search. As highlighted by Mu et al. [9], enabling full-text search engines with NNS paves a coherent way for multi-model searches (e.g., allowing users to express their interests in both visual and textual queries), at which most of other NNS systems fall short.

Ready for Production. Last but not least, as emphasized by Rygl et al. [14], NNS systems built upon full-text search engines are extremely well-prepared for production deployment. Due to the cutting-edge engineering designs from full-text search engines (e.g., Elasticsearch and Solr), important features like horizontal scaling, I/O and cache optimization, security configuration, index and cluster management, real-time monitoring and RESTful APIs are immediately ready to be consumed by such NNS systems, so that engineers can effectively avoid reinventing the wheel themselves.

Blessed by all the above major benefits, we continue this journey to explore specifically effective ways to achieve exact nearest neighbor search in Hamming space (i.e., the set of binary codes) on top of full-text search engines.

Problem Statement. Specifically, with the following dataset of binary codes

$$\mathcal{B} = \{\boldsymbol{b}_1, \boldsymbol{b}_2, \ldots, \boldsymbol{b}_n\} \subset \{0,1\}^m, \tag{1}$$

the goal of our paper is to enable full-text search engines with the capability of efficiently finding all r-neighbors of $\boldsymbol{q}$ in $\mathcal{B}$, namely

$$B_H(\boldsymbol{q}, r) := \{\boldsymbol{b} \in \mathcal{B} \mid d_H(\boldsymbol{b}, \boldsymbol{q}) \le r\}, \tag{2}$$

where $d_H(\boldsymbol{b}, \boldsymbol{q}) := \sum_{i=1}^m \mathbb{1}_{\{b_i \ne q_i\}}$ denotes the Hamming distance between binary code $\boldsymbol{b}$ and $\boldsymbol{q}$.[1] Similar to previous works [1,9,13,14], without the loss of general applicability to other full-text search engines, we elaborate our core ideas concretely using Elasticsearch–one of the most popular full-text search engines built upon Apache Lucene.

Organization. The rest of the paper is organized as follows. In Sect. 2, we first review the term match approach–a technique widely used by nowadays full-text search engines to find nearest neighbors among binary codes. In Sect. 3, we propose a better one for full-text search engines to accomplish this task. Specifically, we implement an Elasticsearch-based solution called FENSHSES (Fast

---

[1] It is worth noting that the r-neighbor search problem studied by the paper can also be easily adapted to conduct k-NN (k-nearest neighbors) search by progressively increasing the Hamming search radius r until k neighbors are found.

Exact Neighbor Search in Hamming Space on Elasticsearch) to conduct nearest neighbor search in Hamming space. We incorporate three techniques into FEN-SHSES: *bit operation*, which enables Elasticsearch to compute Hamming distance with just a few bit operations; *sub-code filtering*, which instructs Elasticsearch to conduct a simple but effective screening process before any Hamming distance calculation and therefore empower FENSHSES with sub-linear search times; *data preprocessing with permutation*, which preprocesses binary codes with appropriate permutation to maximize the effect of sub-code filtering. In Sect. 4, we show that FENSHSES outperforms the term match approach dramatically in terms of search latency.

2   Term Match from LIRE

Based on its definition, Hamming distance is nothing but the number of positions at which two binary codes vary. As a result, full-text search engines can naturally compute this through term match. Specifically, for each binary code b, we can index its positions corresponding to ones and zeros; and when the query binary code q arrives, full-text search engines can simply calculate its Hamming distance to each binary code $b \in \mathcal{B}$ by matching its zero and one positions with $b's$. This term match approach, firstly developed by Lux and Marques [7] in their Java library called LIRE (Lucene Image Retrieval) to find visually similar images (based on their binary visual features), is currently the cutting-edge approach for full-text search engines to find nearest neighbors within binary codes. Some of its variants (e.g., using fuzzy query based on Levenshtein edit distance) are also widely used on full-text search engines nowadays.

3   Proposed Approach: FENSHSES

The term match approach treats each binary digit (i.e., bit) in a textual way, which heavily overlooks the intrinsic and special properties of binary codes. By making better uses of these properties, we introduce a novel approach called FENSHSES (Fast Exact Neighbor Search in Hamming Space on Elasticsearch), whose complete JSON-encoded Elasticsearch request body can be found in JSON 2. In essence, FENSHSES integrates three techniques: bit operation, sub-code filtering and data preprocessing with permutation, which should be generally applicable to other full-text search engines besides Elasticsearch. These three techniques are pervasively used in nearest neighbors search for binary codes; but as far as we know, this is the first-time such techniques are seamlessly integrated into full-text search engines, and thus leads to a novel NNS solution with minimal main memory consumption, full support in multi-modal search and extreme readiness to be deployed in production (per our discussions in Sect. 1).

3.1   Bit Operation

Motivated by the well-known fact that hamming distances between binary codes can be computed extremely fast using bit operations, in this part, we will explore how we can replace term match by natively empowering Elasticsearch to calculate hamming distances through bit operations.

For an m-bit binary code b, we will first segment it into s sub-codes:[2]

$$[\underbrace{b_1,\ldots,b_{\frac{m}{s}}}_{b^1},\underbrace{b_{\frac{m}{s}+1},\ldots,b_{\frac{2m}{s}}}_{b^2},\ldots\ldots,\underbrace{b_{m-\frac{m}{s}+1},\ldots,b_m}_{b^s}]. \tag{3}$$

Since $d_H(q,b)=\sum_{i\in[s]}d_H(q^i,b^i)$, the Hamming distance calculation is reduced into s ones with binary codes of much shorter length. In JSON 1, We re-implement the assembly codes found in the notable HAKMEM memo [3] to compute the Hamming distance between two short binary codes of length 64 or less into Painless–a simple and secure scripting language designed specifically for Elasticsearch. When the query binary code q is issued, we will invoke hmd64bit s times to calculate $\left\{d_H(q^i,b^i)\right\}_{i=1}^{s}$ by specifying q^i and b^i as parameters accordingly and then sum them up. The whole process can be efficiently implemented in Elasticsearch using the *function score query*, where several functions are combined to calculate the score of each document (see lines 15–26 in JSON 2).

JSON 1 Create the script called hmd64bit into Elasticsearch.

```
1   POST _scripts/hmd64bit
2   {
3     "script": {
4       "lang": "painless",
5       "source": """
6         long u = params.subcode^doc[params.field].value;
7         long uCount = u-((u>>>1)&-5270498306774157605L)
8                        -((u>>>2)&-7905747460161236407L);
9         return ((uCount+(uCount>>>3))&8198552921648689607L)%63;
10        """
11  }}
```

3.2   Sub-Code Filtering

So far, regardless of the term match approach or the bit operation one, we have to exhaustively compute the Hamming distance between q and each binary code in $\mathcal{B}$. This expensive linear scan is not desirable for many applications where the number of codes in $\mathcal{B}$ is in the order of millions or even billions [17]. As a remedy, in this part, we will borrow a simple but powerful counting argument from Norouzi et al. [11] to conduct a screening process before any Hamming distance calculation, which successfully empowers our FENSHSES approach with sub-linear search times.

Suppose binary codes are segmented into s sub-codes as in (3). Then for two codes b and q within r Hamming distance, among all their s sub-code pairs

---

[2] For simplicity, we assume s divides m.

$\{(b^i, q^i)\}_{i=1}^{s}$, there must be at least one pair with Hamming distance no larger than $\lfloor \frac{r}{s} \rfloor$, which mathematically implies

$$B_H(q, r) \subseteq \bigcup_{i=1}^{s} \left\{ b \in \mathcal{B} \;\middle|\; b^i \in B_H\left(q^i, \lfloor \tfrac{r}{s} \rfloor\right) \right\}. \qquad (4)$$

This simple counting argument yields great potentials in reducing the number of Hamming distance calculations needed to find all r-neighbors q in $\mathcal{B}$. Specifically, according to relationship (4), it is safe to just consider binary codes belonging to the set on the right side of (4), whose size could be substantially smaller than n for $r \ll m$. It is worth noting that similar ideas have been frequently revisited in many different contexts–e.g., multi-index hashing [11] and string similarity joins [6], and a generalized version of (4) is also derived recently [12].

Due to the inverted-indexing nature of full-text search engines, this sub-code filtering step is extremely suitable and straightforward to be implemented on full-text search engines. Specifically, on Elasticsearch, we can simply leverage the *filter context* (see lines 8–14 in JSON 2), within which each sub-code Hamming ball $B_H\left(q_i, \lfloor \frac{r}{s} \rfloor\right)$ is obtained by the *terms query* (e.g., line 11 in JSON 2), and the union is achieved through a *boolean combination of should clauses*.

3.3   Data Preprocessing with Permutation

The effectiveness of sub-code filtering will be maximized if bits within the same sub-code group are statistically independent. Since hamming distance is invariant to permutation transformation, it is tempting to transform binary codes in $\mathcal{B}$ with appropriate permutation towards this desired group independence property.

For two Bernoulli random variables x and y, they are independent if and only if their correlation coefficient $\rho(x, y) = 0$. Therefore, it is natural to find a permutation $\bar{\pi}$ to minimize correlation effects among each sub-code segment. This immediately leads to the following optimization problem essentially solved by Wan et al. [16] to improve the performance of [11]:

$$\min_{\pi:[m] \to [m]} \quad \langle D, P_\pi M_{\mathcal{B}} P_\pi^\top \rangle \qquad \text{s.t. } \pi \text{ is a permutation.} \qquad (5)$$

Here $D = \mathrm{diag}\,(I_{d \times d}, \ldots, I_{d \times d}) \in \mathbb{R}^{m \times m}$ is a block diagonal matrix with $I_{d \times d}$ as a matrix of ones and $d = m/s$, P_π is the permutation matrix induced by π $P_\pi = \begin{bmatrix} e_{\pi(1)} & e_{\pi(2)} & \cdots & e_{\pi(m)} \end{bmatrix}^T$ and M is a matrix in $\mathbb{R}^{m \times m}$ whose (i, j)-entry is obtained from $\mathcal{B}$ as the absolute value of the correlation between the i-th and the j-th bits.

4   Experiment

We compare search latencies between the term match approach and FENSHSES with semantic binary codes generated from Jet.com's catalog images. To better understand the contribution of each technique involved in FENSHSES, we experiment systematically with four methods: the term match baseline, FEN-SHSES with just bit operation, FENSHSES without data preprocessing and FENSHSES.

JSON 2 Elasticsearch request body of the FENSHSES approach

```
1   {
2     "min_score": m-r,
3     "query": {
4       "function_score":{
5         "query":{
6           "constant_score":{
7             "boost": m,
8             "filter":{
9               "bool":{
10                "should":[
11                  {"terms":{"b1": r/m-neighbor of q1}},
12                  ......,
13                  {"terms":{"bm": r/m-neighbor of qs}}
14                ]}}}},
15         "functions":[
16           {"script_score": {"script":
17                             {"id": "hmd64bit",
18                              "params": {"field": "b1", "subcode": q1}}},
19           "weight": -1},
20           ...
21           {"script_score": {"script":
22                             {"id": "hmd64bit",
23                              "params": {"field": "bm", "subcode": qm}}},
24           "weight": -1}],
25         "boost_mode": "sum",
26         "score_mode": "sum"
27   }}}
```

Settings. Our dataset $\mathcal{B}$ is generated using half a million images selected from Jet.com's furniture catalog through the pretrained INCEPTION-RESNET-V2 model [15] with iterative quantization (ITQ) [4].[3] We choose the length of binary codes to be 128 and 256 respectively. For the setting of FENSHSES, we keep the sub-code length as 64 for bit operation and 16 for sub-code filtering throughout the experiment, since we observe such segmentations consistently yield satisfactory performances. Each Elasticsearch index is created with five shards and zero replica on a single-node Elustersearch cluster deployed on a Microsoft Azure virtual machine with 16 cores and 112 GiB of RAM. We randomly select 1,000 binary codes from $\mathcal{B}$ to act as query codes q. For each q, we compare the search latencies among all four methods with Hamming distance $r \in \{5, 10, 15, 20\}$.

---

[3] Note that the purpose of the experiment is not to compare different embedding models, but to evaluate the performance of FENSHESES, which should be generally applicable to NNS in any Hamming space.

Table 1. Means and standard deviations (in brackets) of search latency (measured in ms) under different scenarios. FENSHSES is dramatically faster than the term match approach, and all of the three techniques involved in FENSHSES contribute substantially to this performance improvement.

m	r	Term match	Bit operation	FENSHSES w/o prep.	FENSHSES
128	5	641.99 (19.01)	41.38 (6.38)	2.80 (3.50)	1.08 (1.25)
	10	638.20 (16.65)	42.24 (7.39)	7.40 (5.07)	3.62 (1.54)
	15	637.63 (16.14)	43.08 (7.90)	7.19 (5.09)	3.45 (1.55)
	20	638.41 (17.41)	42.65 (7.59)	15.51 (5.88)	9.51 (2.18)
256	5	1259.22 (30.66)	75.35 (11.87)	6.24 (6.48)	2.18 (2.02)
	10	1257.04 (20.68)	75.06 (11.27)	6.28 (6.63)	2.13 (1.97)
	15	1270.38 (25.88)	75.81 (12.22)	6.70 (6.93)	2.09 (1.56)
	20	1278.47 (25.56)	75.50 (11.94)	18.02 (10.71)	7.67 (2.85)

Results. As shown in Table 1, FENSHSES is much faster than the term match approach. In the following, we address the contribution of each component of FENSHSES respectively.

- By computing the Hamming distance using bit operation instead of term match, we consistently observe around sixteen times speedup over different m and r.
- The amount of speed-up introduced by sub-code filtering varies with the radius r. Specifically, as $\lfloor \frac{16r}{s} \rfloor$ heavily influence the number of data points to be considered for Hamming distance computation (see (4)), for r's with the same value of $\lfloor \frac{16r}{s} \rfloor$, the search latencies of FENSHSES w/o prep. are quite similar. As $\lfloor \frac{16r}{s} \rfloor$ becomes larger, the sub-code filtering technique will become less effective. In practice, since we most likely care about nearest neighbors within a small radius, the sub-code filtering technique should be capable of greatly reducing the search latency.
- By reshuffling binary codes to reduce their correlations within each sub-code group, the technique of data processing with permutation not only accelerates FENSHSES in terms of the average search latency, but also stabilizes its overall performance with much smaller standard deviation.
- A comprehensive comparison between FENSHSES and FAISS [5] in terms of indexing speed, search latency and RAM consumption is also conducted in [8], where FENSHSES demonstrates competitive performance.

5 Conclusion

It has been recently demonstrated that NNS systems built upon full-text search engines are capable of effectively reducing main memory consumption, coherently supporting multi-model search and being well-prepared for production deployment. Motivated by these clear advantages, in this paper, we explore how to empower full-text search engines to efficiently find nearest neighbors in Hamming space. By revisiting bit operation, sub-code filtering and data preprocessing with permutation, we propose a novel approach to accomplish this task, which is shown empirically to be substantially faster than the term match approach (the state-of-art one for nowadays full-text search engines to find nearest neighbors within binary codes). By implementing the proposed approach non-trivially on the Elasticsearch platform, we delivered a cutting-edge engineering solution called FENSHSES. In the future, we will also explore how to implement our approach efficiently on other full-text search engines (e.g., Solr and Sphinx).

References

1. Amato, G., Bolettieri, P., Carrara, F., Falchi, F., Gennaro, C.: Large-scale image retrieval with elasticsearch. In: SIGIR (2018)
2. Bernhardsson, E.: Annoy: Approximate Nearest Neighbors in C++/Python (2018). Python package version 1.13.0. https://pypi.org/project/annoy/
3. Beeler, M., Gosper, R.W., Schroeppel, R.: Hakmem. MIT Artificial Intelligence Laboratory (1972)
4. Gong, Y., Lazebnik, S.: Iterative quantization: a procrustean approach to learning binary codes. In: CVPR (2011)
5. Johnson, J., Douze, M., Jégou, H.: Billion-scale similarity search with GPUs. arXiv preprint arXiv:1702.08734 (2017)
6. Li, G., Deng, D., Wang, J., Feng, J.: Pass-join: a partition-based method for similarity joins. In: VLDB (2012)
7. Lux, M., Marques, O.: Visual Information Retrieval Using Java and LIRE, vol. 25. Morgan & Claypool Publishers (2013)
8. Mu, C., Yang, B., Yan, Z.: An empirical comparison of FAISS and FENSHSES for nearest neighbor search in hamming space. In: SIGIR eCommerce Workshop (2019)
9. Mu, C., Zhao, J., Yang, G., Zhang, J., Yan, Z.: Towards practical visual search engine within elasticsearch. In: SIGIR eCommerce Workshop (2018)
10. Muja, M., Lowe, D.G.: Scalable nearest neighbor algorithms for high dimensional data. IEEE Trans. Pattern Anal. Mach. Intell. **36**(11), 2227–2240 (2014)
11. Norouzi, M., Punjani, A., Fleet, D.J.: Fast search in hamming space with multi-index hashing. In: CVPR (2012)
12. Qin, J., Wang, Y., Xiao, C., Wang, W., Lin, X., Ishikawa, Y.: GPH: similarity search in hamming space. In: ICDE (2018)
13. Ruzicka, M., Novotny, V., Sojka, P., Pomikalek, J., Rehurek, R.: Flexible similarity search of semantic vectors using fulltext search engines (2018). http://ceur-ws.org/Vol-1923/article-01.pdf
14. Rygl, J., Pomikalek, J., Rehurek, R., Ruzicka, M., Novotny, V., Sojka, P.: Semantic vector encoding and similarity search using fulltext search engines. In: RepL4NLP Workshop (2017)
15. Szegedy, C., Ioffe, S., Vanhoucke, V., Alemi, A.: Inception-v4, Inception-Resnet and the impact of residual connections on learning. In: AAAI (2017)
16. Wan, J., Tang, S., Zhang, Y., Huang, L., Li, J.: Data driven multi-index hashing. In: ICIP (2013)
17. Yang, F., et al.: Visual search at ebay. In: KDD (2017)

k-Distance Approximation
for Memory-Efficient RkNN Retrieval

Max Berrendorf$^{(\boxtimes)}$, Felix Borutta, and Peer Kröger

Lehrstuhl für Datenbanksysteme und Data Mining,
Ludwig-Maximilians-Universität München, Munich, Germany
{berrendorf,borutta,kroeger}@dbs.ifi.lmu.de

Abstract. For a given query object, Reverse k-Nearest Neighbor queries retrieve those objects that have the query object among their k-nearest neighbors. However, computing the k-nearest neighbor sets for all points in a database is expensive in terms of computational costs. Therefore, specific index structures have been invented to apply pruning heuristics which aim at reducing the search space. At time, the state-of-the-art index structure for enabling fast RkNN query processing in general metric spaces is the MRkNNCoP-Tree which uses linear functions to approximate lower and upper bounds on the k-distances to prune the search space. Storing those linear functions results in additional storage costs in $\mathcal{O}(n)$ which might be infeasible in situation where storage space is limited, e.g., on mobile devices. In this work, we present a novel index based on the MRkNNCoP-Tree as well as recent developments in the field of neural indexing. By learning a single neural network model that approximates the k-nearest neighbor distance bounds for all points in a database, the storage complexity of the proposed index structure is reduced to $\mathcal{O}(1)$ while the index is still able to guarantee exact query results. As shown in our experimental evaluations on synthetic and real-world data sets, our approach can significantly reduce the required storage space in trade-off to some growth in terms of refinement sets when relying on exact query processing. We provide our code at www.github.com/mberr/k-distance-prediction.

Keywords: Reverse k-nearest neighbor · k-nearest neighbor ·
Query processing · Neural indexing

1 Introduction

In many applications like resource allocation, targeted marketing or in general decision support systems, it has proven to be very useful to have knowledge about influence sets. Considering the example of opening a new store, one of the decisive points when asking where to open the store obviously is the number of customers that might be attracted by this store. However, such an influence set, i.e. the set of potential customers, not only depends on spatial proximity, but also on the influence of other competitors on the customers. Therefore, the

© Springer Nature Switzerland AG 2019
G. Amato et al. (Eds.): SISAP 2019, LNCS 11807, pp. 57–71, 2019.
https://doi.org/10.1007/978-3-030-32047-8_6

determination of influence sets is not straightforward. An established tool for determining influence sets are *Reverse k-Nearest Neighbor* (RkNN) queries. A Reverse k-Nearest Neighbor query retrieves all objects from a database having a given query object as one of their k nearest neighbors.

In general, the problem of solving RkNN queries has been considered widely under various conditions. However, a natural problem of these queries is that they heavily rely on k nearest neighbor computations and hence are rather complex. Precisely, a naive solution of the RkNN problem requires $O(n^2)$ time since the k-nearest neighbors for all n objects in the database have to be determined. A common approach to overcome this drawback is the usage of index structures. The idea is to store data objects within tree-like data structures such that queries can be processed efficiently by employing certain pruning strategies. At time, the state-of-the-art solution for index structures that support fast RkNN query processing for general metric data is the *MRkNNCoP*-Tree. This index structure makes use of the observation that the distance distributions for data points often follow the power-law in natural datasets. In somewhat more detail, the MRkNNCoP-Tree approximates the k nearest neighbor distances up to some pre-defined value k_{max} for each data object by computing two linear functions that approximate the k nearest neighbor distances conservatively and progressively in log-log-space. Using these two linear functions as upper and lower bound for pruning, the MRkNNCoP-Tree supports fast query processing for RkNN queries with $k \leq k_{max}$. However, beside being limited to $k \leq k_{max}$ at query time, the MRkNNCoP-Tree requires a storage overhead in $\mathcal{O}(n)$ due to storing four additional parameters for each data object (i.e., slope and intercept for both linear functions). The latter might become problematic when considering modern, embedded systems where memory is potentially limited, e.g., on moving sensor devices that for instance may have to frequently compute influence sets to optimize sensor to sensor communication.

Therefore, we present an approach to overcome this issue and tackle the problem of k-distance approximation with focus on memory-efficient RkNN retrieval in this work. Precisely, we state that the coefficients of the linear model functions correlate spatially and based on our findings, we learn a constant space index structure which is related to the field of neural indexing as we aim at learning the distribution of coefficients. By training a single regression model we can approximate the k nearest neighbor distances for all data objects in the data base. However, since using these approximations for RkNN query processing potentially leads to incorrect query answers, we also propose a solution to obtain guaranteed bounds for exact answers. Our experimental evaluations with different regression models on a variety of synthetic and real-world datasets show that our approach can significantly reduce the required storage space in trade-off to some growth in terms of refinement sets when relying on exact query processing.

To summarize, the contribution of this paper are as follows:

1. The coefficients stored in the MRkNNCoP-Tree are analyzed, and the potential for compression is identified.

2. A novel constant size index structure based upon progress made in the field of neural indexing is proposed.
3. Thorough experiments on both, synthetic and real-world, data sets showcase the capabilities of this index structure.

2 Preliminaries

The following section gives some fundamental definitions related to RkNN query processing to establish an unified notation. First, as RkNN query processing generally relies on k nearest neighbor distance calculations we formally define a *distance space* as follows.

Definition 1. *Let $\mathcal{U}$ be an arbitrary set, and* dist $: \mathcal{U} \times \mathcal{U} \rightarrow \mathbb{R}$. *A space $(\mathcal{U}, \text{dist})$ is called distance space with distance d, if the following properties hold*

1. *Non-Negativity:* $\text{dist}(x, y) \geq 0$ *for all* $x, y \in \mathcal{U}$,
2. *Symmetry:* $\text{dist}(x, y) = \text{dist}(y, x)$,
3. $\text{dist}(x, x) = 0$.

Given such a distance space, we define the *k-distance* and subsequently the set of *k-nearest neighbors* with respect to some query object as below.

Definition 2. *For a given data set $D \subseteq \mathcal{U}$, the k-distance of a data point $q \in D$ can be defined as*

$$\text{nndist}(q, k) = \min_{\substack{D' \subseteq D \\ |D'| = k}} \max_{o \in D'} \text{dist}(q, o) \tag{1}$$

Further, the set of k-nearest neighbors of $q \in D$ are given by

$$\text{nn}(q, k) = \{o \in D \mid \text{dist}(q, o) \leq \text{nndist}(q, k)\} \tag{2}$$

Note that the set of k-nearest neighbors is *not* symmetric, i.e. $o \in \text{nn}(q, k)$ does not imply that $q \in \text{nn}(o, k)$. Given the definition of nearest neighbors, we can define the set of *reverse k-nearest neighbors* as the set of those data points that have the query point among their k-nearest neighbors.

Definition 3. *Given a data set $D \subseteq \mathcal{U}$ and some query object $q \in D$, the reverse nearest neighbors of q are given as*

$$\text{rnn}(q, k) = \{o \in D \mid q \in \text{nn}(o, k)\} = \{o \in D \mid \text{dist}(o, q) \leq \text{nndist}(o, k)\} \tag{3}$$

Equation (3) directly shows the necessity of calculating the distances between any data object $o \in D$ and the query object q in order to determine those data object o whose distance to q is less or equal than the k-distance of o. Naively, this requires $\mathcal{O}(n)$ distance calculations per query object. Hence, the ability to determine the validity of $\text{dist}(o, q) \leq \text{nndist}(o, k)$ rapidly is crucial for efficient RkNN query processing.

It is noteworthy that the type of RkNN query defined in Definition 3 only considers data objects stemming from a single set of data objects D. In general, this kind of RkNN queries are called *monochromatic* RkNN queries. However, in most applications RkNN queries are more useful when considering different sets of data entities D and Q for the query object $q \in Q$ and the result set rnn$(q, k) \subseteq D$ (e.g. for the store-customer example given in Sect. 1). Such kind of RkNN queries are called *bichromatic* RkNN queries. Nonetheless, since we are primarily interested in approximating the k-distances in this work, we only focus on monochromatic RkNN queries for the sake of simplicity.

3   Related Work

Since this paper presents an approach which aims at combining methods from neural indexing with RkNN query processing, we briefly review related work from both fields in the following.

3.1   RkNN Query Processing

The problem of RkNN search in Euclidean and general metric spaces has already been investigated extensively in previous work. In general, there are two kinds of existing approaches, called *self-pruning* and *mutual-pruning* approaches. Self pruning approaches are typically characterized by using kNN information of the currently regarded objects to decide whether to prune them or not. Most of these approaches are connected with hierarchically organized tree-like index structures. The first work on RkNN is presented in [11] and proposes the concept of RNN-Tree which is an R-Tree based index structure that stores pre-computed NN spheres and uses them for fast query processing. Hence, the result set for rnn(q, k) is given by the centers of all such spheres, which encompass q. For answering queries in a dynamic setting with insertions and deletions, an additional tree, that is used to resolve kNN queries, is maintained. This idea has been extended by the RdNN-Tree in [22] such that dynamic queries are supported by only using a single index structure. However, though the RdNN-Tree can be applied for metric data, both concepts require a fixed value of k. To overcome this limitation, [1–3] developed the MRkNNCoP-Tree which benefits from progressive and conservative approximations of kNN distances for an arbitrary value of k less or equal a fixed maximal value k_{max}. Assuming that distances in (real-world) data sets often follow a power-law, they approximate the function mapping k to the k-distance of a specific point by two linear functions in log-log space. By storing only the coefficients of the functions, they can process RkNN queries efficiently with an index of size $\mathcal{O}(n)$. Another solution for RkNN query processing in metric spaces which is based on the M-Tree structure can be found in [19]. The approach presented there makes use of the M-Tree structure to identify candidates which are refined subsequently.

In opposite to self-pruning approaches, mutual-pruning approaches generally use other objects of the database to prune a given index object. In [16], the authors present a technique which uses *Voronoi cells* to process R1NN queries in Euclidean space. The work proposed in [18] overcomes the problem of being restricted to $k = 1$ and shows an approach to process RkNN queries with arbitrary values of k. Another mutual pruning based solution for the R1NN problem in dynamic data sets is presented in [17]. However, like the other approaches that follow the mutual pruning scheme, this method relies on Euclidean space properties to partition the space for reducing the search space. This generally limits the applicability of mutual pruning approaches, especially when considering high-dimensional data. Finally, one approach that aims at combining both pruning paradigms with the objective to reduce I/O costs for RkNN query processing in dynamic data bases is presented in [4].

Further, and more specialized, approaches on RkNN query processing tackle the problem on solving RkNN queries in uncertain databases [7,13], in continuous settings [8,20], or in spatial applications, e.g. road networks [5,6]. However, in this work, we aim at solving the RkNN problem in the more generalized setting, i.e. for general metric data.

3.2   Neural Indexing

The field of neural indexing combines indexing structures with the power of Machine Learning models, as the latter have proven to be able to effectively approximate data distributions. The first work coming up with the general idea to replace indexes with machine learning models can be found in [12]. The authors argue that an index, in their case a B-Tree, Hash Map or Bloom Filter, can be interpreted as a regression, resp. classification, model in machine learning terminology, since the goal is to map keys to specific positions with a min- and max-error, resp. to predict the existence of a data record. By using key-position instances for training, the regression model finally is able to predict the position of a data record (just like it is the case for index structures like B-Trees or Hash Maps) at inference time. Kraska et al. also point out that learned indexes are able to manage a major requirement which is to guarantee bounds on the min- and max-error as is given for conventional index structures. For monotonic models in range indices for instance[1], one possibility is to remember the worst over- and underestimate for calculating the guarantee boundaries subsequently. However, the main advantages of such *learned indexes* are the reduced costs for look-up operations (in $\mathcal{O}(1)$ instead of $\mathcal{O}(\log(n))$ in the best case) and the reduced amount of storage costs (in $\mathcal{O}(1)$ instead of $\mathcal{O}(n)$).

The works presented in [9,23] are somehow related to neural indexing as they propose to use associative memories based on neural networks to accelerate (approximate) nearest neighbor query processing. The idea is to partition the data into equi-sized classes and subsequently refine only those classes at query time that exhibit the largest overlap with a given query object. Building upon the ideas presented in [12], the work in [14] investigates the usefulness of learned indexes, i.e. learned Bloom filters, for conjunctive Boolean queries by performing term-search queries on a document database. In particular, they propose and

---

[1] As is given in our case since we replace the linear functions serving as bounds for the MRkNNCoP-Tree.

evaluate various approaches each of which makes a different tradeoff between computational costs and the amount of required storage space. The Pavo index [21] is a learned inverted index for which the underlying hash function is replaced by a hierarchy of recurrent neural network models.

4   Neural Indexing for R*k*NN Retrieval

The contribution of this work is three-fold: in the first subsection, an analysis of the coefficients of the MR*k*NNCoP-Tree reveals their spatial correlation. Next, an approximation framework is proposed where the prediction of the coefficients is posed as a regression task with the data points given as input and the corresponding coefficients given as targets. Finally, for a static setting, a methodology similar to [12] is used to obtain guaranteed bounds that can be used for exact query processing.

4.1   Analysis of MR*k*NNCoP Tree Coefficients

Before we propose our approach to further compress the MRkNNCoP-Tree, we first analyze the coefficients of data points in the vicinity of each other. Precisely, we investigate the slope and offset values of the linear lower, resp. upper, bound approximations of k-distances for each data point with respect to their local neighborhood. Figure 1 exemplary visualizes both MR*k*NNCoP-Tree coefficients of the lower bound ($k_{max} = 256$) for the *Oldenburg* dataset as well as for a synthetic data set of multiple Gaussian blobs. The data points are plotted according to their x and y coordinates (both datasets are two-dimensional) and the color coding indicates the value of the corresponding coefficient. Lighter colors stand for higher values.

For both coefficients we can observe similar values for data points that are in similarly dense areas which indicates that the coefficients are spatially correlated. This can be observed particularly clear for the offset value when considering the synthetic dataset. Data points that are located within sparse areas have higher offset values than data points in dense areas, in order to compensate for the large 1-distance. For the slope parameter we generally observe a converse behaviour, as points in dense areas have many points in close vicinity and hence a slower increasing k-distance.

In general, we could observe the same behaviour for each of the datasets listed in Table 1 and in particular also for the upper bound approximations. Given this spatial correlation, we claim that the values of the coefficients defining the lower and upper bound approximations can be learned from the data distribution. Based on our findings, we therefore see the potential to replace the set of linear functions that are stored within the MR*k*NNCoP-Tree structure (i.e. two linear functions for each data object) with a single trained regression model that is able to predict the coefficients of the bounds given the location of a certain data point.

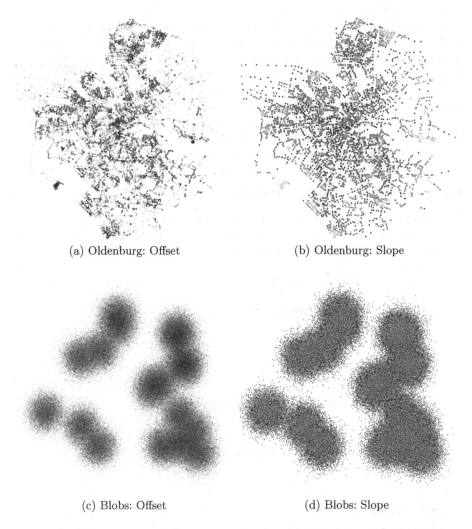

(a) Oldenburg: Offset   (b) Oldenburg: Slope

(c) Blobs: Offset   (d) Blobs: Slope

Fig. 1. MRkNNCoP-Tree coefficients of the lower bound for the real-world *Oldenburg* dataset (upper two images) and a synthetic data set of multiple Gaussian blobs comprising approx. 200,000 points. Lighter colors indicate higher values.

4.2 k-Distance Approximation

Motivated by the correlation of the coefficients of the MRkNNCoP tree bounds of adjacent data points, we propose to solve the regression task

$$\phi(x) : x \mapsto (\alpha_0, \beta_0, \alpha_1, \beta_1) \tag{4}$$

mapping a point to the coefficients of both bounds of the MRkNNCoP tree. For the approximation any kind of machine learning model may be used. If we model ϕ in a parametric form $\phi(x; \theta)$ for some parameters θ, e.g. as regression tree or neural network, we can constrain the number of parameters to achieve a constant-sized approximator. We denote with $\mathrm{low}(x, k) := \exp\left(\left(\phi(x)\right)_0 \cdot \log k + \left(\phi(x)\right)_1\right)$ the predicted lower bound, and analogously with $\mathrm{up}(x, k)$ the upper bound. As the bounds of the k-distance are not approximated directly but through regression of the MRkNNCoP tree coefficients, it is to be expected that we can achieve at most the same quality. However, one might be able to achieve significant reduction in the number of parameters while at the same time not losing too much quality. In Sect. 5 we investigate different choices of regression models as well as the influence of the number of parameters.

4.3   Obtaining Guaranteed Bounds

For a static data set D we can calculate guaranteed lower and upper bounds of the exact k-distance, by computing the maximal deviation in positive and negative direction over the whole dataset,

$$\varepsilon_0^D(k) = \max_{x \in D} \left\{ \mathrm{low}(x, k) - \mathrm{nndist}(x, k) \right\} \tag{5}$$

$$\varepsilon_1^D(k) = \max_{x \in D} \left\{ \mathrm{nndist}(x, k) - \mathrm{up}(x, k) \right\} \tag{6}$$

Now we have guaranteed that the true k-distance is bounded by

$$\mathrm{low}(x, k) - \varepsilon_0^D(k) \leq \mathrm{nndist}(x, k) \leq \mathrm{up}(x, k) + \varepsilon_1^D(k) \tag{7}$$

Notice that the maximum deviation requires storing $\mathcal{O}(k_{\max})$ values. This may be further reduced, e.g. by storing only $\varepsilon_i^D = \max_k \varepsilon_i^D(k)$ for $i \in \{0, 1\}$ at the cost of loosening the bounds. Also the maximum error for a specific k may be predicted using another regression model, in addition to it's maximum error, which is then a scalar.

5   Experiments

The following section evaluates the proposed method on various synthetic and real-world datasets by comparing the performance of our proposed neural index with the MRkNNCoP-Tree by considering both the size of the index and the size of the candidate sets that have to be refined for ensuring exact RkNN retrieval.

5.1   Setup

In order to achieve reproducibility, the following subsection describes the experimental setup in detail.

Table 1. Overview of the datasets used for evaluation. n denotes the number of samples, and d the dimensionality.

Dataset	n	d	Type
blobs.2.12	4,096	2	Synthetic
OL	6,105	2	Real-world
TG	18,263	2	Real-world
cal	21,048	2	Real-world
blobs.2.15	32,768	2	Synthetic
blobs.4.15	32,768	4	Synthetic
blobs.8.15	32,768	8	Synthetic
SF	174,956	2	Real-world

Datasets. The experiments are conducted on synthetic datasets as well as real-world datasets (summarized in Table 1). The synthetic ones are generated by the `make_blobs` routine of scikit-learn [15]. This routine generates a dataset by drawing random samples from a uniform mixture of c Gaussian distributions with standard deviation of one, i.e.

$$x \sim \sum_{i=1}^{c} \frac{1}{c} \mathcal{N}(x \mid \mu_i, 1),$$

where the centers $\mu_i \sim U([-10, 10]^d)$, with U denoting the uniform distribution. We arbitrarily chose $c = \log_2 n$ for all synthetic data sets. The synthetic datasets are as *blobs.x.y* with x being the dimensionality and 2^y indicating the size of the corresponding dataset.

Furthermore, we use the following road networks as real-world datasets: *Oldenburg* (OL), *California* (cal), *City of San Joaquin County* (TG) and *San Francisco* (SF)[2]. For all datasets Euclidean distance is used as distance measure.

Models. The following three model classes are investigated further. For all of them, we use the implementation as provided by scikit-learn.

- *Regression Trees*: We vary the parameters `max_depth`, controlling the maximum depth of the tree, and `min_samples_split`, controlling the minimum number of samples in nodes that shall be considered for splitting.
- Gradient-Boosting with regression trees: Here we vary the parameters `max_depth`, and `learning_rate`, where the latter controls the step size for the gradients.
- Fully-connected neural networks. We consider only sequential networks, i.e. networks that are a sequence of fully-connected layers without any branches of skip-connections. We use the ReLU activation function for the hidden layers

---

[2] Download from https://www.cs.utah.edu/~lifeifei/SpatialDataset.htm.

(i.e. $\max\{0, x\}$). All networks are trained for 1024 epochs and we use the Adam optimizer [10] with a learning rate of 10^{-3}.

Evaluation Protocol. First, we evaluate the quality of the MRkNNCoP-Tree coefficient prediction using mean absolute error (MAE). For true values y_i and predictions $\hat{y}_i$, for $i = 1, \ldots, n$, the MAE is given by $\mathrm{MAE}(y, \hat{y}) = \frac{1}{n} \sum_i |y_i - \hat{y}_i|$ To reduce the computational complexity of the model search we employ a skyline-based pruning procedure. Precisely, we only consider those models for further evaluations that lie on the skyline of model size, in terms of number of parameters, and model quality, in terms of MAE. For these models, guaranteed bounds, required for exact query processing, are derived using the procedure described in Sect. 4.3. Using the guaranteed bounds, the candidate set size is computed as the number of data points between the upper and the lower bound, and subsequently compared against the candidate set retrieved from the MRkNNCoP-Tree. To compare memory consumption, we only consider the models' parameters, and neglect the overhead of storing, e.g. the class structure.

5.2 Results

In the following, we provide detailed results from the conducted experiments. We begin with a qualitative example to further enhance the understanding about our method. In Fig. 2, the approximation bounds are exemplarily shown for the *cal* dataset. The black solid line shows the true k-distances for one selected point and values of $k \in \{1, \ldots, k_{\max}\}$. The solid red and blue line are the optimal bounds computed by the MRkNNCoP-Tree. With dashed line-style we can see the predicted bounds, observing two phenomena: First, the red predicted upper bound is correct, i.e. does not violate the upper bounding property but not as

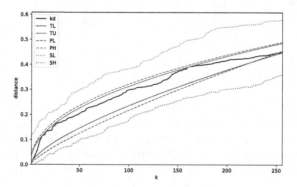

Fig. 2. Comparison of the approximation bounds for the *cal* dataset. The black line is the true k-distance for the given data point. The solid lines are the MRkNNCoP-Tree bounds, the dashed lines the predictions, and the dotted lines the guaranteed predictions.

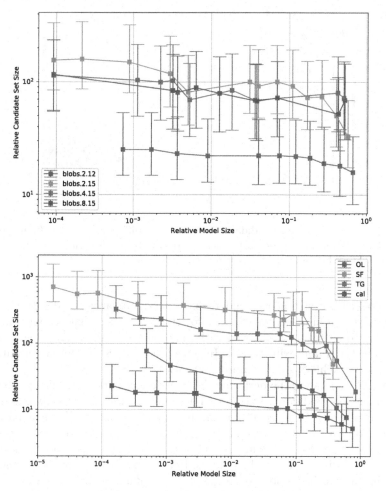

Fig. 3. Comparison of relative model size with respect to the number of parameters of MRkNNCoP-Tree, and the relative candidate set size (upper: synthetic, lower: real-world). The error bars show the inter-quartile range. Notice the log-log scale.

tight as the optimal one from the MRkNNCoP-Tree. Second, the blue predicted lower bound does violate the lower bounding property for very large k. With dotted line style the guaranteed bounds are shown, i.e. the prediction plus the maximum error across the full dataset. The guaranteed bounds encompass a much larger area, corresponding to a potentially larger candidate set.

Figure 3 shows a comparison of all models for all datasets. Each point corresponds to one model, having a relative model size compared to the MRkNNCoP-Tree, and a mean relative candidate set size over all query points and all values of k, also compared to the MRkNNCoP-Tree. Furthermore, the error bar shows the inter-quartile range of the latter value. The results presented in the figure emphasize our expectations: Smaller models are not strong enough to accurately

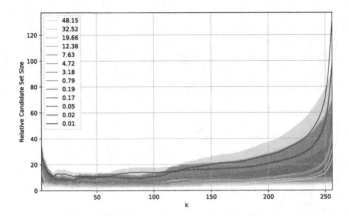

Fig. 4. Analysis of relative candidate set size across different values of k, and different model sizes for the *cal* dataset. Shaded areas indicate inter-quartile range; color corresponds to relative model size (in percentage).

fit the coefficients and hence tend to have a large relative candidate set size. However, we can also see, for instance for the *cal* dataset that strong compression factors of 10^4 can be achieved with only increasing the mean candidate set size by a factor of approximately 12. In general it appears that for larger datasets it is harder to achieve good compression ratios.

In Fig. 4 we analyze the performance for different values of k. The plot shows different models as differently colored lines, where the color corresponds to the relative model size in percent, ranging from 48% to 0.01%. The shaded areas are used to indicate the inter-quartile ranges. We can observe that very high candidate set sizes occur at values of k close the $k_{\max}$. This might be the case, due to the exponential nature of the bounds. Hence, a small error in the coefficients of the linear line in log-log space has a huge influence for large values of k. The additional peak close to $k = 1$ can be explained with the high variety of 1-nearest-neighbor distances, which are not as smooth as the distances for larger k-values.

Figure 5 shows the distribution of errors in the predicted upper bound for the *cal* datasets for different values of k encoded by color. In particular when taking the logarithmic scale of the x-axis into consideration, we can observe that the vast majority of predictions has a relatively small error. Thus, the offset for the guaranteed bound, and thereby the penalty in candidate set size, is dominated by a few points with relatively bad predictions. Hence, for future research it might be worthwhile to consider scenarios with approximate query processing, where the bound can be guaranteed to hold for e.g. 99% of the dataset, but at the same time is much tighter than the currently considered exact bound.

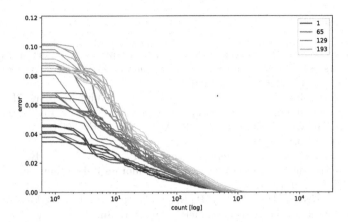

Fig. 5. Distribution of errors for the *cal* dataset. The plot shows the errors of the predicted upper bound ordered in decreasing order with logarithmic x-scale.

6   Conclusion

In this work, we presented a neural indexing approach which replaces the conservative and progressive k-distance approximations of the MRkNNCoP-Tree by a trained regression model, and hence allows for reducing the required storage space from $\mathcal{O}(n)$ to $\mathcal{O}(1)$. Based on our findings that the coefficients of the linear approximation functions are spatially correlated, we propose to substitute the set of approximation functions by a single regression model which has been trained with the objective to predict the offset and slope values of the approximation functions at query time. In our experiments, we discuss the performance of the neural index compared to the original MRkNNCoP-Tree with emphasis on the trade-off between model size and the number of candidates that must be refined to get an exact query result. The results show that the set of candidates increases with decreasing model size. However, the growth of the candidate set is mainly reasoned by the fact that the prediction error of the regression model is only for very few data points significant while it is quite low for the majority of data objects. Therefore, it might be of special interest to investigate the proposed neural index in the context of approximate RkNN query retrieval in future work. Another (and especially in the context of approximate query processing) promising direction is to study the possibility of supporting fast RkNN query processing with a variant of our neural index that predicts k-distances directly.

Acknowledgements. Parts of this work have been funded by the German Research Foundation (DFG) under grant number AC 242/4-2. This work has been developed in cooperation with the Munich Center for Machine Learning (MCML), funded by the German Federal Ministry of Education and Research (BMBF) under Grant No. 01IS18036A. The authors of this work take full responsibilities for its content.

References

1. Achtert, E., Böhm, C., Kröger, P., Kunath, P., Pryakhin, A., Renz, M.: Approximate reverse k-nearest neighbor queries in general metric spaces. In: Proceedings of CIKM, pp. 788–789. Citeseer (2006)
2. Achtert, E., Böhm, C., Kröger, P., Kunath, P., Pryakhin, A., Renz, M.: Efficient reverse k-nearest neighbor search in arbitrary metric spaces. In: Proceedings of SIGMOD, SIGMOD 2006, pp. 515–526. ACM, New York (2006). https://doi.org/10.1145/1142473.1142531
3. Achtert, E., Böhm, C., Kröger, P., Kunath, P., Pryakhin, A., Renz, M.: Efficient reverse k-nearest neighbor estimation. Informatik-Forschung und Entwicklung **21**(3–4), 179–195 (2007)
4. Achtert, E., Kriegel, H.P., Kröger, P., Renz, M., Züfle, A.: Reverse k-nearest neighbor search in dynamic and general metric databases. In: Proceedings of EDBT, pp. 886–897. ACM (2009)
5. Borutta, F., Nascimento, M.A., Niedermayer, J., Kröger, P.: Monochromatic RkNN queries in time-dependent road networks. In: Proceedings of SIGSPATIAL MobiGIS, pp. 26–33. ACM (2014)
6. Borutta, F., Nascimento, M.A., Niedermayer, J., Kröger, P.: Reverse k-nearest neighbour schedules in time-dependent road networks. In: Proceedings of SIGSPATIAL, p. 27. ACM (2015)
7. Cheema, M.A., Lin, X., Wang, W., Zhang, W., Pei, J.: Probabilistic reverse nearest neighbor queries on uncertain data. IEEE TKDE **22**(4), 550–564 (2010)
8. Cheema, M.A., Zhang, W., Lin, X., Zhang, Y., Li, X.: Continuous reverse k nearest neighbors queries in euclidean space and in spatial networks. VLDB J. **21**(1), 69–95 (2012)
9. Gripon, V., Löwe, M., Vermet, F.: Associative memories to accelerate approximate nearest neighbor search. Appl. Sci. **8**(9), 1676 (2018)
10. Kingma, D.P., Ba, J.: Adam: a method for stochastic optimization. In: Proceedings of ICML (2015)
11. Korn, F., Muthukrishnan, S.: Influence sets based on reverse nearest neighbor queries. In: ACM Sigmod Record, vol. 29, pp. 201–212. ACM (2000)
12. Kraska, T., Beutel, A., Chi, E.H., Dean, J., Polyzotis, N.: The case for learned index structures. In: Proceedings of SIGMOD, pp. 489–504. ACM (2018)
13. Lian, X., Chen, L.: Efficient processing of probabilistic reverse nearest neighbor queries over uncertain data. VLDB J. **18**(3), 787–808 (2009)
14. Oosterhuis, H., Culpepper, J.S., de Rijke, M.: The potential of learned index structures for index compression. arXiv preprint arXiv:1811.06678 (2018)
15. Pedregosa, F., et al.: Scikit-learn: machine learning in python. J. Mach. Learn. Res. **12**, 2825–2830 (2011)
16. Singh, A., Ferhatosmanoglu, H., Tosun, A.Ş.: High dimensional reverse nearest neighbor queries. In: Proceedings of CIKM, pp. 91–98. ACM (2003)
17. Stanoi, I., Agrawal, D., El Abbadi, A.: Reverse nearest neighbor queries for dynamic databases. In: SIGMOD Workshop DMKD, pp. 44–53 (2000)
18. Tao, Y., Papadias, D., Lian, X.: Reverse kNN search in arbitrary dimensionality. In: Proceedings of VLDB, pp. 744–755. VLDB Endowment (2004)
19. Tao, Y., Yiu, M.L., Mamoulis, N.: Reverse nearest neighbor search in metric spaces. IEEE TKDE **9**, 1239–1252 (2006)
20. Xia, T., Zhang, D.: Continuous reverse nearest neighbor monitoring. In: 22nd International Conference on Data Engineering (ICDE 2006), p. 77. IEEE (2006)

21. Xiang, W., Zhang, H., Cui, R., Chu, X., Li, K., Zhou, W.: Pavo: a RNN-based learned inverted index, supervised or unsupervised? IEEE Access **7**, 293–303 (2019). https://doi.org/10.1109/ACCESS.2018.2885350
22. Yang, C., Lin, K.I.: An index structure for efficient reverse nearest neighbor queries. In: Proceedings of ICDE, pp. 485–492. IEEE (2001)
23. Yu, C., Gripon, V., Jiang, X., Jégou, H.: Neural associative memories as accelerators for binary vector search. In: Proceedings of Cognitive, pp. 85–89 (2015)

Pruning Algorithms for Low-Dimensional Non-metric k-NN Search: A Case Study

Leonid Boytsov$^{(\boxtimes)}$ and Eric Nyberg

Carnegie Mellon University, Pittsburgh, PA, USA
{srchvrs,enh}@cs.cmu.edu

Abstract. We focus on low-dimensional non-metric search, where tree-based approaches permit efficient and accurate retrieval while having short indexing time. These methods rely on space partitioning and require a pruning rule to avoid visiting unpromising parts. We consider two known data-driven approaches to extend these rules to non-metric spaces: TriGen and a piece-wise linear approximation of the pruning rule. We propose and evaluate two adaptations of TriGen to non-symmetric similarities (TriGen does not support non-symmetric distances). We also evaluate a hybrid of TriGen and the piece-wise linear approximation pruning. We find that this hybrid approach is often more effective than either of the pruning rules. We make our software publicly available.

Keywords: k-NN search · Non-metric distance · VP-tree · TriGen

1 Introduction and Problem Definition

We consider a k nearest neighbor (k-NN) search, which is a popular technology used in many domains including, machine learning (ML), data mining, information retrieval, and natural language processing. Informally, k-NN search is a task of retrieving k data set entries closest to a query point with respect to some distance or similarity function. This problem originated from the real-world spatial search. In particular, Knuth famously formulated k-NN search as the (nearest) post-office problem [14]. With subsequent developments of the vector-space abstraction, the problem was generalized to searching in a multi-dimensional vector and/or generic metric space, where the latter may lack the structure of the vector space [10,21]. Motivated by emergence of useful non-metric distances—such as Bregman divergences [7]—the problem was recently generalized to more challenging domains [5,8,23,27].

Formally, we assume to have a possibly infinite domain containing objects x, y, z, ..., which are commonly called data points or simply points. The domain—sometimes called a *space*—is equipped with a *distance function* $d(x, y)$, which is used to measure dissimilarity of objects x and y. The value of $d(x, y)$ is interpreted as a degree of dissimilarity. The larger is $d(x, y)$, the more dissimilar points x and y are. Some distances

---

Authors gratefully acknowledge the support by the NSF grant #1618159: "Matching and Ranking via Proximity Graphs: Applications to Question Answering and Beyond".

G. Amato et al. (Eds.): SISAP 2019, LNCS 11807, pp. 72–85, 2019.
https://doi.org/10.1007/978-3-030-32047-8_7

Table 1. Distance functions

Denotation/Name	d(x, y)
Euclidean distance (L_2)	$\|x - y\|_2 = \left[\sum_i (x_i - y_i)^2\right]^{1/2}$
L_p ($p > 0$)	$\left[\sum_{i=1}^{m} (x_i - y_i)^p\right]^{1/p}$
Squared euclidean (L_2^2)	$\|x - y\|_2^2 = \sum_i (x_i - y^i)^2$
Cosine distance	$1 - \dfrac{\sum_i x_i y_i}{\|x\|_2 \|y\|_2}$
Kullback-Leibler diverg. (KL-div.) [15]	$\sum_{i=1}^{m} x_i \log \dfrac{x_i}{y_i}$
Itakura-Saito distance [13]	$\sum_{i=1}^{m} \left[\dfrac{x_i}{y_i} - \log \dfrac{x_i}{y_i} - 1\right]$
Rényi diverg. [20]	$\dfrac{1}{\alpha-1} \log \left[\sum_{i=1}^{m} x_i^\alpha y_i^{1-\alpha}\right]$, $\alpha > 0$ and $\alpha \neq 0$

are non-negative and become zero only when x and y have the highest possible degree of similarity. The *metric* distances are additionally symmetric and satisfy the triangle inequality. However, in general, we do not impose any restrictions on the value of the distance function (except that smaller values represent more similar objects).

We further assume that there is a data set D containing a *finite* number of domain points and a set of queries that belong to the domain but not to D. We then consider a standard top-k retrieval problem. Given a query q, a retrieval task consists in finding k data set points $\{x_i\}$ with smallest values of distances to the query among all data set points (ties are broken arbitrarily). Data points $\{x_i\}$ are called *nearest neighbors*. A search should return $\{x_i\}$ in the order of increasing distance to the query. If the distance is not symmetric, two types of queries can be considered: *left* and *right* queries. In a *left* query, a data point compared to the query is always the first (i.e., the left) argument of $d(x, y)$. For simplicity of exposition we consider only the case of left queries.

We employ a space-partitioning method VP-tree [19, 24, 26], but many other space-partitioning approaches can be used. Importantly, applying space-partitioning methods to non-metric data of even moderate dimensionality entails two problems. First, exact space-partitioning methods can degenerate to a brute-force search for just a dozen of dimensions [1, 25]. Second, many generic space-partitioning methods incorporate pruning rules that crucially rely on the triangle inequality, which does not generally hold in non-metric spaces. Most existing non-metric space-partitioning methods employ specialized extensions specific to a concrete class of distances, e.g., to Bregman divergences [8, 27] or Ptolemaic distances [12]. However, in a more general case we clearly need to resort to empirically derived analogs of the triangle inequality, which are inferred from data with a certain degree of approximation.

For these reasons, we focus only on *approximate* search methods. We also restrict our attention to low- and moderate-dimensional methods, because even approximate pruning methods are not effective in truly high dimensions. There has been a tremendous effort put into design of metric space-partitioning algorithms [10, 21], but many

Table 2. Data sets

Name	Max. # of rec.	Dimensionality	Source
RandHist-d	0.5×10^6	$d = 8$	Histograms sampled uniformly from a simplex
RCV-d	0.5×10^6	$d \in \{8, 32, 128\}$	d-topic LDA [2] RCV1 [16] histograms
Wiki-d	2×10^6	$d \in \{8, 32, 128\}$	d-topic LDA [2] Wikipedia histograms

fewer methods are designed for non-metric domains. We aim to fill this gap by making the following contribution, which we detail in the rest of the paper:

- We carry out the first experimental comparison of two existing generic pruning algorithms, which include the piecewise linear approximation of the pruning rule [5] and TriGen [22].
- Unlike most prior work, many of our distances are non-symmetric. To deal with non-symmetry, we propose two adaptation of TriGen to non-symmetric distances and demonstrate that the choice of the symmetrization algorithm can be quite important.
- In our comprehensive evaluation, which includes 40 combinations of data sets and distances, we demonstrate the feasibility of accurate non-metric k-NN search for data of moderate dimensionality.
- We demonstrate that often best results can be achieved by combining these pruning methods.
- We find that on data of moderate dimensionality, the pruning algorithm needs to be quite efficient.

2 Methods and Materials

2.1 Data Sets and Distances

In our experiments, we use the following non-metric distances: L_2^2 (squared Euclidean) L_p distance, cosine distance, KL-divergence, the Itakura-Saito distance, and the family of Rényi divergence distances. The first three distances are symmetric. The remaining distances are statistical distances defined over probability distributions. For expository purposes, we also use the Euclidean metric distance L_2. Distances are listed in Table 1.

Statistical distances in general and, KL divergence in particular, play an important role in ML [8, 17]. They are typically non-symmetric. Both the KL-divergence and the Itakura-Saito distances were used in prior work [8]. The Rényi divergence is a single-parameter family of distances, which are not symmetric when the parameter $\alpha \neq 0.5$. By changing the parameter we can vary the degree of symmetry. In particular, large values of α and close-to-zero values result in highly non-symmetric distances. This flexibility allows us to "stress-test" retrieval methods on challenging non-symmetric distances.

The data sets are listed in Table 2. Wiki-d and RCV-d data sets consist of dense vectors of topic histograms with d topics. RCV-d set are created by Cayton [8] by applying the latent Dirichlet allocation (LDA) method [2] to the RCV1 collection [16]. These data sets have only 500K entries. Thus, we created larger sets from Wikipedia following a similar methodology. RandHist-d is a synthetic set of topics sampled uniformly from a d-dimensional simplex.

2.2 Pruning Algorithms for Space-Partitioning Methods

We employ a simple approach called a *vantage-point* tree (VP-tree) [19,24,26]. There are two reasons for this choice: for low- and moderate-dimensional data, it is often a hard-to-beat method. For example, in a preliminary experiment with L_2 on Wiki-8 data set for exact 10-NN search using NMSLIB [4], SA-tree [18], GH-tree [24], MVP-tree (binary version) [6], and VP-tree are respectively 70×, 210×, 1200×, 1600× faster than the brute-force search. This comparison was done using the leaf bucket of size 50 for all methods (except SA-tree, which does not easily support bucketing) and without using any specific optimizations for any of the methods. We can see that VP-tree can outperform fancier alternatives including MVP-tree, which carries out 3× fewer distance computations in this experiment.

VP-tree is a hierarchical space-partitioning method, which divides the space using hyperspheres. The output of an indexing algorithm is a hierarchical partitioning of the data set represented by a binary tree. This algorithm is a *recursive* procedure that operates on a subset of data—which we call an *active subset*—and on a partially built tree. At each step of recursion, the indexing algorithm checks if the number of active data points is below a certain threshold called the *bucket* size. If this is the case, the active data points are simply stored as a single bucket. Otherwise, the algorithm divides the active subset into two nearly equal parts, each of which is further processed recursively.

Division of the active subset starts with selecting a pivot π (e.g., randomly) and computing the distance from π to every other data point in the active subset. Assume that R is the median distance. Then, the active subset is divided into two subsets by the hypersphere with radius R and center π. Two subtrees are created. Points inside the pivot-centered hypersphere are placed into the left subtree. Points outside the pivot-centered hypersphere are placed into the right subtree. Points on the separating hypersphere may be placed arbitrarily. Because R is the median distance, each of the subtrees contains approximately half of active points.

In VP-tree k-NN search can be seen as a range search with a shrinking radius. The search algorithm is a *best-first* traversal procedure that starts from the root of the tree and proceeds recursively. It updates the search radius r as it encounters new close data points. Let us consider one step of recursion. If the search algorithm reaches a leaf of the tree, i.e., a bucket, all bucket elements are compared against the query. In other words, elements in the buckets are searched via brute-force search.

If the algorithm reaches an internal node X, there are exactly two subtrees representing two spaces partitions. The query belongs to exactly one partition. This is the "best" partition and the search algorithm always explores this partition recursively before deciding whether to explore the other partition. While exploring the best partition, we may encounter new close data points (pivots or bucket points) and further shrink the search radius. On completing the sub-recursion and returning to node X, we make a decision about pruning or exploring the other partition.

Piecewise-Linear Approximation of the Decision Rule. An essential part of this process is a decision function, which identifies situations when pruning is possible without sacrificing accuracy. Let us review the decision process. Recall that each internal node keeps pivot π and radius R, which define the division of the space into two subspaces. Although there are many ways to place a query ball, all locations can be divided into three categories, which are illustrated by Fig. 1. The red query ball "sits" inside the inner partition. Note that it does not intersect with the outer partition. For this reason,

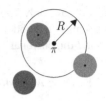

Fig. 1. Three types of query balls in VP-tree.

the outer partition cannot have sufficiently close data points, i.e., points with radius r from the query. Hence, this partition can be safely pruned. The blue query ball is located in the outer partition. Likewise, it does not intersect the other, inner, partition. Thus, this inner partition can be safely pruned. Finally, the gray query ball intersects both partitions. In this situation, sufficiently close points may be located in both partitions and no safe pruning is possible.

The pruning algorithm can be seen as the *binary classification problem*, which tells us whether we should visit both partitions or only the partition that contains the query. As we show previously [5], the problem can be solved by collecting training data and building a non-parametric model, but a simple two-parameter approach—described below—delivers better results. Let us first consider the case of a metric distance. From the triangle inequality it follows that the VP-tree search algorithm visits:

- *only* the left subtree if $d(\pi, q) < R - r$;
- *only* the right subtree if $d(\pi, q) > R + r$;
- both subtrees if $R - r \leq d(\pi, q) \leq R + r$.

Let us rewrite these rules using notation $D_{\pi,R}(x) = |R - x|$. It is easy to see that the search algorithm has to visit both partitions if and only if $r \geq D_{\pi,R}(d(\pi, q))$. If $r < D_{\pi,R}(d(\pi, q))$, we need to visit only one partition that contains the query point whereas the other partition can be safely pruned.

In other words, the pruning decision is made by comparing the query radius r with the value of the function $D_{\pi,R}(x)$, whose only argument is the distance from the query to the pivot $d(\pi, q)$.[1] This basic rule can also be learned from data for non-metric distances. Our initial approach to learn $D_{\pi,R}(x)$ employed a stratified sampling procedure (see §2 of the supplemental materials of our publication [5]). However, it was expensive and not very accurate. For this reason, we also implemented a simple *parametric* approximation whose parameters are selected to optimize efficiency at a given value of recall.

To choose the right parametric representation, we examine the (*approximate*) functions $D_{\pi,R}(x)$ learned by the sampling algorithm. Plots of functions $D_{\pi,R}(x)$ learned from data are shown in Fig. 2. Small dots in these plots represent function values obtained by sampling. Blue curves are fit to these dots. In these plots, we use only topic histogram data RCV-d, where $d \in \{8, 32\}$ and random 8-dimensional histograms (RandHist-8).

---

[1] Recall that k-NN search is executed as a best-first range search with a shrinking radius.

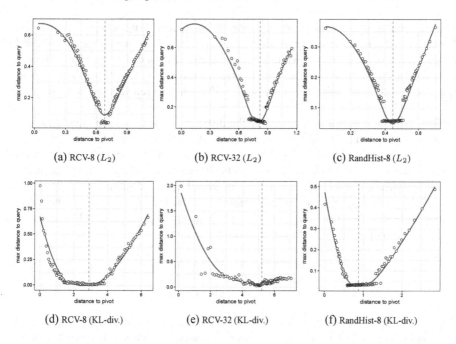

(a) RCV-8 (L_2)   (b) RCV-32 (L_2)   (c) RandHist-8 (L_2)

(d) RCV-8 (KL-div.)   (e) RCV-32 (KL-div.)   (f) RandHist-8 (KL-div.)

Fig. 2. The empirically obtained (*approximate*) pruning decision function $D_{\pi,R}(x)$

For the Euclidean data (Panels 2a–2c in Fig. 2), $D_{\pi,R}(x)$ resembles a piecewise linear function close to the exact metric pruning function $|R - x|$. For the KL-divergence data (Panels 2d–2f in Fig. 2), $D_{\pi,R}(x)$ looks like either a U-shape or a hockey-stick curve. These observations originally motivated the use of a piecewise *polynomial* decision function, which is formally defined as:

$$D_{\pi,R}(x) = \begin{cases} \alpha_{left}|x - R|^{\beta_{left}}, & \text{if } x \leq R \\ \alpha_{right}|x - R|^{\beta_{right}}, & \text{if } x \geq R \end{cases}, \tag{1}$$

where β_i are positive integers. However, preliminary experiments convinced us to switch to a simple piece-wise linear variant. First, we learned that using different β_i did not make our pruning function sufficiently more accurate. However, it made the optimization problem harder due to additional parameters (so we set $\beta = \beta_1 = \beta_2$). Second, we found that in many cases a polynomial approximation was not better than a piecewise linear one, especially when dimensionality was high.

This is not very surprising: Due to the concentration of measure, for most data points the distance to the pivot π is close to the median distance R (which corresponds to the boundary between two VP-tree partitions). If we explore the shape of $D_{\pi,R}(x)$ in Panels 2a and 2e around the median, we can see that a piecewise linear shape approximation is quite reasonable. To sum up, we ended up using the piecewise linear parametric decision rule defined as:

$$D_{\pi,R}(x) = \begin{cases} \alpha_{left}|x - R|, & \text{if } x \leq R \\ \alpha_{right}|x - R|, & \text{if } x \geq R \end{cases} \tag{2}$$

This is similar to stretching of the triangle inequality proposed by Chávez and Navarro [9]. There are two crucial differences, however. First, we utilize different values of α_i, i.e., $\alpha_{left} \neq \alpha_{right}$, while Chávez and Navarro used $\alpha_{left} = \alpha_{right}$. Second, we devise a simple training procedure to obtain values of α_i that maximize efficiency at a given recall value. For details, we address the reader to relevant publications [3,5].

TriGen. TriGen consists in "stretching" the distance function using a monotonic concave transformation [22] that reduces non-metricity of the distance. TriGen is designed only for *bounded, semimetric* distances, which are crucially *symmetric*, non-negative, and become zero only for identical data points. We are not aware of any prior extensions to non-symmetric distances except a straightforward filter-and-refine approach.

Let x, y, z be an arbitrary ordered triple of points such that $d(x, y)$ is the largest among three pairwise distances, i.e., $d(x, y) \geq \max(d(x, z), d(z, y))$. If $d(x, y)$ is a metric distance, the following conditions should all be true:

$$d(x, y) \leq d(x, z) + d(z, y)$$
$$d(y, z) \leq d(y, x) + d(x, z) \qquad (3)$$
$$d(x, z) \leq d(x, y) + d(y, z)$$

Because $d(x, y) \geq \max(d(x, z), d(z, y))$, the second and the third inequalities in (3) are trivially satisfied for (not necessarily metric) *symmetric* and *non-negative* distances. However, the first condition can be violated if the distance is non-metric. The closer is the distance to the metric distance, the less frequently we encounter such violations. Thus, it is quite reasonable to assess the degree of deviation from metricity by estimating a probability that the triangle inequality is violated (for a randomly selected triple), which is exactly what is suggested by Skopal [22].

Skopal proposes a clever way to decrease non-metricity by constructing a new distance $f(d(x, y))$, where $f()$ is a monotonically increasing concave function. The concave function "stretches" the distance and makes it more similar to a true metric compared to the original distance $d(x, y)$. At the same time, due to the monotonicity of such a transformation, the k-NN search using the modified distance produces the same result as the k-NN search using the original distance. Thus, the TriGen strategy to dealing with non-metric data consists in (1) employing a monotonic transformation that makes a distance approximately metric while preserving the original set of nearest neighbors, and (2) indexing data using an exact metric-space access method.

A TriGen mapping $f(x)$—defined for $0 \leq x \leq 1$—is selected from the union of two parametric families of concave functions, which are termed as bases:

- A fractional power base $FP(x, w) = x^{\frac{1}{1+w}}$;
- A Rational Bézier Quadratic (RBQ) base $RBQ_{(a,b)}(x, w)$, $0 \leq a < b \leq 1$. The exact functional form of RBQ is not relevant to this discussion (see [22] for details).

Note that parameters w, a, and b are treated as constants, which define a specific functional form. By varying these parameters we can design a necessary stretching function. The larger is the value of w the more concave is the transformation and the more "metric" is the transformed distance. In particular, as $w \to \infty$, both RBQ and FP converge to one minus the Dirac delta function. This limit function of all bases is equal to zero for

$x = 0$ and to one for $0 < x \leq 1$. As noted by Skopal [22], applying such a degenerate transformation produces a *trivial* metric space where $d(x, x) = 0$ and $d(x, y) = C$ for some constant $C > 0$ and all $x \neq y$.

A learning objective of TriGen, however, is to select a *single* concave function that satisfies the accuracy requirements while allowing for efficient retrieval. The fraction of violations is computed for a set of `trigenSampleTripletQty` ordered data point triples sampled from a set of `trigenSampleQty` data points, which are, in turn, selected randomly from the data set (uniformly and without replacement). The fraction of violations is required to be above the threshold `trigenAcc`. Values `trigenSampleTripletQty`, `trigenSampleQty`, and `trigenAcc` are all parameters in our implementation of TriGen. To assess efficiency Skopal uses the value of an intrinsic dimensionality as a proxy metric (see [22] for details). The idea is that the modification of the distance with the lowest intrinsic dimensionality should result in the fast retrieval method.

Because it is not feasible to optimize over the infinite set of transformation functions, TriGen employs a finite pool of bases, which includes FB and multiple RBQ bases for all possible combinations of parameters a and b such that $0 \leq a < b \leq 1$. For each base, TriGen uses a binary search to find the minimum parameter w such that the transformed distance deviates from a metric distance within specified limits. Then the base with minimum intrinsic dimensionality is selected.

TriGen has two major limitations: In addition to be non-negative, the distance should be symmetric and bounded. Bounding can be provided by using $\min(d(x, y)/D_{\max}, 1)$ instead of the original distance.[2] Note that $D_{\max}$ is an empirically estimated maximum distance (obtained by computing $d(x, y)$ for a sample of data set point pairs).

As noted by Skopal [22], searching with a non-symmetric distance can be partially provided by a filter-and-refine approach where a fully *min*-symmetrized distance $\min(d(x, y), d(y, x))$ is used during the filtering step. However, as we learn from our prior work §§ 2.3.2.3–2.3.2.4 [3], the filtering step has to carry out a k_c-NN search with k_c (sometimes substantially) larger than k. This is required to compensate for the lack of accuracy due to replacing the original distance with the symmetrized one. In that, using $k_c > k$ leads to reduced efficiency. Thus, instead of the complete filter-and-refine symmetrization, we consider two simple alternatives. In both cases we first apply the TriGen algorithm to the min-symmetrized distance. As a result, we obtain a mapping that makes this min-symmetrized distance to be closer to a metric distance. However, this mapping is used differently in the two modifications of TriGen.

Recall that in a typical space-partitioning method, we divide the data into reasonably large buckets (50 in our experiments). The k-NN search is simulated as a range search with a shrinking radius. In the case the first modification of TriGen, while we traverse the tree, we compute the original and the min-symmetrized distance for two purposes:

- shrinking the dynamic radius of the query using the *symmetrized* distance;
- checking if the *original* distance is small enough to update the current set of k nearest neighbors.

---

[2] For efficiency reasons this is simulated via multiplication by inverse maximum distance.

Table 3. Efficiency-effectiveness results for metric VP-tree on non-metric data for 10-NN search (using complete data sets).

	RCV-8		Wiki-8		RandHist-8		Wiki-128	
	Recall	Impr. in eff.	Recall	Impr. in eff.	Recall	Impr. in eff.	Recall	Impr. in eff.
$L_p(p = 0.125)$	0.41	1065	0.66	15799	0.45	136	0.07	14845
$L_p(p = 0.25)$	0.61	517	0.78	14364	0.66	115	0.09	396
$L_p(p = 0.5)$	0.91	926	0.94	14296	0.92	174	0.50	33
L_2^2	0.69	1607	0.78	5605	0.56	1261	0.55	114
Cosine dist	0.67	1825	0.62	3503	0.58	758	0.73	55
Rényi div. ($\alpha = 0.25$)	0.66	5096	0.70	24246	0.50	3048	0.48	1277
Rényi div. ($\alpha = 0.75$)	0.61	9587	0.66	35940	0.50	4673	0.50	468
Rényi div. ($\alpha = 2$)	0.40	22777	0.66	46122	0.38	11762	0.71	55
KL-div.	0.52	1639	0.67	5271	0.46	610	0.56	41
Itakura-Saito	0.46	706	0.69	4434	0.41	1172	0.14	384

When we reach a bucket, for every data point in the bucket, we can compute both the original and the symmetrized distance. The symmetrized distance is used to update the query radius, while the original distance is used to update the set of k nearest neighbors. This is our first modification of TriGen which we refer to as *TriGen 0*.

In the second variant of TriGen, which we refer to as *TriGen 1*, we use *only* the original distance to compute the distance from the query to bucket data points. When we compute the distance to the pivots, we compute the min-symmetrized distance and apply a metrizing transformation. However, when we process bucket data points, we compute only the original distance. Consequently, we shrink the dynamic query radius using values of $f(d(x, y))$ instead of $\min(f(d(x, y)), f(d(y, x)))$, In TriGen 1, we expect the query radius to shrink somewhat slower compared to TriGen 0, which, in turn, can reduce the effectiveness of pruning. However, we hope that nearly halving the number of distance computations would have a larger effect on overall retrieval time.

3   Experiments

3.1   Experimental Setup and Preliminary Experiments

We compare TriGen and the piecewise-linear pruning approach using the NMSLIB [4] implementation of the VP-tree (method `vptree_trigen`)[3]. Experiments are run on a laptop (i7-4700MQ @ 2.40 GHz with 16 GB of memory). The accuracy of retrieval is measured via recall (equal to the average fraction of neighbors found).

We use two variants of TriGen (TriGen 0 and TriGen 1), but for symmetric distances, we use only TriGen 1. The TriGen algorithm that finds an optimal mapping function was downloaded from the author's website[4] and incorporated into NMSLIB.

---

[3] https://github.com/nmslib/nmslib/tree/nmslib4a_bigger_reruns.

[4] http://siret.ms.mff.cuni.cz/skopal/download.htm.

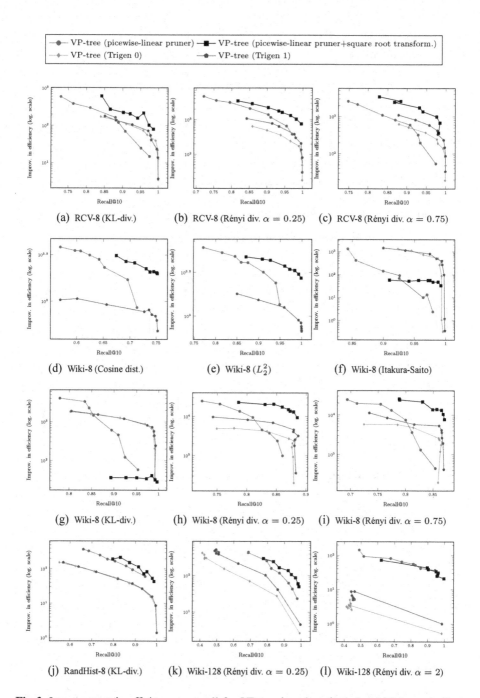

Fig. 3. Improvement in efficiency vs recall for VP-tree based methods in 10-NN search. Best viewed in color.

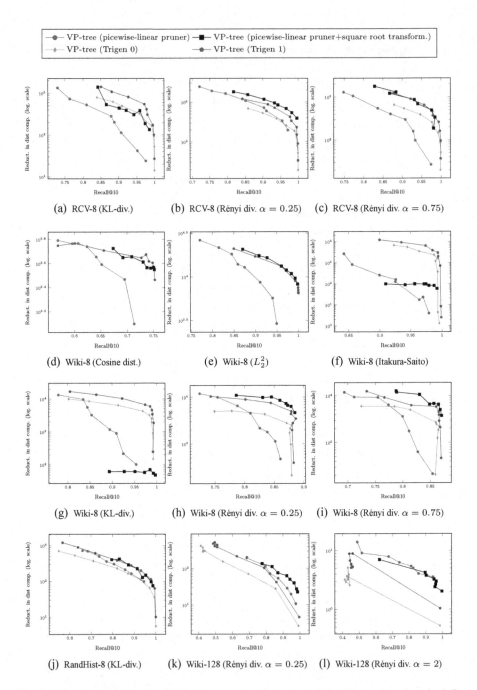

Fig. 4. Reduction in the number of distance computations vs recall for VP-tree based methods in 10-NN search. Best viewed in color.

The optimization procedure employs a combination of parameters a and b, where a are multiples of 0.01, b are multiples of 0.05, and $0 \leq a < b \leq 1$. The sampling parameters are set as follows: `trigenSampleTripletQty` = 10000, `trigenSampleQty` = 5000.

TriGen is compared against two variants of NMSLIB VP-tree, which rely on the piecewise-linear pruner. The second variant uses a clever TriGen idea of applying a concave mapping to make the distance more similar to a metric one. However, unlike TriGen [22], we do not carry an extensive search for an optimal transformation but rather apply, perhaps, the simplest and fastest *monotonic concave* transformation possible, which consists in taking a square root. On Intel the square root is computed the instruction `sqrtss`, which typically takes less than 10 cycles [11].

In our main experiments, we employ 40 combinations of data sets and distances. All distances are non-metric: We experiment with both symmetric and non-symmetric ones. Due to space limitations, we do not present all the results here and refer the reader to our unpublished technical report for the complete set of results (§ 2.3.3 [3]).

Before we proceed, we must answer the following question: "How difficult are these data sets and distances"? To ensure we do not deal with mildly non-metric data, we attempted to index this data using a metric variant VP-tree without adapting the pruning rule to non-metric distances. Results for randomly selected 1K queries are presented in Table 3 (for a subset of distances and data sets), where we show improvement in efficiency and respective recall.

It can be seen that nearly all the combinations of data and distance functions are substantially non-metric: Searching using a metric VP-tree is usually fast, but the accuracy is *unacceptably* low. In particular, this is true for Wiki-8 and Wiki-128 data sets with KL-divergence (which are also used in our main experiments). One exception, is the L_p distance for $p = 0.5$, where recall of about 90% is achieved for three low-dimensional data sets. However, as p decreases, the recall decreases sharply, i.e., the distance function becomes "less" metric. To summarize, we clearly deal with challenging non-metric data sets, where both accurate and efficient retrieval is not possible to achieve by a straightforward application of metric-space search methods.

3.2 Main Experiments

Experimental results for 16 out of 40 cases are presented in Figs. 3 and 4. The remaining results can be found in the technical report (§ 2.3.3 [3]). In Fig. 3, we measure efficiency directly in terms of wall-clock time improvement over the brute-force search. In Fig. 4, we show improvement in the number of distance computations (again compared to the brute-force search).

First and foremost, we can see that VP-tree with a data-adapted pruning rule can enable accurate non-metric k-NN search for data of moderate dimensionality. When comparing TriGen against the piecewise linear pruner in terms of pure efficiency, the results are a bit of the mixed bag. Yet, the piecewise linear pruner is typically better (in 23 cases out of 40 on the full set, see § 2.3.3 [3]).

However, the piecewise linear pruner combined with the square-root distance transform is nearly always better than the basic piecewise linear pruner. In Panels 3d, 3e, 3a, 3b, 3c the improvement is up to one order of magnitude. The combination of the piecewise linear pruner with the square root transform outperforms TriGen in all but two cases, sometimes by an order of magnitude. In Panels 3g and 3f, however, TriGen can also be an order of magnitude faster than the piecewise linear pruner.

It is important to note, however, that there is often little to no difference between the hybrid pruning approach and TriGen in terms of the reduction in the number of distance computations (see Fig. 4). The most likely explanation for this discrepancy is that the transformation functions used in the adopted TriGen implementation are quite expensive to compute.

Finally, we can see that TriGen 1 is never less efficient than TriGen 0. Furthermore, TriGen 1 is up two times more efficient in four cases (see Panels 3h, 3i, 3k, 3l). This is somewhat unsurprising, because TriGen 0 computes both $d(x, q)$ and $d(q, x)$ for every data point visited by the search. Although this may permit a more effective pruning, the cost of extra distance computations outweigh the benefits (at least on our data).

4   Conclusion

We carry out the first comparison of two generic pruning approaches for non-metric data. Our approach is comprehensive and involves 40 combinations of data sets and distances, which cannot be handled by a classic metric-space access method. We extend TriGen to the case of non-symmetric distances and demonstrate that VP-tree with a data-adapted pruning rule can enable accurate non-metric k-NN search for data of moderate dimensionality by using the modified TriGen, the piecewise linear approximation of the metric pruning rule, or by the hybrid approach. In that, we find that this hybrid approach is often more effective than either of the pruning rules. Our software is publicly available: NMSLIB branch `nmslib4a_bigger_reruns`, search method `vptree_trigen`.[5]

Acknowledgments. This work was done while Leonid Boytsov was a PhD student at CMU. Authors gratefully acknowledge the support by the NSF grant #1618159.

References

1. Beyer, K.S., Goldstein, J., Ramakrishnan, R., Shaft, U.: When is "nearest neighbor" meaningful? In: Beeri, C., Buneman, P. (eds.) ICDT 1999. LNCS, vol. 1540, pp. 217–235. Springer, Heidelberg (1999). https://doi.org/10.1007/3-540-49257-7_15
2. Blei, D.M., Ng, A.Y., Jordan, M.I.: Latent dirichlet allocation. J. Mach. Learn. Res. **3**, 993–1022 (2003)
3. Boytsov, L.: Efficient and accurate non-metric k-NN search with applications to text matching. Ph.D. thesis, Carnegie Mellon University (2017)
4. Boytsov, L., Naidan, B.: Engineering efficient and effective non-metric space library. In: Brisaboa, N., Pedreira, O., Zezula, P. (eds.) SISAP 2013. LNCS, vol. 8199, pp. 280–293. Springer, Heidelberg (2013). https://doi.org/10.1007/978-3-642-41062-8_28
5. Boytsov, L., Naidan, B.: Learning to prune in metric and non-metric spaces. In: Proceedings of NIPS 2013, pp. 1574–1582 (2013)

---

[5] https://github.com/nmslib/nmslib/tree/nmslib4a_bigger_reruns.

6. Bozkaya, T., Özsoyoglu, Z.M.: Indexing large metric spaces for similarity search queries. ACM Trans. Database Syst. **24**(3), 361–404 (1999). https://doi.org/10.1145/328939.328959
7. Bregman, L.: The relaxation method of finding the common point of convex sets and its application to the solution of problems in convex programming. USSR Comput. Math. Math. Phys. **7**(3), 200–217 (1967)
8. Cayton, L.: Fast nearest neighbor retrieval for bregman divergences. In: Proceedings of the 25th International Conference on Machine Learning, pp. 112–119. ACM (2008)
9. Chávez, E., Navarro, G.: Probabilistic proximity search: fighting the curse of dimensionality in metric spaces. Inf. Process. Lett. **85**(1), 39–46 (2003)
10. Chávez, E., Navarro, G., Baeza-Yates, R.A., Marroquín, J.L.: Searching in metric spaces. ACM Comput. Surv. **33**(3), 273–321 (2001)
11. Fog, A.: Instruction tables: Lists of instruction latencies, throughputs and micro-operation breakdowns for intel, AMD and VIA CPUs (2011)
12. Hetland, M.L., Skopal, T., Lokoč, J., Beecks, C.: Ptolemaic access methods: challenging the reign of the metric space model. Inf. Syst. **38**(7), 989–1006 (2013)
13. Itakura, F., Saito, S.: Analysis synthesis telephony based on the maximum likelihood method. In: Proceedings of the 6th International Congress on Acoustics, pp. C17–C20 (1968)
14. Knuth, D.E.: The Art of Computer Programming: Volume 3: Sorting and Searching. Atmospheric Chemistry & Physics (1973)
15. Kullback, S., Leibler, R.A.: On information and sufficiency. Ann. Math. Stat. **22**(1), 79–86 (1951)
16. Lewis, D.D., Yang, Y., Rose, T.G., Li, F.: RCV1: a new benchmark collection for text categorization research. J. Mach. Learn. Res. **5**, 361–397 (2004)
17. Markatou, M., Chen, Y., Afendras, G., Lindsay, B.G.: Statistical distances and their role in robustness. In: Chen, D.-G., Jin, Z., Li, G., Li, Y., Liu, A., Zhao, Y. (eds.) New Advances in Statistics and Data Science. IBSS, pp. 3–26. Springer, Cham (2017). https://doi.org/10.1007/978-3-319-69416-0_1
18. Navarro, G.: Searching in metric spaces by spatial approximation. VLDB J. **11**(1), 28–46 (2002)
19. Omohundro, S.M.: Five balltree construction algorithms (1989). iCSI Technical Report TR-89-063. http://www.icsi.berkeley.edu/icsi/publication_details?ID=000562
20. Rényi, A.: On measures of entropy and information. In: Proceedings of the Fourth Berkeley Symposium on Mathematical Statistics and Probability, vol. 1, pp. 547–561 (1961)
21. Samet, H.: Foundations of Multidimensional and Metric Data Structures. Morgan Kaufmann Publishers Inc., San Francisco (2005)
22. Skopal, T.: Unified framework for fast exact and approximate search in dissimilarity spaces. ACM Trans. Database Syst. **32**(4), 29 (2007)
23. Skopal, T., Bustos, B.: On nonmetric similarity search problems in complex domains. ACM Comput. Surv. **43**(4), 34 (2011)
24. Uhlmann, J.K.: Satisfying general proximity/similarity queries with metric trees. Inf. Process. Lett. **40**(4), 175–179 (1991)
25. Weber, R., Schek, H.J., Blott, S.: A quantitative analysis and performance study for similarity-search methods in high-dimensional spaces. In: VLDB, vol. 98, pp. 194–205 (1998)
26. Yianilos, P.N.: Data structures and algorithms for nearest neighbor search in general metric spaces. In: Proceedings of ACM/SIGACT-SIAM 1993, pp. 311–321 (1993)
27. Zhang, Z., Ooi, B.C., Parthasarathy, S., Tung, A.K.H.: Similarity search on Bregman divergence: towards non-metric indexing. PVLDB **2**(1), 13–24 (2009)

Non-metric Similarity Search Using Genetic TriGen

David Bernhauer[1,2] and Tomáš Skopal[2(✉)]

[1] Faculty of Information Technology,
Czech Technical University in Prague, Prague, Czech Republic
bernhdav@fit.cvut.cz
[2] Faculty of Mathematics and Physics,
Charles University, Prague, Czech Republic
skopal@ksi.mff.cuni.cz

Abstract. The metric space model is a popular and extensible model for indexing data for fast similarity search. However, there is often need for broader concepts of similarities (beyond the metric space model) while these cannot directly benefit from metric indexing. This paper focuses on approximate search in semi-metric spaces using a genetic variant of the TriGen algorithm. The original TriGen algorithm generates metric modifications of semi-metric distance functions, thus allowing metric indexes to index non-metric models. However, "analytic" modifications provided by TriGen are not stable in predicting the retrieval error. In our approach, the genetic variant of TriGen – the TriGenGA – uses genetically learned semi-metric modifiers (piecewise linear functions) that lead to better estimates of the retrieval error. Additionally, the TriGenGA modifiers result in better overall performance than original TriGen modifiers.

Keywords: Approximate similarity search · Semi-metric space · Genetic TriGen

1 Introduction

The similarity search models stand in the center of methods for content-based retrieval in datasets of multimedia and other unstructured data. For decades, the metric space model [7] has been widely accepted as the standard model for similarity search applications. The metric space model is both extensible (supporting black-box descriptors and similarities) as well as indexable (due to metric properties), thus providing efficient similarity search by metric access methods (MAMs) [2,7].

In the era of Big data – with the increasing diversity and complexity of data and algorithms for entity matching – there is a need for more general schemes of similarity modeling. The metric space properties could be too restrictive in many fields [6], for example in bioinformatics/cheminformatics. At the same time, the datasets grow to sizes that are not possible to query without indexing.

© Springer Nature Switzerland AG 2019
G. Amato et al. (Eds.): SISAP 2019, LNCS 11807, pp. 86–93, 2019.
https://doi.org/10.1007/978-3-030-32047-8_8

Hence, it is extremely challenging to provide scalability in both retrieval aspects – the effectiveness (retrieval quality) and efficiency (system performance). There have been many approaches developed, trading effectiveness for efficiency, e.g., approximate search methods. However, most of the results were based on the metric space model. Only a few approaches considered a more general non-metric case, one of which is the TriGen algorithm [5]. In this paper, we build on the idea of TriGen-based modification of non-metric space into approximate metric space, that enables metric indexing of (initially) non-metric data models for approximate search. As a contribution, we introduce a variant of TriGen based on genetic algorithm, that produces more robust semimetric-to-metric modifiers and better efficiency-effectiveness tradeoffs.

2   Non-metric/Approximate Similarity Search by TriGen

When talking about non-metric similarities, we usually consider distance functions that do not satisfy some of the metric axioms (reflexivity, non-negativity, symmetry, triangle inequality). Most of the practical non-metric distances actually miss just one of the axioms, like pseudo-metrics (reflexivity), quasi-metrics (symmetry), or semi-metrics (triangle inequality). As the lack of reflexivity and symmetry can be solved easily in the design of indexing/query algorithms, the real challenging problem is the semi-metric case; the lack of triangle inequality.

The *TriGen* algorithm [5] was developed to transform a semi-metric space (dataset- and distance-specific) into an equivalent (approximate) metric space. The idea behind TriGen is to use an increasing modifying function $f : \mathbb{R} \to \mathbb{R}$ that preserves query-induced similarity ordering when applied to a semi-metric distance function δ. Having a query object $q \in \mathbb{U}$ and database objects $x_i \in \mathbb{S} \subset \mathbb{U}$, then ordering/ranking of the database objects based on $\delta(q, x_i)$ is the same as when based on $f(\delta(q, x_i))$. Whereas all modifying functions behave the same with regard to the similarity ordering (thus to sequential similarity search), they are dramatically different in terms of the degree of triangle inequality exhibited by $f(\delta(\cdot, \cdot))$. Consider three objects $x_1, x_2, x_3 \in \mathbb{U}$ and the distances $\delta(x_i, x_j)$ among them – a distance triplet $\delta(x_1, x_2), \delta(x_2, x_3), \delta(x_1, x_3)$. In semi-metric spaces, some triplets form triangles and some do not (one distance is larger than the sum of the others). It is easy to show that concave modifiers increase the degree of triangle inequality by turning more distance triplets into triangles. A concave function magnifies short distances more than large distances, so that any distance triplet can be f-modified to a triangle if f is concave enough. On the other hand, convex modifiers do the opposite (break triangles).

From practical point of view, concave modifiers increase the degree of triangle inequality, hence eventually turn the semi-metric space into a metric space (indexable by MAMs). Unfortunately, they also increase the intrinsic dimensionality [2] of the space by decreasing the variance of distance distribution (up to equilateral triangles). Convex modifiers decrease the intrinsic dimensionality but also decrease the degree of triangle inequality, increasing thus retrieval error when such a semi-metric space is indexed by a MAM.

The TriGen algorithm finds the optimal level of concavity/convexity of a modifier in order to minimize intrinsic dimensionality of the resulting space, while keeping the expected retrieval error (degree of triangle inequality violation) below user-defined threshold. Hence, TriGen not only provides a solution for efficient (approximate) search in semi-metric spaces, but also fast approximate search in metric spaces. Specifically, TriGen utilizes a set of similarity-preserving T-base modifiers with convexity/concavity parameter w (see Fig. 1). Using binary search on w, such T-base f and w is found that exhibits the lowest intrinsic dimensionality for a given T-error threshold (where T-error is the proportion of non-triangles in all sampled distance triplets).

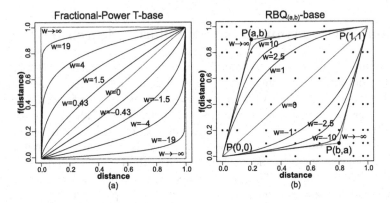

Fig. 1. T-bases of TriGen [5], parameterized by covexity/concavity weights w.

3   Genetic TriGen

In this paper, we present the TriGenGA (developed from an experiment [1]), a variant of TriGen that replaces the binary-search algorithm of finding modifiers by a genetic algorithm. The original TriGen algorithm finds just one T-base parameter w determining the concavity/convexity weight of the respective modifier function. In the genetic variant, we have implemented a new modifier $g_v : \langle 0,1 \rangle \mapsto \langle 0,1 \rangle$ represented by a piecewise linear function (Eq. 1, Fig. 3). As in TriGen, the modifier is a strictly increasing continuous function with $g_v(0) = 0$, $g_v(1) = 1$. However, instead of a predefined pool of T-base functions (i.e., FP-base, RBQ-bases), each parameterized by w, the genetic modifier g_v is composed by n linear segments, given $n - 1$ parameters (n is defined by user). The starting/ending points (x,y) of the i-th/$(i-1)$-th segment are defined as $(i/n, v_i)$, where vector $v = [v_1, v_2, ..., v_{n-1}]$ stores the parameters. The genetic algorithm is then used to find the n parameters of the modifier, given a dataset $\mathbb{S}$ and a distance function δ. Unlike the original TriGen where modifiers are strictly concave or convex (due to the single-parameter w optimization), the multi-parameter optimization provided by genetic TriGen is able to generate locally convex/concave modifiers. We anticipate such modifiers could better control the degree of triangle inequality (achieving lower T-error for the same or lower intrinsic dimensionalities), resulting in better precision/efficiency tradeoff exhibited by MAMs when searching.

$$g_v(x) = \begin{cases} v_1(n-1)x & 0 \le x \le \frac{1}{n} \\ v_1 + (v_2 - v_1)((n-1)x - 1) & \frac{1}{n} \le x \le \frac{2}{n} \\ \quad \vdots \\ v_i + (v_{i+1} - v_i)((n-1)x - i) & \frac{i}{n} \le x \le \frac{i+1}{n} \\ v_{n-1} + (1 - v_{n-1})(n-1)(x-1) & \frac{n-1}{n} \le x \le 1 \end{cases} \tag{1}$$

3.1 The Algorithm

The genetic algorithm (GA) consists of an evolution cycle described by Algorithm 1. The population is a set of vectors representing the modifiers (piecewise linear functions). For the selection of potentially successful individuals, we implemented the Tournament selector with variable size of the tournament k, because it performed better than other selectors. The selector randomly samples k individuals, and the best individual (with the highest fitness) is chosen in the selection. The one-point crossover of two parents (potentially successful individuals v and u) randomly generates a breakpoint b_p. The new individual (child) is generated as

$$child_i = \begin{cases} v_i & \text{for } 1 \le i < b_p, \\ u_i & \text{for } b_p \le i \le n - 1. \end{cases}$$

Mutation with probability p_M randomly chooses one parameter and moves it down or up randomly (still keeping the modifier g_v monotonous).

As there were several cases when the fitness of the population degenerated quickly, we have implemented the catastrophic scenario. When the best fitness score has not changed for several generations, part of the generation is replaced by randomly generated. This process should bring a different kind of genomes into the population. If the catastrophic scenario is repeatedly not successful (the best fitness score is the same) or the maximum number of generations is reached, the genetic algorithm terminates. Such modifier g_v is selected from the final population for which the intrinsic dimensionality of $(\mathbb{S}, g_v(\delta))$ is minimal.

3.2 The Fitness Function

The most important part of TriGenGA is the fitness function, which is the optimization criterion. As it is not possible to optimize $(\mathbb{U}, \delta)$ space globally, we have taken over the idea of triplet sampling from TriGen where a random subset of the dataset $\mathbb{S}^* \subset \mathbb{S}$ is used. However, we sample the triplets (x, y, z) in a different way. First, a distance matrix on $\mathbb{S}^*$ is computed, i.e., all distances $\delta(x, y) \; x, y \in \mathbb{S}^*$. Then, a fraction of pairs are selected for construction of the

---

Algorithm 1. TriGenGA

Result: modifier g_v with the best fitness score

$Population \longleftarrow$ GenerateRandom(pop_size)

while $UnsuccessfulCatastrophe() \leq max. \#$ of catastrophes **do**

    $Parents \longleftarrow$ TournamentSelector$_k$($Population$)

    for $\forall pair$ of $Parents$ **do**

        /* combine modifiers   */

        $Child \longleftarrow$ Crossover($pair$)

        if $mutation$ $probability$ $succeeded$ **then**

            /* modify modifier   */

            Mutate($Child$)

        $Population \longleftarrow Population \cup \{Child\}$

    keep only pop_size best individuals in $Population$

    if $best$ $fitness$ $score$ **does not change** for $last$ l $iterations$ **then**

        CatastropheScenario($Population$)

return best individual in $Population$

---

sample. For every selected pair (x, y), the third object $z \in \mathbb{S}^*$ is found that maximizes $\frac{\delta(x,z)}{\delta(x,y)+\delta(y,z)}$ (with $\delta(x, z)$ maximal). This way we obtain as many non-triangle triplets (x, y, z) in the sample as possible.

The fitness function $fit(\boldsymbol{v})$ consists of two parts, it takes into account the T-error (the proportion of non-triangle triplets in all triplets) as well as the index-ability (e.g., intrinsic dimensionality). In preliminary experiments, we found that modifiers g_v with small number of alternating concave and convex segments perform better. Based on that observation, we have proposed the ConFactor indicator (Eq. 2) which is part of the fitness function. The number of concave segments is defined as $c_v^+ = |\{i|2\boldsymbol{v}_i > \boldsymbol{v}_{i-1} + \boldsymbol{v}_{i+1}\}|$, and number of convex segments is defined $c_v^- = |\{i|2\boldsymbol{v}_i < \boldsymbol{v}_{i-1} + \boldsymbol{v}_{i+1}\}|$, for both $1 < i < n - 1$. We utilized the idea of TriGen, which assumes more triangle-preserving modifiers imply worse indexability (higher intrinsic dimensionality), so the current implementation of fit is described by Eq. 3, where ϵ_T is the T-error and $\epsilon_{threshold}$ is the T-error threshold (expected retrieval error specified by user at query time).

$$\text{ConFactor}(\boldsymbol{v}) = \frac{|c_v^+ - c_v^-|}{c_v^+ + c_v^-} \tag{2}$$

$$fit(\boldsymbol{v}) = \begin{cases} 1 - \epsilon_T(\boldsymbol{v}) & \text{for } \epsilon_T(\boldsymbol{v}) > \epsilon_{threshold}, \\ 1 + \epsilon_T(\boldsymbol{v}) \cdot \text{ConFactor}(\boldsymbol{v}) & \text{otherwise} \end{cases} \tag{3}$$

4   Experimental Results

We have experimented with kNN queries, where pivot tables (LAESA [4]) were used with TriGen and TriGenGA, as well as the sequential search. There were randomly sampled 400 10NN queries from the respective dataset and efficiency and retrieval error was measured. The efficiency was defined as the proportion of distance computations needed with respect to sequential search. The real retrieval error was defined as $\frac{|E \cap O|}{\max\{|E|,|O|\}}$, where E were the expected objects (a result of sequential search) and O were observed objects (result of LAESA).

For all experiments, we have used the same configuration. The size of the database subset used by TriGen/TriGenGA was $|\mathbb{S}^*| = 1000$. Triplet sample size was 25000. 15%, 30%, 45%, 60%, 75%, and 90% triplets were selected by the algorithm maximizing the ratio of erroneous triplets; the other part was sampled randomly. The size of the GA population was 150 individuals. Probability of mutation was set to 5%. The catastrophe scenario was invoked after 10 iterations without the best fitness score improvement. The algorithm terminates after 1000 iterations or ten catastrophe scenarios, without improvement of fitness score. The dimensions $n = 5$ and $n = 7$ were used for piecewise linear modifiers.

4.1 Datasets

SISAP NASA dataset [3] (40150 objects with 20 dimensions) was used for vector-based descriptors with metric Minkowski L_p and semi-metric Fractional L_p distances (L_3, L_2, $L_{0.75}$, $L_{0.5}$, $L_{0.25}$ and $L_{0.125}$), SISAP English dictionary [3] (69069 English words) for string-based descriptors with metric Levenshtein distance, and an industrial dataset of ATM withdrawal time series (5985 ATM's time-series with 168 dimensions) with semi-metric dynamic time warping (DTW) distance bounded by Sakoe-Chiba band of size 5. We have tested the T-error threshold $\epsilon_{threshold}$ in five different ranges (0.0, 0.025, 0.05, 0.1 and 0.2).

4.2 Results

Figure 2 summarizes the results. Note that TriGen with FP-modifier has a smooth progress as only one parameter defines the modifier. In contrast, the non-linear and non-deterministic optimization in TriGenGA can generate two modifiers with the same retrieval error but different efficiency. However, the standard deviation of repeated experiments with different random seed was 1% for both the retrieval error and distance computations. TriGenGA performs better than original TriGen for DTW distance on ATM dataset, as well as for metric Minkowski and most of the semi-metric Fractional distance measures (NASA dataset). On the other hand, for the Levenshtein distance and $L_{0.125}$ the TriGen algorithm dominates TriGenGA.

We evaluated not only the error/efficiency tradeoff but also the T-error threshold vs. real retrieval error dependency. Figure 3 shows this difference – the x coordinate of a circle shows the T-error threshold, while left endpoint of the connected line shows the real retrieval error (y coordinate is the efficiency). Hence, TriGenGA behaves better in terms of real retrieval error estimation.

The distance computation cost is the same for both approaches. The distance matrix on $\mathbb{S}^*$ is precomputed (see Sect. 3.2). In terms of complexity, both

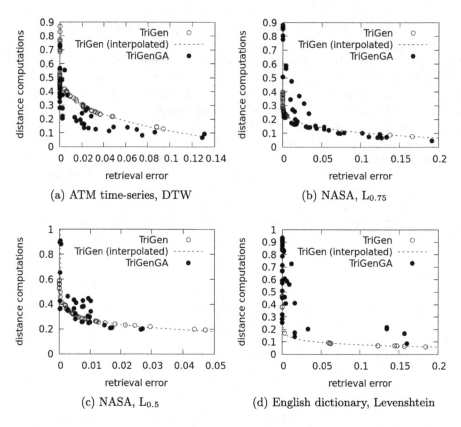

Fig. 2. Comparison of TriGen and TriGenGA on various datasets/distances

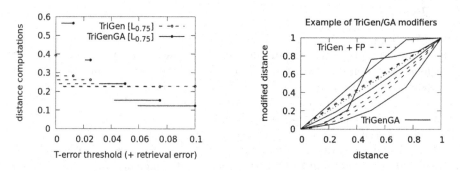

Fig. 3. Left figure shows difference between T-error threshold and retrieval error. Right figure illustrates examples of modifiers generated by TriGen/GA.

approaches can limit the number of iterations. The difference is only in the type of algorithm, the binary search versus the genetic algorithm. Intuitively, the genetic variant has larger requirements for learning parameters because the binary search has only one parameter with only one optimum. However, these

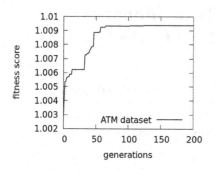 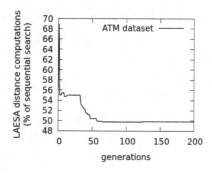

Fig. 4. Effects of TriGenGA early termination (limited number of generations).

computations are done only once before the querying. In Fig. 4, see the evolution convergence and the effects of sub-optimal modifier obtained by early termination (i.e., fitness achieved and distance computations of LAESA search).

5   Conclusion

We proposed a genetic variant of the TriGen algorithm for approximate similarity search in semi-metric spaces. Experiments proved that the piecewise linear modifier generated by the genetic algorithm can outperform the original Tri-Gen algorithm with FP-modifier, as TriGenGA in some cases provides modifiers exhibiting both lower retrieval error and lower number of distance computations (LAESA-based search). The TriGenGA-generated modifiers also provide better retrieval error estimation given a user-specified T-error threshold.

Acknowledgments. This research has been supported in part by the Czech Science Foundation (GAČR) project Nr. 17-22224S.

References

1. Bernhauer, D., Skopal, T.: Approximate search in dissimilarity spaces using GA. In: GECCO, pp. 279–280. ACM (2019)
2. Chávez, E., Navarro, G., Baeza-Yates, R., Marroquín, J.L.: Searching in metric spaces. ACM Comput. Surv. **33**(3), 273–321 (2001)
3. Figueroa, K., Navarro, G., Chávez, E.: Metric spaces library (2007). http://www.sisap.org/Metric_Space_Library.html
4. Mico, L., Oncina, J., Vidal, E.: A new version of the nearest-neighbour approximating and eliminating search algorithm (AESA) with linear preprocessing time and memory requirements. Pattern Recogn. Lett. **15**, 9–17 (1994)
5. Skopal, T.: Unified framework for fast exact and approximate search in dissimilarity spaces. ACM Trans. Database Syst. **32**(4), 29 (2007)
6. Skopal, T., Bustos, B.: On nonmetric similarity search problems in complex domains. ACM Comput. Surv. **43**(4), 34:1–34:50 (2011)
7. Zezula, P., Amato, G., Dohnal, V., Batko, M.: Similarity Search: The Metric Space Approach. Springer, New York (2005). https://doi.org/10.1007/0-387-29151-2

Privacy–Preserving Text Similarity via Non-Prefix-Free Codes

M. Oğuzhan Külekci[(✉)], Ismail Habib, and Amir Aghabaiglou

Informatics Institute, Istanbul Technical University, Istanbul, Turkey
{kulekci,habib19,aghabaiglou}@itu.edu.tr

Abstract. Many methods have been proposed to compute the similarity score $\alpha \leftarrow S(\mathcal{A}, \mathcal{B})$ in between two *plain* documents $\mathcal{A}$ and $\mathcal{B}$. However, when their contents are confidential, special processing is required to protect privacy. A great extent of the solutions offered to date is mostly based on homomorphic encryption or secure multi-party computation techniques, where their computational cost inhibits the practical usage, especially on massive sets. In this study we propose an alternative by encoding the documents with non-prefix-free (NPF) coding before applying the preferred similarity metric $S()$. The NPF coding simply represents the symbols with variable-length codewords, where the codeword set is generated without the prefix-free restriction. Thus, a codeword may be a prefix of another, and without the explicit codeword boundary information, retrieving the original data from the encoded stream becomes hard due to the lack of unique decodability in non-prefix-free codes. We provide the combinatorial analysis of this hardness, and experimentally compare the similarity scores obtained on NPF encoded documents and on original plain text versions. We have considered normalized compression distance (NCD) and Jaccard coefficient (JC) for the similarity metric $S()$. When $\mathcal{A}'$ and $\mathcal{B}'$ denote the NPF-encoded documents, experiments conducted on METER corpus revealed that the difference between $\alpha' \leftarrow S(\mathcal{A}', \mathcal{B}')$ and $\alpha \leftarrow S(\mathcal{A}, \mathcal{B})$ lie in the range of 0.5% and 3% for both NCD and JC.

1 Introduction

Similarity computation between documents [18] has many applications such as plagiarism detection, copyright management, duplicate report detection, document classification/clustering, and search engines, just to list a few. Many solutions [2,4,6,12,17] have appeared assuming that the contents of the documents are public, where the general tendency here is to first compute a feature vector that is mainly based on the number of occurrences of words or n-gram frequencies in the document. Those features, or the fingerprint values extracted from them, are then compared to compute the similarity score. The necessity for the privacy preserving similarity detection appears when the owners of the

This work has been supported by the TÜBİTAK grant number 117E865.

documents want to keep their contents secret. Some example scenarios include duplicate submission control between the related academic venues, information sharing between the insurance companies or intelligence agencies, and etc. The methods [3,5,14,20] proposed to date for that problem mostly relied on multi-party homomorphic encryption schemes [10] such that the features extracted from the documents are encrypted, and then the comparison in between them are achieved by using the homomorphic properties. The practical difficulties of the homomorphic schemes [21] unfortunately inhibit the usage of these solutions particularly on large document collections.

In this work, we propose an alternative approach by using non-prefix-free (NPF) coding of the input documents as a privacy-preserving transformation with the observation that the NPF encoded documents are hard to be decoded in absence of the codeword boundaries, while the syntactic structures are preserved, giving us the opportunity to run some standard similarity calculations on the encoded versions of the documents. It had been previously shown that the cipher-text only attacks on non-prefix-free codes [19] and even on prefix codes [9,11,22] are hard.

The performance of the privacy preserving similarity score calculation can be measured by observing how much the similarity scores obtained between the encoded documents deviate from the ones obtained on plain versions. We considered normalized compression distance (NCD) [7] and Jaccard coefficient (JC) [13,15] for the similarity measurements. When $\mathcal{A}'$ and $\mathcal{B}'$ denote the NPF-encoded documents, experiments conducted on METER corpus [8] revealed that the difference between $\alpha' \leftarrow S(\mathcal{A}', \mathcal{B}')$ and $\alpha \leftarrow S(\mathcal{A}, \mathcal{B})$ lie in the range of 0.5% and 3% for both NCD and JC.

2   Preliminaries and Primitives

Let $\mathcal{A} = a_1 a_2 \ldots a_p$ and $\mathcal{B} = b_1 b_2 \ldots b_q$ represent two documents of respectively p and q symbols long, where each character a_i and b_j is drawn from the alphabet $\Sigma = \{\epsilon_1, \epsilon_2, \ldots, \epsilon_\sigma\}$ for $1 \leq i \leq p$ and $1 \leq j \leq q$. The similarity score computed with the method $S()$ is denoted by $S(\mathcal{A}, \mathcal{B})$.

Non-Prefix-Free (NPF) Coding: We encode the documents with a NPF coding scheme. Assume $W = \{w_1, w_2, \ldots, w_\sigma\}$ is a code-word set generated for the alphabet Σ. Each w_i is a bit sequence of arbitrary length, which can be a prefix of another code-word w_j from W. Given a sequence $\mathcal{T} = t_1 t_2 \ldots t_n$ for $t_i \in \Sigma$, the $NPF : \Sigma \rightarrow W$ coding replaces all occurrences of ϵ_i with its corresponding w_i on $\mathcal{T}$. Figure 1 demonstrates a sample NPF coding on a given $\mathcal{T} =$ NONPREFIXFREE.

Since the code-words in W are not prefix free, the code-word boundaries on the encoded stream cannot be determined. Therefore, a NPF encoded stream can have many possible parses according to the code-word set W, e.g., the initial bits 0011111 of the sample in Fig. 1 can also represent the sequence PFXX, and many others as well.

$$\mathcal{T} = \texttt{NONPREFIXFREE}$$
$$\Sigma = \{ \text{ E, R, F, N, I, O, P, X } \}$$
$$W = \{ \text{ 01, 0, 111, 001, 010, 1111, 00, 1 } \}$$
$$NPF(\mathcal{T}) = \underline{00}\underline{111}\underline{1}\underline{1}\underline{00}\underline{1}\underline{00}\underline{0}\underline{111}\underline{0}\underline{10}\underline{111}\underline{00}\underline{101}$$

Fig. 1. An example non-prefix-free coding of the sequence $\mathcal{T} = \texttt{NONPREFIXFREE}$.

Due to that lack of unique decodability, the non-prefix-free codes have not been found interesting in terms of data compression. Though, there had been some recent efforts [1,16] to solve this problem efficiently.

Although the NPF codes has not made much sense in terms of data compression, they provide an opportunity in terms of security. It had been previously shown that the prefix-free codes, such as the Huffman, are difficult to cryptanalyze [11]. In prefix-free coding, none of the codewords can be a prefix of another, and this property indeed can be used in crypt-analysis of a sequence that is known to be variable-length prefix-free encoded. Even in that case it is hard to extract the original sequence from the prefix-free encoded version [9,22]. In case of non-prefix-free codes there is no such restriction, and actually every parse of the input bit stream defines a valid possible source. Recently, [19] has considered using NPF as a substitution cipher, and has shown that a cipher-text only attack is difficult. The difficulty stems from the fact that the codeword boundaries in the non-prefix-free encoded bit-stream cannot be determined analytically, and any arbitrarily selected boundary actually maps to a symbol assuming that the NPF codeword set is designed appropriately.

Despite the difficulty for extracting the original sequence, the NPF coding scheme actually preserves the syntactic structures of the documents. The portions that are equal to each other in distinct documents are mapped to the same bit stream. Thus, the shared information content is preserved after the NPF transform.

This observation has led us to the motivating idea of this study as the NCD and the Jaccard Coefficient (JC) between the documents should also be preserved in their NPF transformed versions. More formally, $|S(\mathcal{A}, \mathcal{B}) - S(NPF(\mathcal{A}), NPF(\mathcal{B}))|$ is expected to be small. We now provide the definitions of NCD and JC below.

Normalized Compression Distance (NCD): The NCD similarity metric introduced in [7] is based on the fact that if two documents are similar then each of them should be able to be compressed well with the model extracted from the other. In other words, when one of them is concatenated to the end of the other document, and send to a compressor, the compression ratio obtained should be reflecting their similarity. Normalizing the compression ratio by respecting possible different lengths of the documents, the NCD similarity score is represented by the formula

$$NCD(\mathcal{A}, \mathcal{B}) = \frac{C(\mathcal{A}\mathcal{B}) - min\big(C(\mathcal{A}), C(\mathcal{B})\big)}{max\big(C(\mathcal{A}), C(\mathcal{B})\big).}$$

Here, $C()$ is an arbitrary compressor and $\mathcal{AB}$ is the concatenation of the documents $\mathcal{A}$ and $\mathcal{B}$. The score is between 0 and 1. The similarity increases towards 0 and decreases towards 1.

Jaccard Coefficient (JC): Jaccard Coefficient is used to measure similarity between finite sample sets. It is also known as the intersection over union which basically assumes that the more elements two sets have in common the closer they are. JC is measured using the below formula [15], where $\mathcal{W_A}$ and $\mathcal{W_B}$ are two sets whose elements are the n-grams of documents $\mathcal{A}$ and $\mathcal{B}$ respectively. The similarity according to JC between documents $\mathcal{A}$ and $\mathcal{B}$ increases towards 1 and decreases towards 0.

$$J(\mathcal{A}, \mathcal{B}) = \frac{\mid \mathcal{W_A} \cap \mathcal{W_B} \mid}{\mid \mathcal{W_A} \cup \mathcal{W_B} \mid}$$

3   The Method

Assume Alice has document $\mathcal{A}$ and Bob has $\mathcal{B}$, where both have the same source alphabet Σ. Protocol 1 describes the proposed schema that computes the similarity of $\mathcal{A}$ and $\mathcal{B}$ without revealing their contents by the help of a trusted third party. The formal definition of the NPF encoding is given below.

Protocol 1. Privacy-Preserving Text Similarity Protocol

Required:
Alice and Bob agree on seed and random function to use;
Alice,
(I) generates $P = \{p_1, p_2, \ldots, p_\sigma\}$ based on random(seed),
(II) shuffles the alphabet $\Sigma = \{\epsilon_1, \epsilon_2, \ldots \epsilon_\sigma\}$ using P,
(III) generates $W = \{w_1, w_2, \ldots, w_\sigma\}$ such that $w_i = MBR(p_i)$,
(IV) generates $NPF(\mathcal{A})$ by substituting p_i with w_i,
(V) sends $NPF(\mathcal{A})$ to a trusted entity for similarity calculation.
Bob,
(I) generates $P = \{p_1, p_2, \ldots, p_\sigma\}$ based on random(seed),
(II) shuffles the alphabet $\Sigma = \{\epsilon_1, \epsilon_2, \ldots \epsilon_\sigma\}$ using P,
(III) generates $W = \{w_1, w_2, \ldots, w_\sigma\}$ such that $w_i = MBR(p_i)$,
(IV) generates $NPF(\mathcal{B})$ by substituting p_i with w_i,
(V) sends $NPF(\mathcal{B})$ to a trusted entity for similarity calculation.
Third Party,
(I) receives $NPF(\mathcal{A})$ and $NPF(\mathcal{B})$,
(II) computes the similarity score S between both documents by using either the NCD or the Jaccard distance metrics,
(III) shares the similarity score S with Alice and Bob.

Definition 1. *Let $P = \{p_1, p_2, \ldots, p_\sigma\}$ be a random permutation of $\{2, \ldots, \sigma + 1\}$, where σ denotes the number of symbols in the source alphabet $\Sigma = \{\epsilon_1, \epsilon_2, \ldots \epsilon_\sigma\}$. $W = \{w_1, w_2, \ldots, w_\sigma\}$ is a codeword set such that the codeword that represents ϵ_i is $NPF(\epsilon_i) = w_i = MBR(p_i)$, where $MBR(i)$ is the binary representation of integer i omitting its leftmost 1 bit (e.g., $MBR(5) = 01$ as $5 = (101)_2$).*

Assuming an input sequence $\mathcal{T} = t_1 t_2 \ldots t_n$, where $t_i \in \Sigma$ for all $1 \leq i \leq n$, the **non-prefix-free encoding** of $\mathcal{T}$ by codeword set W is denoted by $NPF(\mathcal{T}) = c_1 c_2 \ldots c_n$, where $c_i = NPF(t_i)$, for $c_i \in W$ and $t_i \in \Sigma$.

The pseudo-random permutation P can be obtained by shuffling the sequence $X = \langle 2, \ldots, \sigma + 1 \rangle$ with the help of a pseudo-random number generator PRNG. The seed of the PRNG in that case becomes the secret key of the scheme, since the permutations, hence the codeword set W, obtained by using the same seed will be identical. Thus, the Alice and Bob agrees on the PRNG and share the same seed in Protocol 1.

4   The Hardness of Decoding NPF

We consider the hardness of decoding the original $\mathcal{T}$ from its NPF encoded version $NPF(\mathcal{T})$. There appears two difficulties based on the lack of information regarding the codeword set W and the codeword boundaries on $NPF(\mathcal{T})$. Due to the Definition 1, the lengths of the codewords in W vary between 1 and $\lceil \log(\sigma + 1) \rceil$. Let f_ℓ denotes the number of symbols that are represented by ℓ-bits long codewords, for $1 \leq \ell \leq \lceil \log(\sigma + 1) \rceil$, in $NPF(\mathcal{T})$. Assuming that an attacker has the knowledge of the input alphabet Σ, and correctly guessed $f_1, f_2, \ldots f_{\lceil \log(\sigma+1) \rceil}$, the number of possible parses of the $NPF(\mathcal{T})$ is,

$$\frac{n!}{f_1! f_2! \cdots f_{\lceil \log(\sigma+1) \rceil}!}, \tag{1}$$

where $n = f_1 + f_2 + \ldots + f_{\lceil \log(\sigma+1) \rceil}$. This is akin to counting the anagrams of a word with repeated letters[1]. In $NPF(\mathcal{T})$ sequence, there are n codewords that are of $\lceil \log(\sigma + 1) \rceil$ different lengths. When f_i is the number of length-i codewords, then the number of different permutations of such a set can be computed with Eq. 1.

Since the correct codeword boundaries that are used in the encoding phase are not available, it will be hard for the attacker to generate all possible candidates. Excluding the exaggerated cases that set the Eq. 1 value becomes small enough for a brute-force analysis (e.g., small input size n), the number of possible parses is large, and thus, computationally hard to enumerate.

Assuming that the statistical distribution of the input data is available, the attacker will consider evaluating the appropriateness of any generated candidate by observing how close it is to that distribution, such as considering the expected k-gram counts. However, there appears yet another difficulty here since the codeword set W is not public, and it is not straight forward to reconstruct the symbol sequence from the parsed codewords. There are $\sigma!$ possible codeword set W as each permutation defines a unique W, where the secret seed used in the PRNG determines the correct one. Thus, besides the huge number of candidate NPF

---

[1] For example, the number of anagrams of word MISSISSIPPI is $\frac{11!}{4!1!2!4!}$ as letters i, m, p, s appears 4, 1, 2, and 4 times respectively.

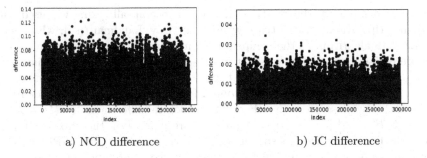

a) NCD difference   b) JC difference

Fig. 2. Absolute difference of the similarity scores on plain and NPF-encoded files computed with NCD and JC metrics.

stream parses, mapping codewords to symbols introduce an additional difficulty of around $\sigma!$.

It may seem more feasible to try sequential decoding of the NPF stream symbol by symbol. In this case the attacker still suffers from the unknown codeword assignments as well as unknown codeword boundaries. The codeword assigned to a symbol can be of $\lceil \log(\sigma + 1) \rceil$ different lengths, and any of the σ symbols may map to these codewords. For example, on a byte alphabet with $\sigma = 256$, the lengths of the codewords vary between 1 to 8, and each encoded symbol then has $8 \cdot 256 = 2^{11}$ possible assignments. In such a sample case, consecutive r characters would bring $2^{r \cdot 11}$ possibilities to analyze. Notice also the fact that since the codeword stream is variable-length, wrong decision on a symbol effects all the remaining ones. Therefore, one-by-one incremental decoding of the NPF stream is still difficult to achieve.

5   Experimental Results

We observed the absolute differences $|NCD(\mathcal{A}, \mathcal{B}) - NCD(\mathcal{A}', \mathcal{B}')|$ and $|JC(\mathcal{A}, \mathcal{B}) - JC(\mathcal{A}', \mathcal{B}')|$, where $\mathcal{A}' = NPF(\mathcal{A})$ and $\mathcal{B}' = NPF(\mathcal{B})$, on a set of file pairs obtained from the METER corpus [8] to evaluate the performance of the proposed scheme. The METER corpus was originally designed to evaluate the schemes for text reuse detection and includes news text appeared in the newspapers and also the source of those news published by the news agencies. We have sampled 770 documents from the METER corpus and computed the similarity of one document with all others both in plain and NPF encoded versions. Thus, in total we have $\binom{770}{2} \approx 300K$ file pairs.

In the implementation of our scheme in C language, we simply used the `rand()` function as the PRNG, which is seeded by an arbitrarily selected seed value x via the `srand(x)`. The permutation P is simply computed by shuffling the number from 2 to 257 via this PRNG assuming a one-byte-alphabet consisting of 256 symbols. The NPF transform of each document is accordingly achieved and stored as binary strings. Before applying the NPF encoding on a file, the punctuation symbols are removed and all letters are converted to lowercase.

While computing the NCD similarity score on all pairs we used the Python script available at https://github.com/DavyLandman/ncd, which uses the LZMA2 algorithm as the compression function. For the JC calculation, we considered 5-grams on plain versions and 30-grams (which is roughly 5 times the average NPF codeword length on text files) on NPF encoded bit streams.

We observed that the scores computed for the plain and NPF-encoded versions are very close to each other. The absolute differences $|NCD(\mathcal{A},\mathcal{B}) - NCD(\mathcal{A}',\mathcal{B}')|$ and $|JC(\mathcal{A},\mathcal{B}) - JC(\mathcal{A}',\mathcal{B}')|$ are plotted on the y-axis in Fig. 2, where the file pair indices are shown on the x-axis. On average the NPF-encoded scores are only 0.0356 points different than the scores obtained on plain documents. The standard deviation and the maximum are 0.0189 and 0.124 respectively. For the JC case, we observed that the JC difference between the plain and NPF encoded documents are within the range of 0 and 0.0343 with an average of 0.005 and standard deviation 0.0032.

Overall, for both the JC and NCD metrics, the similarity scores computed on NPF encoded files deviate around 3 percent from the scores computed on original plain versions.

Table 1. Average results of the experiment

	Class	0–0.25	0.25–0.5	0.5–0.75	0.75–1
NCD	Avg. plain score	0.0793	0.3017	0.575	0.984
	Avg. NPF score	0.1144	0.2853	0.5622	0.9595
	Difference	**0.0351**	**0.0164**	**0.0098**	**0.0245**
	Count	297489	862	24	770
JC	Avg. plain score	0.0258	0.3168	0.594	1
	Avg. NPF score	0.0308	0.3358	0.6113	1
	Difference	**0.005**	**0.019**	**0.0173**	**0**
	Count	294967	248	24	770

In Table 1 we have also summarized the results by classifying the file pairs into 4 clusters according to their plain text similarities. For instance, on the set of 862 file pairs whose plain similarity scores with NCD are between 0.25 and 0.5, the absolute difference of NCD scores on NPF encoded versions is 0.0164 on the average.

6   Conclusion

We have presented a novel privacy preserving document similarity detection method. The privacy in the proposed method is based on the lack of unique decodability in non-prefix-free codes, which is different from the solutions offered to date that rely on mostly the homomorphic schemes and multi-party secure computation methods. Besides the similarity computation, the intrinsic decodability problem of non-prefix-free codes might also provide further results on secure data search, distributed data storage on the cloud, and compressed secure text indexes.

References

1. Adaş, B., Bayraktar, E., Külekci, M.O.: Huffman codes versus augmented non-prefix-free codes. In: Bampis, E. (ed.) SEA 2015. LNCS, vol. 9125, pp. 315–326. Springer, Cham (2015). https://doi.org/10.1007/978-3-319-20086-6_24
2. Bennett, C.H., Gács, P., Li, M., Vitányi, P.M., Zurek, W.H.: Information distance. IEEE Trans. Inf. Theory **44**(4), 1407–1423 (1998)
3. Blundo, C., De Cristofaro, E., Gasti, P.: Espresso: efficient privacy-preserving evaluation of sample set similarity. J. Comput. Secur. **22**(3), 355–381 (2014)
4. Buttler, D.: A short survey of document structure similarity algorithms. In: International Conference on Internet Computing, pp. 3–9 (2004)
5. Buyrukbilen, S., Bakiras, S.: Secure similar document detection with simhash. In: Jonker, W., Petković, M. (eds.) SDM 2013. LNCS, vol. 8425, pp. 61–75. Springer, Cham (2014). https://doi.org/10.1007/978-3-319-06811-4_12
6. Chen, X., Francia, B., Li, M., Mckinnon, B., Seker, A.: Shared information and program plagiarism detection. IEEE Trans. Inf. Theory **50**(7), 1545–1551 (2004)
7. Cilibrasi, R.L., Vitanyi, P.: The google similarity distance. IEEE Trans. Knowl. Data Eng. **19**(3), 370–383 (2007)
8. Clough, P., Gaizauskas, R., Piao, S.S., Wilks, Y.: Meter: measuring text reuse. In: Proceedings of the 40th Annual Meeting of ACL, pp. 152–159 (2002)
9. Fraenkel, A.S., Klein, S.T.: Complexity aspects of guessing prefix codes. Algorithmica **12**(4–5), 409–419 (1994)
10. Gentry, C.: Fully homomorphic encryption using ideal lattices. In: STOC, vol. 9, pp. 169–178 (2009)
11. Gillman, D.W., Mohtashemi, M., Rivest, R.L.: On breaking a Huffman code. IEEE Trans. Inf. Theory **42**(3), 972–976 (1996)
12. Hammouda, K.M., Kamel, M.S.: Efficient phrase-based document indexing for web document clustering. IEEE Trans. Knowl. Data Eng. **16**(10), 1279–1296 (2004)
13. Jaccard, P.: Étude comparative de la distribution florale dans une portion des alpes et des jura. Bull. Soc. Vaudoise Sci. Nat. **37**, 547–579 (1901)
14. Jiang, W., Murugesan, M., Clifton, C., Si, L.: Similar document detection with limited information disclosure. In: IEEE 24th International Conference on Data Engineering, pp. 735–743 (2008)
15. Jiang, W., Samanthula, B.K.: N-gram based secure similar document detection. In: Li, Y. (ed.) DBSec 2011. LNCS, vol. 6818, pp. 239–246. Springer, Heidelberg (2011). https://doi.org/10.1007/978-3-642-22348-8_19
16. Külekci, M.O.: Uniquely decodable and directly accessible non-prefix-free codes via wavelet trees. In: IEEE International Symposium on Information Theory, pp. 1969–1973 (2013)
17. Li, M., Chen, X., Li, X., Ma, B., Vitányi, P.: The similarity metric. IEEE Trans. Inf. Theory **50**(12), 3250–3264 (2004)
18. Manber, U., et al.: Finding similar files in a large file system. Usenix Winter **94**, 1–10 (1994)

19. Muralidhar, R.B.: Substitution cipher with nonprefix codes. Master's thesis, San Jose State University (2011)
20. Murugesan, M., Jiang, W., Clifton, C., Si, L., Vaidya, J.: Efficient privacy-preserving similar document detection. VLDB J. **19**(4), 457–475 (2010)
21. Naehrig, M., Lauter, K., Vaikuntanathan, V.: Can homomorphic encryption be practical? In: Proceedings of the 3rd ACM Workshop on Cloud Computing Security Workshop, pp. 113–124. ACM (2011)
22. Rubin, F.: Cryptographic aspects of data compression codes. Cryptologia **3**(4), 202–205 (1979)

Explainable Similarity of Datasets Using Knowledge Graph

Petr Škoda, Jakub Klímek, Martin Nečaský, and Tomáš Skopal[✉]

Department of Software Engineering, Faculty of Mathematics and Physics,
Charles University, Malostranské náměstí 25, 118 00 Praha 1, Czech Republic
{skoda,klimek,necasky,skopal}@ksi.mff.cuni.cz

Abstract. There is a large quantity of datasets available as Open Data on the Web. However, it is challenging for users to find datasets relevant to their needs, even though the datasets are registered in catalogs such as the European Data Portal. This is because the available metadata such as keywords or textual description is not descriptive enough. At the same time, datasets exist in various types of contexts not expressed in the metadata. These may include information about the dataset publisher, the legislation related to dataset publication, language and cultural specifics, etc. In this paper we introduce a similarity model for matching datasets. The model assumes an ontology/knowledge graph, such as Wikidata.org, that serves as a graph-based context to which individual datasets are mapped based on their metadata. A similarity of the datasets is then computed as an aggregation over paths among nodes in the graph. The proposed similarity aims at addressing the problem of explainability of similarity, i.e., providing the user a structured explanation of the match which, in a broader sense, is nowadays a hot topic in the field of artificial intelligence.

Keywords: Similarity · Datasets · Search · Graph

1 Introduction and Motivation

There is an enormous volume of datasets available on the Web as Open Data. Open Data is data which can be freely re-used by anyone without any restrictions. The key prerequisite is that datasets can be easily found by their potential consumers. There are so-called Open Data catalogs[1] available today which enable data consumers to search for datasets. However, they provide only basic search features based on descriptive metadata recorded in a catalog such as full text search in dataset titles, descriptions and keywords. These classical search methods presume that potential consumers know exactly what they search for

---

[1] E.g., European Data Portal https://www.europeandataportal.eu.

This research has been supported by Czech Science Foundation (GAČR) project Nr. 19-01641S.

G. Amato et al. (Eds.): SISAP 2019, LNCS 11807, pp. 103–110, 2019.
https://doi.org/10.1007/978-3-030-32047-8_10

and what search query leads to the expected search result. However, consumers are able to formulate only some search queries which return only a subset of the datasets of their interest. There are usually other useful datasets but consumers cannot find them because they do not know the necessary search query (i.e., leading to low recall).

A possible solution is to search for datasets similar to datasets a consumer found interesting. In general, we can measure the similarity of two datasets by measuring the similarity of their content as it is common in the world of documents. Documents are homogeneous as they can all be considered as sequences of words which makes content-based similarity a well-defined problem. In this sense, datasets are very heterogeneous. Datasets exist in different formats and are structured differently. Their content is mostly formed by primitive data values such as numbers, dates and short strings (e.g., names). These values have their meaning which is however hidden in the semantics of schemas of the datasets. This makes measuring content-based dataset similarity harder and the general findability of datasets in Open data catalogs is seriously limited.

To demonstrate the problem, consider data consumers who want to show sports grounds in European cities on a map. There is no central database of sports grounds in Europe and, therefore, the consumers need to find datasets published by different publishers. For example, they can find a dataset with sports grounds in Prague 8[2] or Northern Ireland[3]. The first dataset can be found only with a search query containing the words *sports grounds* in Czech. The other can be found with English words *active places* or *sport facilities*. Both datasets encode data about sports grounds in different tabular structures and there are no direct or indirect links between entities encoded in the datasets. Also titles and descriptions of the datasets are available in different languages. Moreover, even when we translate all textual information to a single language we get different terms (e.g., sports ground vs. facility or active place). It is clear that finding all datasets about sports grounds in Europe is extremely difficult in such heterogeneous environment with only existing search capabilities of the current catalogs. It would be helpful to let consumers find an initial set of datasets about sports grounds and then find similar datasets even though they are called, described and structured differently.

1.1 Paper Contribution

In this paper, we present an approach based on considering so-called dataset contexts. We consider datasets which are registered in Open Data catalogs. Catalogs provide records, i.e., metadata descriptions comprising dataset title, description, keywords, etc. Such descriptions are homogeneous in their structure across different datasets. However, they are not so rich compared to content of documents

---

[2] https://data.gov.cz/zdroj/datové-sady/http—opendata.praha.eu-api-3-action-package_show-id-praha8-sportoviste.

[3] https://www.opendatani.gov.uk/dataset/active-places-ni-sports-facilities-database.

used for measuring document similarity. Therefore, we enrich the model with semantic contexts of entities found in the metadata descriptions encoded in a given knowledge graph, e.g., Wikidata. We then compute the similarity of two datasets as an aggregation of similarities of the entities. Similarity of two entities is based on the paths connecting them via their lowest common ancestor in a certain hierarchy encoded in the knowledge graph.

Moreover, we claim that in addition to the similarity itself consumers also need an explanation of that similarity. They need to further process (i.e., integrate, cleanse and transform) the datasets in their consumption process and therefore, they need to know the details why the datasets are similar. Therefore, we not only compute the similarity but we also extract the structure from the knowledge graph which explains the similarity. This structure includes the connecting paths but also information related to the entities and edges on the path, e.g., their labels, their semantic definitions in the knowledge graph or sample instances of entities which represent category concepts (i.e., concepts which categorize other concepts).

2   Related Work

The attributed graphs [4,5] were developed as a graph-theoretic model, where vertices are equipped by a set of attributes. Although in literature the attributed graphs were mainly used for clustering and segmentation of attribute/entity-augmented graphs (e.g., social networks), we consider this model as suitable also for representation of knowledge graphs and for similarity search of entities represented as graph vertices. The attributes in vertices allow to map vertices of the graphs to non-graph elements, such as features of datasets we aim at.

The techniques for enriching entities found in unstructured texts such as dataset titles or descriptions with their semantic contexts in a given knowledge graph are generally called entity linking techniques. In [3] the authors provide a comprehensive survey of entity linking techniques.

In [2], the authors propose a method for measuring document similarity based on semantic relationships between document annotations. Documents are annotated with entities from DBPedia. The similarity of two documents is aggregated similarity of pairs of annotating entities. The similarity of two entities e_1 and e_2 comprises so called *hierarchical* and *transversal similarity*. The hierarchical similarity builds on the hierarchical structure of entity categories encoded in DBPedia. It combines the distance of e_1 and e_2 from their *lowest common ancestor* category and their depth in the hierarchy while the traversal similarity represents a weighted distance of e_1 and e_2 where only non-hierarchical paths connecting e_1 and e_2 are considered. Similarity according to [2] is therefore based on distances where no other characteristics of paths between e_1 and e_2 (e.g. edge semantics, etc.) are considered. This simple approach has a reason in [2] as the authors aim at performance of their technique. Our approach presented in this paper introduces a generic framework for measuring similarities of datasets. On one hand, the similarity techniques introduced in [2] can be adopted as concrete techniques for our framework for measuring similarities between annotations (with a change that instead of documents we annotate datasets). On the other hand, other similarity techniques may be used in our framework as well. There are also other works which propose techniques for measuring similarity of entities from a knowledge graph which can be incorporated to our framework as well. E.g., [6] derives similarity on the base of a shortest path between two entities and considers also semantic labels of the edges on the path.

3   Model Framework for Explainable Dataset Similarity

In our framework, we consider notions from the area of similarity search. X is the dataset universe, where element $x \in X$ is a dataset. F is the feature universe, where a feature $f \in F$ could be, e.g. a vector, a number or a string/keyword. $D = 2^F$ is the descriptor universe, where $d \in D$ is descriptor, i.e. a feature descriptor associated to a dataset. Finally, $e : X \rightarrow D$ is the feature extraction procedure transforming a dataset into a descriptor.

3.1   Knowledge Graph Model

For the model of a knowledge graph we adopt the concept of attributed graphs from the graph theory (see Sect. 2) restricted to directed acyclic graphs for our purposes. Hence, we consider a directed attributed acyclic graph (DAAG) $G(V, E, A)$, where V is a set of vertices, E is a set of directed edges without cycles and A is a set of attributes associated with vertices in V describing vertex properties. The DAAG concept was chosen to provide edges of generalization/specialization hierarchy, i.e. the subClassOf relation. In future, we plan to extend the concept to more general knowledge (multi)graph models where semantic and other types of edges will also take place.

A crucial part of the model is mapping descriptors to the graph vertices. In particular, we define mapping of individual features into vertices as $map : F \rightarrow 2^V$, and mapping of descriptors as $mapD : D \rightarrow 2^V$, where

$$mapD(s) = \bigcup_{f_i \in s} map(f_i)$$

and $s \in D$. The mapping definition itself is left to domain-specific implementations, however, using the DAAG model we assume the dataset features are to be mapped by means of attributes in DAAG vertices using an entity linking method (see Sect. 2).

3.2   Navigational Similarity

As the basis for measuring the similarity between datasets, we consider their mapping to DAAG and evaluating an aggregated distance between the mapped vertices. In particular, we define the set of common ancestors for a pair of vertices

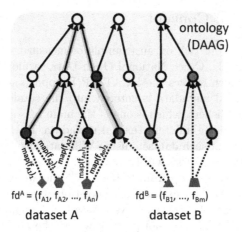

Fig. 1. Explainable similarity between datasets based on aggregation of paths in the knowledge graph. (Color Figure online)

v_1, v_2 as $ca : V \times V \rightarrow 2^V$, such that $ca(v_1, v_2) = W$ while it holds $\forall u \in W : path(v_1, u) \neq \emptyset \wedge path(v_2, u) \neq \emptyset$, where $path(v, u)$ returns a set of edges connecting the vertices v, u, starting in v and ending in u. A set of lowest common ancestors is then defined as $lca : V \times V \rightarrow 2^V$ such that

$$lca(v_1, v_2) = \underset{w \in ca(v_1, v_2)}{\arg\min} \left(|path(v_1, w)| + |path(v_2, w)| \right)$$

A navigational distance between descriptors is defined as $\delta_{nav} : D \times D \rightarrow \mathbb{R}$, where

$$\delta_{nav}(d_1, d_2) = \underset{v_i \in mapD(d_1), v_j \in mapD(d_2)}{\text{AGG}} (lca(v_i, v_j)).distance$$

The AGG function acts as an aggregation operator over the set of lowest common ancestors among datasets' mapped features. Where the AGG component *.distance* refers to the numerical distance score, the component *.structure* refers to a subgraph structure providing a visualization/explanation of the numeric distance as motivated in Subsect. 1.1. The definition of AGG is left to domain-specific applications of the model. It could be a simple max/sum/average of path lengths or a combined distance also taking into account the vertex attributes.

For mapping the feature descriptors of some datasets A and B to the DAAG see Fig. 1. Also note the blue and yellow paths to lowest common ancestors of f_{A1}, f_{B1} and f_{A2}, f_{B1}, respectively, that illustrate the visualization of the similarity evaluated by a domain-specific function AGG (as AGG.*structure*).

4   Early Proof of Concept

We demonstrate our approach on an example of measuring the similarity of a pair of datasets from the Czech National Open Data catalog [1] based on their textual metadata, which follows the DCAT-AP European standard for dataset metadata[4]. The textual metadata is entered into the catalog manually by the individual dataset publishers, which is one of the main reasons why semantically similar datasets often have different textual metadata. In this proof of concept we show the similarity of such datasets using Wikidata[5] as the knowledge graph to which the dataset descriptors will be mapped.

We utilized Wikidata dump from 2018-12-17[6] and we extracted labels and aliases for all entities. Also, we extracted the position of every entity in a category hierarchy. The hierarchy of an entity in Wikidata is represented using the P279[7] (subclass of) property. When it is not present, we use the P31[8] (instanceof) property instead. Note that we use Czech datasets with Czech metadata and Czech Wikidata labels and aliases. For the purpose of this paper we translated the labels into English using Google Translate or, when available, used the English Wikidata labels.

The first dataset $D_{Bohumin}$[9] is called "What, when, where". It covers cultural, sports and free-time events in the Bohumín city in Moravian-Silesian Region. The second dataset $D_{Theater}$[10] is called "Program of the National Moravian-Silesian Theater" and contains the program of the National Moravian-Silesian Theater. The datasets are related because they both contain information about cultural events. At the same time, their textual metadata does not share common words. Therefore, they would not be found by traditional full-text search methods.

In the following paragraphs we demonstrate how we utilized the model framework defined in Sect. 3 to evaluate the dataset similarity. The feature space F is represented by the dataset title, description and keywords. Function e transforms each feature into a bag of words using the following steps: remove punctuation and acute accents, make all characters lowercase, tokenize the string and remove all stop-words.

The mapping function map maps a bag of words to a set of Wikidata entities. Before mapping, we apply e to the Wikidata labels and aliases as well, getting multiple bags of words. We map the dataset to an entity when the entity's bag of words is a subset of any dataset's bag of words. This mapping method is simple, based on the exact match of strings only, while they obviously can have different meanings. The mapping is, however, not the main focus of this paper,

---

[4] https://joinup.ec.europa.eu/release/dcat-ap/121.
[5] https://wikidata.org.
[6] https://dumps.wikimedia.org/other/wikidata/20181217.json.gz.
[7] https://www.wikidata.org/wiki/Property:P279.
[8] https://www.wikidata.org/wiki/Property:P31.
[9] https://data.gov.cz/zdroj/datové-sady/Bohumin/3384768.
[10] https://data.gov.cz/zdroj/datové-sady/https—opendata.ostrava.cz-api-3-action-pac kage_show-id-program-narodniho-divadla-moravskoslezskeho.

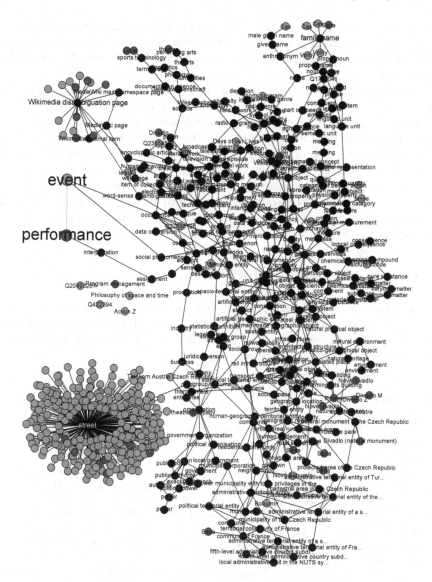

Fig. 2. Example of mapping two datasets $D_{Theater}$ and $D_{Bohumin}$ into the Wikidata hierarchy. Nodes in the displayed graph represent Wikidata entities and edges represent *subClassOf* or *instanceOf* relationships between the entities. The colors present the mapping of the datasets to the entities. $D_{Theater}$ and $D_{Bohumin}$ are mapped to blue and green nodes, respectively. (Color Figure online)

and we are going to improve the matching using commonly used NLP methods in future.

$D_{Bohumin}$ and $D_{Theater}$ map to 335 Wikidata entities which we show on Fig. 2. For the purpose of distance computation we implemented AGG as a max/average/sum/min of distances between the closest pair of mapped entities (v_i, v_j). The computed δ is equal to 10, 2.31, 762 and 1 respectively. The numerical distance does not explain the distance, but the mapping into the knowledge graph (see Fig. 2) does.

We can see from the mapping that the 1 min distance is caused by the entities "event" and "performance" that are connected with an edge. Not only does this relation causes the distance, the relation also explains the distance. The explanation aligns perfectly with our explanation of why the datasets are related.

Due to the simple approach used for entity linking we got clusters of mapped entities around *street*, *Wikimedia disambiguation page* and *family name*. These have high influence on the results of avg/max implementation of *AGG*. This is clearly wrong as in all three examples the mapped entities do not represent any semantic concept in the datasets. Thanks to the explainability of the model this issue is easy to discover.

5   Conclusions and Future Work

In this paper we introduced a framework for measuring similarity of open datasets by means of knowledge graph model in an explainable way. In future work, we plan to develop domain-specific instances of the model to provide semantic search in Open Data catalogs.

References

1. Klímek, J.: DCAT-AP representation of Czech National Open Data Catalog and its impact. J. Web Semant. **55**, 69–85 (2019). https://doi.org/10.1016/j.websem.2018.11.001
2. Paul, C., Rettinger, A., Mogadala, A., Knoblock, C.A., Szekely, P.: Efficient graph-based document similarity. In: Sack, H., Blomqvist, E., d'Aquin, M., Ghidini, C., Ponzetto, S.P., Lange, C. (eds.) ESWC 2016. LNCS, vol. 9678, pp. 334–349. Springer, Cham (2016). https://doi.org/10.1007/978-3-319-34129-3_21
3. Shen, W., Wang, J., Han, J.: Entity linking with a knowledge base: issues, techniques, and solutions. IEEE Trans. Knowl. Data Eng. **27**(2), 443–460 (2015). https://doi.org/10.1109/TKDE.2014.2327028
4. Xu, Z., Ke, Y., Wang, Y., Cheng, H., Cheng, J.: A model-based approach to attributed graph clustering. In: Candan, K.S., Chen, Y., Snodgrass, R.T., Gravano, L., Fuxman, A. (eds.) Proceedings of the ACM SIGMOD International Conference on Management of Data, SIGMOD 2012, Scottsdale, AZ, USA, 20–24 May 2012, pp. 505–516. ACM (2012). https://doi.org/10.1145/2213836.2213894
5. Zhou, Y., Cheng, H., Yu, J.X.: Graph clustering based on structural/attribute similarities. PVLDB **2**(1), 718–729 (2009). https://doi.org/10.14778/1687627.1687709
6. Zhu, G., Iglesias, C.A.: Computing semantic similarity of concepts in knowledge graphs. IEEE Trans. Knowl. Data Eng. **29**(1), 72–85 (2017). https://doi.org/10.1109/TKDE.2016.2610428

The Curse of Dimensionality

The Role of Local Intrinsic Dimensionality in Benchmarking Nearest Neighbor Search

Martin Aumüller$^{(\boxtimes)}$ⓘ and Matteo Ceccarelloⓘ

IT University of Copenhagen, Copenhagen, Denmark
{maau,mcec}@itu.dk

Abstract. This paper reconsiders common benchmarking approaches to nearest neighbor search. It is shown that the concept of local intrinsic dimensionality (LID) allows to choose query sets of a wide range of difficulty for real-world datasets. Moreover, the effect of different LID distributions on the running time performance of implementations is empirically studied. To this end, different visualization concepts are introduced that allow to get a more fine-grained overview of the inner workings of nearest neighbor search principles. The paper closes with remarks about the diversity of datasets commonly used for nearest neighbor search benchmarking. It is shown that such real-world datasets are not diverse: results on a single dataset predict results on all other datasets well.

1 Introduction

Nearest neighbor (NN) search is a key primitive in many computer science applications, such as data mining, machine learning and image processing. For example, Spring and Shrivastava very recently showed in [25] how nearest neighbor search methods can yield large speed-ups when training neural network models. In this paper, we study the classical k-NN problem. Given a dataset $S \subseteq \mathbb{R}^d$, the task is to build an index on S to support the following type of query: For a query point $\mathbf{x} \in \mathbb{R}^d$, return the k closest points in S under some distance measure D.

In many practical settings, a dataset consists of points represented as high-dimensional vectors. For example, word representations generated by the `glove` algorithm [23] associate with each word in a corpus a d-dimensional real-valued vector. Common choices for d are between 50 and 300 dimensions. Finding the true nearest neighbors in such a high-dimensional space is difficult, a phenomenon often referred to as the "curse of dimensionality" [8]. In practice, it means that finding the true nearest neighbors, in general, cannot be solved much more efficiently than by a linear scan through the dataset (requiring time $O(n)$ for n data points) or in space that is exponential in the dimensionality d, which is impractical for large values of d.

While we cannot avoid these general hardness results [1], most datasets that are used in applications are not *truly* high-dimensional. This means that the

© Springer Nature Switzerland AG 2019
G. Amato et al. (Eds.): SISAP 2019, LNCS 11807, pp. 113–127, 2019.
https://doi.org/10.1007/978-3-030-32047-8_11

dataset can be embedded onto a lower-dimensional space without too much distortion. Intuitively, the intrinsic dimensionality (ID) of the dataset is the minimum number of dimensions that allows for such a representation [11]. There exist many explicit ways of finding good embeddings for a given dataset. For example, the Johnson-Lindenstrauss transformation [16] allows us to embed n data points in $\mathbb{R}^d$ into $\Theta((\log n)/\varepsilon^2)$ dimensions such that all pairwise distances are preserved up to a $(1+\varepsilon)$ factor with high probability. Another classical embedding often employed in practice is given by principal component analysis (PCA), see [17].

In this paper, we put our focus on "local intrinsic dimensionality" (LID), a measure introduced by Houle in [11]. We defer a detailed discussion of this measure and its estimation to Sect. 2. Intuitively, the LID of a data point $\mathbf{x}$ at a distance threshold $r > 0$ measures how difficult it is to distinguish between points at distance r and distance $(1+\varepsilon)r$ in a dataset. Most importantly for this study, LID is a *local* measure that can be associated with a single query. It was stated in [12] that the LID might serve as a characterization of the difficulty of k-NN queries. One purpose of this paper is to shed light on this statement.

A focus of this paper is an empirical study of how the LID influences the performance of NN algorithms. To be precise, we will benchmark four different implementations [18] which employ different approaches to NN search. Three of them (HNSW [21], FAISS-IVF [15], Annoy [6]) stood out as most performant in the empirical study conducted by Aumüller et al. in [4]. Another one (ONNG) was proposed very recently [13] and shown to be competitive to these approaches. We base our experiments on [4] and describe their benchmarking approach and the changes we made to their system in Sect. 3. We analyze the LID distribution of real-world datasets in Sect. 4. We will see that there is a substantial difference between the LID distributions among datasets. We will next conduct two experiments: First, we fix a dataset and choose as query set the set of points with smallest, medium, and largest estimated LIDs. In addition, we choose a set of "diverse" query points w.r.t. their LID estimates. As we will see, there is a clear tendency such that the larger the LID, the more difficult the query for all implementations. Next, we will study how the different LID distributions between datasets influence the running time performance. In a nutshell, it cannot be concluded that LID by itself is a good indicator for the relative performance of a fixed implementation over datasets. These statements will be made precise in the evaluation that is discussed in Sect. 5.

In the first part of our evaluation, we work in the "classical evaluation setting of nearest neighbor search". This means that we relate a performance measure (such as the achieved throughput measured in queries per second) to a quality measure (such as the average fraction of true nearest neighbors found over all queries). While this is the most commonly employed evaluation method, we reason that this way of representing results in fact hides interesting details about the inner workings of an implementation. Using non-traditional visualization techniques provide new insights into their query behavior on real-world datasets. As one example, we see that reporting average recall on the graph-based approaches from [13,21] hides an important detail: For a given query, they either find all true nearest neighbors or not a single one. This behavior is not shared by the two other approaches that we consider; both yield a continuous transition from "finding no nearest neighbors" to "finding all of them".

As a final point, we want, ideally, to benchmark on a collection of "interesting" datasets that show the strengths and weaknesses of individual approaches [24]. We will conclude that there is little diversity among the considered real-word datasets: While the individual performance observations change from dataset to dataset, the relative performance between implementations stays the same.

Our Contributions. The main contributions of this paper are

- a detailed evaluation of the LID distribution of many real-world datasets used in benchmarking frameworks,
- an evaluation of the influence of the LID on the performance of NN search implementations,
- considerations about the result diversity, and
- an exploration of different visualization techniques that shed light on individual properties of certain implementation principles.

A preliminary workshop version of this paper appeared as [5]. In this paper we expand the experimental study with the correlation between LID and recall; we also consider different ways of generating synthetic datasets to investigate the relationship between LID and performance.

Related Work on Benchmarking Frameworks for NN. We use the benchmarking system described in [4] as the starting point for our study. Different approaches to benchmarking nearest neighbor search are described in [9,10,20]. We refer to [4] for a detailed comparison between the frameworks.

2 Local Intrinsic Dimensionality

We consider a distance-space $(\mathbb{R}^d, D)$ with a distance function $D \colon \mathbb{R}^d \times \mathbb{R}^d \to \mathbb{R}$. As described in [2], we consider the distribution of distances within this space with respect to a reference point $\mathbf{x}$. Such a distribution is induced by sampling n points from the space $\mathbb{R}^d$ under a certain probability distribution. We let $F \colon \mathbb{R} \to [0, 1]$ be the cumulative distribution function of distances to the reference point $\mathbf{x}$.

Definition 1 ([11]). *The local continuous intrinsic dimension of F at distance r is given by*

$$ID_F(r) = \lim_{\varepsilon \to 0} \frac{\ln(F((1 + \varepsilon)r)/F(r))}{\ln((1 + \varepsilon)r/r)},$$

whenever this limit exists.

The measure relates the increase in distance to the increase in probability mass (the fraction of points that are within the ball of radius r and $(1+\varepsilon)r$ around the query point). Intuitively, the larger the LID, the more difficult it is to distinguish true nearest neighbors at distance r from the rest of the dataset. As described in [12], in the context of k-NN search we set r as the distance of the k-th nearest neighbor to the reference point $\mathbf{x}$.

Estimating LID. We use the Maximum-Likelihood estimator (MLE) described in [2,19] to estimate the LID of $\mathbf{x}$ at distance r. Let $r_1 \leq \ldots \leq r_k$ be the sequence of distances of the k-NN of $\mathbf{x}$. The MLE $\hat{\text{ID}}_\mathbf{x}$ is then

$$\hat{\text{ID}}_\mathbf{x} = -\left(\frac{1}{k} \sum_{i=1}^{k} \ln \frac{r_i}{r_k} \right)^{-1}. \tag{1}$$

Amsaleg et al. showed in [2] that MLE estimates the LID well. We remark that in very recent work, Amsaleg et al. proposed in [3] a new MLE-based estimator that works with smaller k values than (1).

3   Overview over the Benchmarking Framework

We use the `ann-benchmarks` system described in [4] to conduct our experimental study. Ann-benchmarks is a framework for benchmarking NN search algorithms. It covers dataset creation, performing the actual experiment, and storing the results of these experiments in a transparent and easy-to-share way. Moreover, results can be explored through various plotting functionalities, e.g., by creating a website containing interactive plots for all experimental runs.

Ann-benchmarks interfaces with a NN search implementation by calling its preprocess (index building) and search (query) methods with certain parameter choices. Implementations are tested on a large set of parameters usually provided by the original authors of an implementation. The answers to queries are recorded as the indices of the points returned. Ann-benchmarks stores these parameters together with further statistics such as individual query times, index size, and auxiliary information provided by the implementation. See [4] for more details.

Compared to the system described in [4], we added tools to estimate the LID based on Eq. (1), pick "challenging query sets" according to the LID of individual points, and added new datasets and implementations. Moreover, we implemented a mechanism that allows an implementation to provide further query statistics after answering a query. To showcase this feature, all implementations in this study report the number of distance computations performed to answer a query.[1]

---

[1] We thank the authors of the implementations for their help and responsiveness in adding this feature to their library.

Table 1. Datasets under consideration with their average local intrinsic dimensionality (LID) computed by MLE [2] from the 100-NN of all the data points.

Dataset	Data points	Dimensions	LID		Metric
			avg	median	
SIFT [14]	1 000 000	128	21.9	19.2	Euclidean
MNIST	65 000	784	14.0	13.2	Euclidean
Fashion-MNIST [26]	65 000	784	15.6	13.9	Euclidean
GLOVE [23]	1 183 514	100	18.0	17.8	Angular/Cosine
GLOVE-2M [23]	2 196 018	300	26.1	23.4	Angular/Cosine
GNEWS [22]	3 000 000	300	21.1	20.1	Angular/Cosine

4   Algorithms and Datasets

4.1   Algorithms

Nearest neighbor search algorithms for high dimensions are usually graph-, tree-, or hashing-based. We refer the reader to [4] for an overview over these principles and available implementations. In this study, we concentrate on the three implementations considered most performant in [4], namely HNSW [21], Annoy [6] and FAISS-IVF [15] (IVF from now on). We consider the very recent graph-based approach ONNG [13] in this study as well.

HNSW and ONNG are graph-based approaches. This means that they build a k-NN graph during the preprocessing step. In this graph, each vertex is a data point and a directed edge (u, v) means that the data point associated with v is "close" to the data point associated with u in the dataset. At query time, the graph is traversed to generate candidate points. Algorithms differ in details of the graph construction, how they build a navigation structure on top of the graph, and how the graph is traversed.

Annoy is an implementation of a random projection forest, which is a collection of random projection trees. Each node in a tree is associated with a set of data points. It splits these points into two subsets according to a chosen hyperplane. If the dataset in a node is small enough, it is stored directly and the node is a leaf. Annoy employs a data-dependent splitting mechanism in which a splitting hyperplane is chosen as the one splitting two "average points" by repeatedly sampling dataset points. In the query phase, trees are traversed using a priority queue until a predefined number of points is found.

IVF builds an inverted file based on clustering the dataset around a predefined number of centroids. It splits the dataset based on these centroids by associating each point with its closest centroid. During query it finds the closest centroids and checks points in the dataset associated with those.

We remark we used both IVF and HNSW implementations from FAISS[2].

---

[2] https://github.com/facebookresearch/faiss.

4.2   Datasets

Table 1 presents an overview over the datasets that we consider in this study. We restrict our attention to datasets that are usually used in connection with Euclidean distance and Angular/Cosine distance. For each dataset, we compute the LID distribution with respect to the 100-NN as discussed in Sect. 2. SIFT, MNIST, and GLOVE are among the most-widely used datasets for benchmarking nearest neighbor search algorithms. Fashion-MNIST is considered as a replacement for MNIST, which is usually considered too easy for machine learning tasks [26].

Figure 1 provides a visual representation of the estimated LID distribution of each dataset, for $k = 100$. While the datasets differ widely in their original dimensionality, the median LID ranges from around 13 for MNIST to about 23 for GLOVE-2M. The distribution of LID values is asymmetric and shows a long tail behavior. MNIST, Fashion-MNIST, SIFT, and GNEWS are much more concentrated around the median compared to the two `glove`-based datasets.

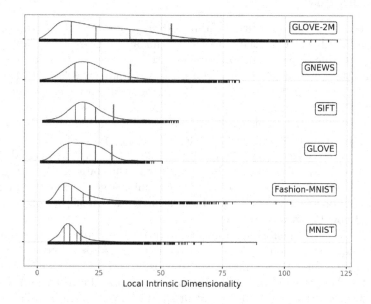

Fig. 1. LID distribution for each dataset. Ticks below the distribution curves represent single queries. Lines within each distribution curve correspond to the 25, 50 and 75 percentile. The red line marks the 10 000 largest estimated LID, which we use as a threshold value to define *hard* query sets. (Color figure online)

5   Evaluation

This section reports on the results of our experiments. Due to space constraints, we only present some selected results. More results and plots can be explored via interactive plots at http://ann-benchmarks.com/sisap19/, which also contains a link to the source code repository. For a fixed implementation, the plots presented here consider the Pareto frontier over all parameter choices [4]. Tested parameter choices and the associated plots are available on the website.

Experimental Setup. Experiments were run on 2x 14-core Intel Xeon E5-2690v4 (2.60 GHz) with 512 GB RAM using Ubuntu 16.10 (kernel 4.4.0). Index building was multi-threaded, queries where answered in a single thread.

Quality and Performance Metrics. As quality metric we measure the individual recall of each query, i.e., the fraction of points reported by the implementation that are among the true k-NN. As performance metric, we record individual query times and the total number of distance computations needed to answer all queries. We usually report on the throughput (the average number of queries that can be answered in one second, in the plots denoted as QPS for *queries per second*), but we will also inspect individual query times.

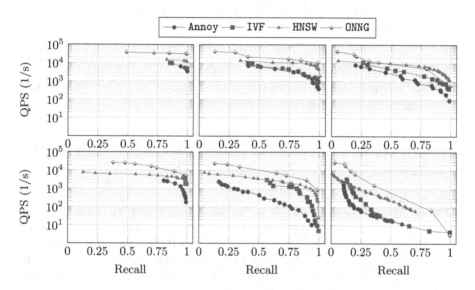

Fig. 2. Recall-QPS (1/s) tradeoff – up and to the right is better; top: SIFT, bottom: GLOVE-2M; left: easy, middle: middle, right: hard.

Objectives of the Experiments. Our experiments are tailored to answer the following questions:

(Q1) How does the LID of a query set influence the running time performance? (Sect. 5.1)
(Q2) How diverse are measurements obtained on datasets? Do relative differences between the performance of different implementations stay the same over multiple datasets? (Sect. 5.2)

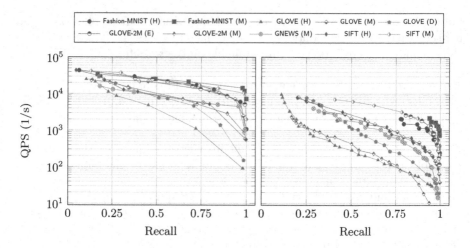

Fig. 3. Recall-QPS (1/s) tradeoff – up and to the right is better; left: ONNG, right: Annoy; (E) — easy, (M) — medium, (H) — hard, (D) — diverse.

(Q3) How well does the number of distance computations reflect the relative running time performance of the tested implementations? (Sect. 5.2)

(Q4) How concentrated are quality and performance measures around their mean for the tested implementations? (Sect. 5.3)

Choosing Query Sets. For each dataset, we select four different query sets: The query set that contains the 10 000 points with the lowest estimated LID (which we denote *easy*), 10 000 points around the data point with median estimated LID (denoted *medium*), 10 000 points with the largest estimated LID (dubbed *hard*), and 5 000 points chosen uniformly according to (integer) LID values (denoted *diverse*). For the latter, we split all data points up into buckets, where bucket i represents all data points that have an estimated LID of i (rounded down). For each query, we pick a non-empty bucket uniformly at random, and inside the bucket we pick a random point (with repetition). Figure 1 marks with a red line the LID used as a threshold to build the *hard* queryset.

5.1 Influence of LID on Performance

Figure 2 shows results for the influence of using only points with low, middle, and large estimated LID as query points, in SIFT and GLOVE-2M. We observe a clear influence of the LID of the query set on the performance: the larger the LID, the more down and to the left the graphs move. This means that, for higher LID, it is more expensive, in terms of time, to answer queries with good recall. For all datasets except GLOVE-2M, all implementations were still able to achieve close to perfect recall with the parameters set. This means that all but one of the tested datasets do not contain too many "noisy queries". Already the queries

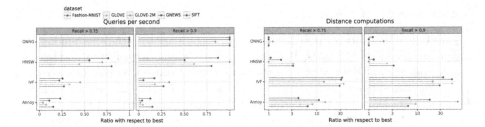

Fig. 4. Ranking of algorithms on five different datasets, according to recall ≥ 0.75 and ≥ 0.9, and according to two different performance measures: number of queries per second (left) and number of distance computations (right). Both plots report the ratio with the best performing algorithm on each dataset: for the queries per second metric a larger ratio is better, for the number of distance computations metric, a smaller ratio is better.

around the median prove challenging for most implementations. For the most difficult queries (according to LID), only IVF and ONNG achieve close to perfect recall on GLOVE-2M.

Figure 3 reports on the results of ONNG and Annoy on selected datasets. Comparing results to the LID measurements depicted in Fig. 1, the estimated median LID gives a good estimate on the relative performance of the algorithms on the data sets. As an exception, SIFT (M) is much easier than predicted by its LID distribution. In particular for Annoy, the hard SIFT instance is as challenging as the medium GLOVE version. The easy version of GLOVE-2M turns out to be efficiently solvable by both implementations (taking about the same time as it takes to answer the hard instance of Fashion-MNIST, which has a much higher LID). From this, we cannot conclude that LID as a single indicator explains performance differences of an implementation across different datasets. However, more careful experimentation is need before drawing a final conclusion. In our setting, the LID estimation is conducted for $k = 100$, while queries are only searching for the 10 nearest neighbors. Moreover, the estimation using MLE might not be accurate enough on these datasets, since it is very dependent on the parameter k being used. We leave the investigation of these two directions as future work.

In general, the diverse query set is more difficult than the medium query set. In particular, at high recall it generally becomes nearly as difficult as the difficult dataset. The reason for this behavior is that none of the implementations can adapt to the difficulty of a query. They only achieve high average recall when they can solve sufficiently many queries with high LID. The parameter settings that allow for such guarantees slow down answering the easy queries by a lot. We believe that the "diverse" query sets thus allow for challenging benchmarking datasets for adaptive query algorithms.

As a side note, we remark that Fashion-MNIST is as difficult to solve as MNIST for all implementations, and is by far the easiest dataset for all implementations. Thus, while there is a big difference in the difficulty of solving the classification task [26], there is no measurable difference between these two datasets in the context of NN search.

5.2 Diversity of Results

Figure 4 gives an overview over how algorithms compare to each other among all "medium difficulty" datasets. We consider two metrics, namely the number of queries per second (left plot), and the number of distance computations (right plot). For two different average recall thresholds (0.75 and 0.9) we then select, for each algorithm, the best performing parameter configuration that attains at least that recall. For each dataset, the plots report the ratio with the best performing algorithm on that dataset, therefore the best performer is reported with ratio 1. Considering different dataset, we see that there is little variation in the ranking of the algorithms. Only the two graph-based approaches trade ranks, all other rankings are stable. Interestingly, Annoy makes much fewer distance computations but is consistently outperformed by IVF.[3]

Comparing the number of distance computations to running time performance, we see that an increase in the number of distance computations is not reflected in a proportional decrease in the number of queries per second. This means that the candidate set generation is in general more expensive for graph-based approaches, but the resulting candidate set is of much higher quality and fewer distance computations have to be carried out. Generally, both graph-based algorithms are within a factor 2 from each other, whereas the other two need much larger candidate lists to achieve a certain recall. The relative difference usually ranges from 5x to 30x more distance computations for the non-graph based approaches, in particular at high recall. This translates well into the performance differences we see in this setting: consider for instance Fig. 2, where the lines corresponding to HNSW and ONNG upper bound the lines relative to the other two algorithms.

5.3 Reporting the Distribution of Performance

In the previous sections, we made extensive use of recall/queries per second plots, where each configuration of each algorithm results in a single point, namely the average recall and the inverse of the average query time. As we shall see in this section, concentrating on averages can hide interesting information in the context of k-NN queries. In fact, not all queries are equally difficult to answer. Consider the plots in Fig. 5, which report performance of the four algorithms[4] on the GLOVE-2M dataset, medium difficulty. The top four plots report the recall versus the number of queries per second, and black dots correspond to the averages. Additionally, for each configuration, we report the distribution of the recall scores: the baseline of each recall curve is positioned at the corresponding queries per second performance. Similarly, the bottom plots report on the inverse of the

---

[3] We note that IVF counts the initial comparisons to find the closest centroids as distance computations, whereas Annoy did not count the inner product computations during tree traversal.

[4] In order not to clutter the plots, we fixed parameters as follows: IVF — number of lists 8192; Annoy — number of trees 100; HNSW — efConstruction 500, M 8; ONNG — edge 100, outdegree 10, indegree 120.

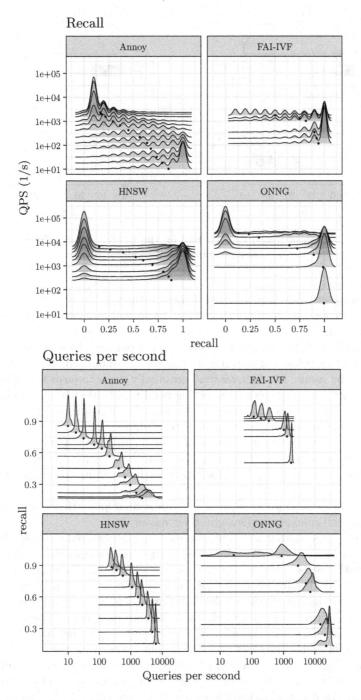

Fig. 5. Distribution of performance for queries on the GLOVE-2M (medium difficulty) dataset. Looking just at the average performance can hide interesting behaviour.

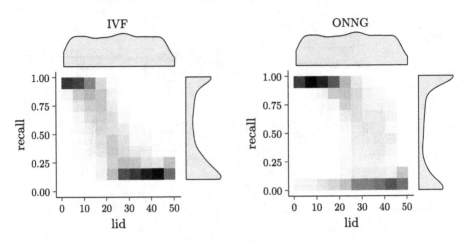

Fig. 6. Distribution of Recall vs. LID plot on the GLOVE dataset. Intensity reflects number of queries that achieve a combination of recall vs. LID.

individual query times (the average of these is the QPS in the left plot) against the average recall. In both plots, the best performance is achieved towards the top-right corner.

Plotting the distributions, instead of just reporting the averages, uncovers some interesting behaviour that might otherwise go unnoticed, in particular with respect to the recall. The average recall gradually shifts towards the right as the effect of more and more queries achieving good recalls. Perhaps surprisingly, for graph-based algorithms this shift is very sudden: most queries go from having recall 0 to having recall 1, taking no intermediate values. Taking the average recall as a performance metric is convenient in that it is a single number to compare algorithms with. However, the same average recall can be attained with very different distributions: looking at such distributions can provide more insight.

For the bottom plots, we observe that individual query times of all the algorithms are well concentrated around their mean.

Figure 6 gives another distributional view on the achieved result quality. The plots shows two runs of IVF and ONNG with fixed parameters on the GLOVE dataset with diverse queries. On the top we see the distribution of estimated LID values for the diverse query set, on the right we see the distribution of recall values achieved by the implementation. Each of the queries corresponds to a single data point in the recall/LID plot and data points are summarized through hexagons, where the color intensity of a hexagon indicates the number of data points falling into this region. The plots show that the higher the LID of a query, there is a clear tendency for the query to achieve lower recall.

For space reasons, we do not report other parameter configurations and datasets, which nonetheless show similar behaviours. All of them can be accessed at the website.

6   Summary

In this paper we studied the influence of LID to the performance of nearest neighbor search algorithms. We showed that LID allows to choose query sets of a wide range of difficulty from a given dataset. We also showed how different LID distributions influence the running time performance of the algorithms. In this respect, we could not conclude that the LID alone can predict running time differences well. In particular, SIFT is usually easier for the algorithms, while GLOVE's LID distribution would predict it to be the easier dataset of the two.

With regard to challenging query workloads, we described a way to choose diverse query sets. They have the property that for most implementations it is easy to perform well for most of the query points, but they contain many more easy and difficult queries than query workloads chosen randomly from the dataset. We believe this is a very interesting benchmarking workload for approaches that try to adapt to the difficulty of an individual query.

We introduced novel visualization techniques to show the uncertainty within the answer to a set of queries, which made it possible to show a clear difference between the graph-based algorithms and the other approaches.

We hope that this study initiates the search for more diverse datasets, or for theoretical reasoning why certain algorithmic principles are generally better suited for nearest neighbor search. On a more practical side, Casanova et al. showed in [7] how dimensionality testing can be used to speed up reverse k-NN queries. We would be interested in seeing whether the LID can be used at other places in the design of NN algorithms to guide the search process or the parameter selection. While we know from [2] that the LID estimation of MLE with $k = 100$ works well on their datasets, it would be interesting to see whether the other estimators mentioned there are also able to characterize the relative performance of queries.

Acknowledgements. The authors would like to thank the anonymous reviewers for their useful suggestions, which helped to improve the presentation of the paper. The research leading to these results has received funding from the European Research Council under the European Union's 7th Framework Programme (FP7/2007-2013)/ERC grant agreement no. 614331.

References

1. Alman, J., Williams, R.: Probabilistic polynomials and hamming nearest neighbors. In: FOCS 2015, pp. 136–150 (2015)
2. Amsaleg, L., et al.: Estimating local intrinsic dimensionality. In: KDD 2015, pp. 29–38. ACM (2015)
3. Amsaleg, L., Chelly, O., Houle, M.E., Kawarabayashi, K.I., Radovanović, M., Treeratanajaru, W.: Intrinsic dimensionality estimation within tight localities. In: Proceedings of the 2019 SIAM International Conference on Data Mining, pp. 181–189. SIAM (2019)

4. Aumüller, M., Bernhardsson, E., Faithfull, A.: ANN-benchmarks: a benchmarking tool for approximate nearest neighbor algorithms. In: Beecks, C., Borutta, F., Kröger, P., Seidl, T. (eds.) SISAP 2017. LNCS, vol. 10609, pp. 34–49. Springer, Heidelberg (2017). https://doi.org/10.1007/978-3-319-68474-1_3
5. Aumüller, M., Ceccarello, M.: Benchmarking nearest neighbor search: influence of local intrinsic dimensionality and result diversity in real-world datasets. In: 1st Workshop on Evaluation and Experimental Design in Data Mining and Machine Learning (EDML 2019) (2019). https://imada.sdu.dk/Research/EDML/
6. Bernhardsson, E.: Annoy. https://github.com/spotify/annoy
7. Casanova, G., et al.: Dimensional testing for reverse k-nearest neighbor search. PVLDB 10(7), 769–780 (2017)
8. Chávez, E., Navarro, G., Baeza-Yates, R., Marroquín, J.L.: Searching in metric spaces. ACM Comput. Surv. 33(3), 273–321 (2001). https://doi.org/10.1145/502807.502808
9. Curtin, R.R., et al.: MLPACK: a scalable C++ machine learning library. J. Mach. Learn. Res. 14, 801–805 (2013)
10. Edel, M., Soni, A., Curtin, R.R.: An automatic benchmarking system. In: NIPS 2014 Workshop on Software Engineering for Machine Learning (2014)
11. Houle, M.E.: Dimensionality, discriminability, density and distance distributions. In: Data Mining Workshops (ICDMW), pp. 468–473. IEEE (2013)
12. Houle, M.E., Schubert, E., Zimek, A.: On the correlation between local intrinsic dimensionality and outlierness. In: Marchand-Maillet, S., Silva, Y.N., Chávez, E. (eds.) SISAP 2018. LNCS, vol. 11223, pp. 177–191. Springer, Cham (2018). https://doi.org/10.1007/978-3-030-02224-2_14
13. Iwasaki, M., Miyazaki, D.: Optimization of Indexing Based on k-Nearest Neighbor Graph for Proximity Search in High-dimensional Data. ArXiv e-prints, October 2018
14. Jégou, H., Douze, M., Schmid, C.: Product quantization for nearest neighbor search. IEEE Trans. Pattern Anal. Mach. Intell. 33(1), 117–128 (2011). https://doi.org/10.1109/TPAMI.2010.57
15. Johnson, J., Douze, M., Jégou, H.: Billion-scale similarity search with GPUs. CoRR abs/1702.08734 (2017)
16. Johnson, W.B., Lindenstrauss, J., Schechtman, G.: Extensions of Lipschitz maps into Banach spaces. Israel J. Math. 54(2), 129–138 (1986)
17. Jolliffe, I.: Principal Component Analysis. Springer, Berlin (2011)
18. Kriegel, H., Schubert, E., Zimek, A.: The (black) art of runtime evaluation: are we comparing algorithms or implementations? Knowl. Inf. Syst. 52(2), 341–378 (2017)
19. Levina, E., Bickel, P.J.: Maximum likelihood estimation of intrinsic dimension. In: NIPS, pp. 777–784 (2005)
20. Li, W., Zhang, Y., Sun, Y., Wang, W., Zhang, W., Lin, X.: Approximate nearest neighbor search on high dimensional data - experiments, analyses, and improvement (v1.0). CoRR abs/1610.02455 (2016)
21. Malkov, Y.A., Yashunin, D.A.: Efficient and robust approximate nearest neighbor search using Hierarchical Navigable Small World graphs. ArXiv e-prints, March 2016
22. Mikolov, T., Chen, K., Corrado, G., Dean, J.: Efficient estimation of word representations in vector space. CoRR abs/1301.3781 (2013)
23. Pennington, J., Socher, R., Manning, C.D.: Glove: global vectors for word representation. In: Empirical Methods in Natural Language Processing (EMNLP), pp. 1532–1543 (2014)

24. Smith-Miles, K., Baatar, D., Wreford, B., Lewis, R.: Towards objective measures of algorithm performance across instance space. Comput. Oper. Res. **45**, 12–24 (2014)
25. Spring, R., Shrivastava, A.: Scalable and sustainable deep learning via randomized hashing. In: KDD 2017, pp. 445–454 (2017). https://doi.org/10.1145/3097983.3098035
26. Xiao, H., Rasul, K., Vollgraf, R.: Fashion-MNIST: a novel image dataset for benchmarking machine learning algorithms. CoRR abs/1708.07747 (2017)

Accurate and Fast Retrieval for Complex Non-metric Data via Neighborhood Graphs

Leonid Boytsov[✉] and Eric Nyberg

Carnegie Mellon University, Pittsburgh, PA, USA
{srchvrs,enh}@cs.cmu.edu

Abstract. We demonstrate that a graph-based search algorithm—relying on the construction of an approximate neighborhood graph—can directly work with challenging *non-metric* and/or *non-symmetric* distances without resorting to metric-space mapping and/or distance symmetrization, which, in turn, lead to substantial performance degradation. Although the straightforward metrization and symmetrization is usually ineffective, we find that constructing an index using a modified, e.g., symmetrized, distance can improve performance. This observation paves a way to a new line of research of designing index-specific graph-construction distance functions.

Keywords: k-NN search · Non-metric distance · Neighborhood graph

1 Introduction and Problem Definition

In this paper we focus on k nearest neighbor (k-NN) search, which is a widely used computer technology with applications in machine learning, data mining, information retrieval, and natural language processing. Formally, we assume to have a possibly *infinite* domain containing objects x, y, z, ..., which are commonly called data points or simply points. The domain—sometimes called a *space*—is equipped with a *distance function* $d(x, y)$, which is used to measure dissimilarity of objects x and y. The value of $d(x, y)$ is interpreted as a degree of dissimilarity. The larger is $d(x, y)$, the more dissimilar points x and y are.

Some distances are non-negative and become zero only when x and y have the highest possible degree of similarity. The *metric* distances are additionally symmetric and satisfy the triangle inequality. However, in general, we do not impose any restrictions on the value of the distance function (except that smaller values represent more similar objects). Specifically, the value of the distance

---

This work was accomplished while Leonid Boytsov was a PhD student at CMU. Authors gratefully acknowledge the support by the NSF grant #1618159: "Matching and Ranking via Proximity Graphs: Applications to Question Answering and Beyond".

© Springer Nature Switzerland AG 2019
G. Amato et al. (Eds.): SISAP 2019, LNCS 11807, pp. 128–142, 2019.
https://doi.org/10.1007/978-3-030-32047-8_12

function can be negative and negative distance values indicate higher similarity than positive ones.

We further assume that there is a data *subset* D containing a *finite* number of domain points and a set of queries that belongs to the domain but not to D. We then consider a standard top-k retrieval problem. Given a query q it consists in finding k data set points $\{x_i\}$ with smallest values of distances to the query among all data set points (ties are broken arbitrarily). Data points $\{x_i\}$ are called *nearest neighbors*. A search should return $\{x_i\}$ in the order of increasing distance to the query. If the distance is not symmetric, two types of queries can be considered: *left* and *right* queries. In a *left* query, a data point compared to the query is always the first (i.e., the left) argument of $d(x, y)$. Henceforth, for simplicity of exposition we consider only the case of left queries.

Exact methods degenerate to a brute-force search for just a dozen of dimensions [35]. Due to diversity of properties, non-metric spaces lack *common* and *easily identifiable* structural properties such as the triangle inequality. There is, therefore, little hope that *fully* generic exact search methods can be devised. Thus, we focus on the *approximate* version of the problem where the search may miss some of the neighbors, but it may not change the order. The accuracy of retrieval is measured via recall (equal to the average fraction of neighbors found). We cannot realistically devise fast exact methods, but we still hope that our approximate methods are quite *accurate* having a recall close to 100%.

There has been a staggering amount of effort invested in designing new and improving existing k-NN search algorithms (see e.g., [8, 29, 30, 34]). This effort has been placed disproportionately on techniques for *symmetric metric* distances, in particular, on search methods for the Euclidean space. Yet, search methods for challenging non-symmetric and non-metric spaces received very *little* attention. A *filter-and-refine* approach is a common way to deal with an unconventional distance. To this end one would map data to a low-dimensional Euclidean space. The goal is to find a mapping without large distortion of the original similarity measure [14, 17]. Jacobs et al. [17] review various projection methods and argue that such a coercion is often against the nature of a similarity measure, which can be, e.g., intrinsically non-symmetric. Yet, they do not provide experimental evidence. We fill this gap and demonstrate that both metric learning and distance symmetrization are, indeed, suboptimal approaches.

Alternatively the metric distance can be learned from scratch [3]. In that, Chechik et al. [9] contended that in the task of distance learning enforcing symmetry and metricity is useful only as a means to prevent overfitting to a small training set. However, when training data is abundant, it can be more efficient and more accurate to learn the distance function in an unconstrained bilinear form. Yet, this approach does not necessarily results in a symmetric metric distance [9]. We, in turn, demonstrate that a graph-based retrieval algorithm—relying on the construction of approximate neighborhood/proximity graphs—can deal with challenging non-metric distances directly without resorting to a low-dimensional mapping or full symmetrization. In that, unlike prior work [24, 25], as we show in Sect. 3, several of our distances are substantially non-symmetric.

Whereas the filter-and-refine symmetrization approach is detrimental, we find that constructing an index using the symmetrized distance can improve results. Furthermore, we show that the index construction algorithm can be quite sensitive to the order of distance function arguments. In most cases, changing the argument order is detrimental. However, this is not a universal truth: Quite surprisingly, we observe small improvements in some cases by building the graph using the argument-reversed distance function. We believe this observations motivates the line of research to design indexing distance functions—different from original distance functions—that result in better performance. The remaining paper contains the description of employed retrieval algorithms and related experimental results.

2   Methods and Materials

2.1   Retrieval Algorithms

We consider two types of retrieval approaches: the filter-and-refine method using brute-force search and indexing using the graph-based retrieval method Small World Graph (SW-graph) [22]. In the filter-and-refine approach, we use a proxy distance to generate a list of k_c candidate entries (closest to the query with respect to the proxy distance) via the brute-force, i.e., exhaustive, search. For k_c candidate entries x_i we compute the true distance values $d(x_i, q)$—or $d(q, x_i)$ for right queries—and select k closest entries.

The filter-and-refine approach can be slow even if the proxy distance is quite cheap [24], whereas indexing can dramatically speed up retrieval. In particular, state-of-the-art performance can be achieved by using graph-based retrieval methods, which rely on the construction of an exact or approximate *neighborhood* graph (see, e.g., [2,24]). The neighborhood graph is a data structure in which data points are associated with graph nodes and sufficiently close nodes are connected by edges. A search algorithm is a graph-traversal routine exploiting a property "the closest neighbor of my closest neighbor is my neighbor as well." The neighborhood graph is often defined as a directed graph [11,12], where the edges go from a vertex to its neighbors (or vice versa), but undirected edges have been used too [20,22] (undirected nodes were also quietly introduced in kgraph[1]). In a recent study, the use of undirected neighborhood graphs lead to a better performance [20].

Constructing an *exact* neighborhood graph is hardly feasible for a large high-dimensional data set, because, in the worst case, the number of distance computations is $O(n^2)$, where n in the number of data points. An *approximate* neighborhood graph can be constructed substantially more efficiently [11,22]. To improve performance, one can use various graph pruning methods [13,20,23]: In particular, it is not useful to keep neighbors that are close to each other [13,20].

Neighborhood graphs have a long history. Toussaint published a pioneering paper where he introduced neighborhood graphs on the plane in 1980 [33]. Arya

---

[1] https://github.com/aaalgo/kgraph.

Table 1. Data sets

Name	Max. # of rec.	Dimensionality	Source
RandHist-d	0.5×10^6	$d \in \{8, 32\}$	Histograms sampled uniformly from a simplex
RCV-d	0.5×10^6	$d \in \{8, 128\}$	d-topic LDA [4] RCV1 [19] histograms
Wiki-d	2×10^6	$d \in \{8, 128\}$	d-topic LDA [4] Wikipedia histograms
Manner	1.46×10^5	1.23×10^5	Question and answers from L5 collection in Yahoo WebScope

Table 2. Distance functions

Denotation/Name	d(x,y)	Notes
Kullback-Leibler diverg. (KL-div.) [18]	$\sum_{i=1}^{m} x_i \log \frac{x_i}{y_i}$	
Itakura-Saito distance [16]	$\sum_{i=1}^{m} \left[\frac{x_i}{y_i} - \log \frac{x_i}{y_i} - 1 \right]$	
Rényi diverg. [27]	$\frac{1}{\alpha-1} \log \left[\sum_{i=1}^{m} x_i^\alpha y_i^{1-\alpha} \right]$, $0 < \alpha < \infty$	We use $\alpha \in 0.25, 0.75, 2$
BM25 similarity [28]	$-\sum_{x_i = y_i} \mathrm{TF}_q(x_i) \cdot \mathrm{TF}_d(y_i) \cdot \mathrm{IDF}(y_i)$	$\mathrm{TF}_q(x)$ and $\mathrm{TF}_d(y)$ are (possibly scaled) term frequencies in a query and document

and Mount were first to apply neighborhood graphs to the problem of k-NN search in a high-dimensional space [1]. Houle and Sakuma proposed the first hierarchical, i.e., multi-layer, variant of the neighborhood graph called SASH, where data points at layer i are connected only to the nodes at layer $i + 1$ [15]. Malkov and Yashunin proposed an efficient multi-layer neighborhood-graph method called a Hierarchical Navigable Small World (HNSW) [23]. It is a generalization and improvement of the previously proposed method navigable Small World (SW-graph) [22], which has been shown to be quite efficient in the past [22,24].

Although there are different approaches to construct a neighborhood graphs, all retrieval strategies known to us rely on a simple semi-greedy graph-traversal algorithm with (possibly) multiple restarts. Such an algorithm keeps a priority queue of elements, which ranks candidates in the order of increasing distance to the query. At each step, the search retrieves one or more elements from the queue that are closest to the query and explores their neighborhoods. Previously unseen elements may be added to the queue. For a recent experimental comparison of several retrieval approaches see [32].

Although, HNSW is possibly the best retrieval method for generic distances [20,23], in our work we use a modified variant of SW-graph, where retrieval starts from a single point (which is considerably more efficient compared to multiple starting points). The main advantage of HNSW over the older version of SW-graph is due to (1) introduction of pruning heuristics, (2) using a single starting point during retrieval. We want to emphasize that comparison of HNSW against SW-graph in [23] is not completely fair, because it basically uses an undertuned SW-graph. Furthermore, gains from using a hierarchy of layers are quite small: see Figs. 3–5 from [23]. At the same time pruning heuristics introduce another confounding factor in measuring the effect of distance symmetrization (and proxying), because symmetrization method used in the pruning approach can be different from the symmetrization method used by k-NN search employed at index time. Thus—as we care primarily about demonstrating usefulness (or lack thereof) of different distance modifications during construction of the graph rather than merely achieving maximum retrieval efficiency—we experiment with a simpler retrieval algorithm SW-graph. The employed algorithm has three main parameters. Parameter NN influences (but does not define directly) the number of neighbors in the graph. Parameters efConstruction and efSearch define the depth of the priority queue used during index and retrieval stages, respectively.

2.2   Data Sets and Distances

In our experiments, we use the following distances (see Table 2): KL-divergence, the Itakura-Saito distance, the Rényi divergence, and BM25 similarity [28]. The first three distances are statistical distances defined over probability distributions. Statistical distances in general and, KL divergence in particular, play an important role in ML [7,31]. Both the KL-divergence and the Itakura-Saito distances were used in prior work [7]. BM25 similarity is a popular and effective similarity metric commonly used in information retrieval. It is a variant of a TF×IDF similarity computed as

$$\sum_{x_i = y_i} \text{TF}_q(x_i) \cdot \text{TF}_d(y_i) \cdot \text{IDF}(y_i), \tag{1}$$

where $\text{TF}_q(x)$ and $\text{TF}_d(y)$ are term frequencies of terms x and y in a query and a document, respectively. IDF is an inverse document frequency (see [28] for more details). When we use BM25 as a distance, we take the negative value of this similarity function. Although BM25 is expressed as an inner product between query and document TF×IDF vectors, this distance is *not* symmetric. Term frequencies are computed differently for queries and documents and the value of the similarity normally changes when we swap function arguments.

The Rényi divergence is a single-parameter family of distances, which are not symmetric when the parameter $\alpha \neq 0.5$. By changing the parameter we can vary the degree of symmetry. In particular, large values of α as well as close-to-zero values result in highly non-symmetric distances. This flexibility allows us to stress-test retrieval methods by applying them to challenging non-symmetric distances.

The data sets are listed in Table 1. Wiki-d and RCV-d data sets consists of dense vectors of topic histograms with d topics. RCV-d set are created by

Cayton [7] from the RCV1 newswire collection [19] using the latent Dirichlet allocation (LDA) method [4]. These data sets have only 500 K entries. Thus, we created larger sets from Wikipedia following a similar methodology. RandHist-d is a synthetic set of topics sampled uniformly from a d-dimensional simplex.

The Manner data set is a collection of TF×IDF vectors generated from data set L5 in Yahoo WebScope[2]. L5 is a set of manner, e.g., how-to, questions posted on the Yahoo answers webite together with respective answers. Note that we keep only a single best answer—as selected by a community member—for each question.

3 Experiments

We carry out two experimental series. In the first series, we test the efficacy of the filter-and-refine approach (using collection subsets) where the distance function is obtained via metrization or symmetrization of the original distance. One of the important objectives of this experimental series is to demonstrate that unlike some prior work [24,25] we deal with *substantially* non-symmetric data. In the second series, we carry out a fully-fledged retrieval experiment using SW-graph [22] with different index- and query-time symmetrization approaches. Overall, we have 31 combination of data sets and distance functions (see Sect. 2.2). However, due to space limitations, we had to omit some experimental results and minor setup details. A fuller description is available in §2.3.2 of the unpublished tech report [5].

Proxying Distance via Metrization and Symmetrization. In this section, we use a proxy distance function to generate a list of k_c candidates, which are compared directly to the query. The candidate generation step employs an *exact* brute-force k-NN search with the proxy distance. On one hand, the larger is k_c, the more likely we find all true nearest neighbors. On the other hand, increasing k_c entails a higher computational cost. We consider two types of proxy distances: a learned distance (which is a metric in four out of five cases), and a symmetrized version of the original non-symmetric distance.

Distance Learning. We considered five approaches to learn a distance and a pseudo-learning approach where we simply use the Euclidean L_2 distance as a proxy. Computing L_2 between data points is a strong baseline, which sometimes outperforms true distance learning methods, especially for high-dimensional data. Four of the distance-learning methods [10,21,26,36] learn a global linear transformation of the data, which is commonly referred to as the Mahalanobis metric learning. The value of the L_2 distance between transformed vectors is used as a proxy distance function. The learned distance, is clearly a metric. We also use a non-linear Random Forest Distance (RFD) method that employs a random-forest classifier [37] and produces generally non-metric, but symmetric,

---

[2] https://webscope.sandbox.yahoo.com.

Table 3. Loss of effectiveness due to symmetrization and distance learning for 10-NN search (using at most 200K points for distance learning and at most 500K points for symmetrization)

Data set	Distance	Symmetrization		Distance learning	
		k_c (cand. k)	Recall reached	k_c (cand. k)	Recall reached
Wiki-8	Itakura-Saito	20	99	2560	99
Wiki-8	KL-div	40	99	640	99
Wiki-8	Rényi div. $\alpha = 0.25$	20	100	640	100
Wiki-8	Rényi div. $\alpha = 2$	20	99	640	99
RCV-128	Itakura-Saito	80	99	20480	58
RCV-128	KL-div	40	100	20480	94
RCV-128	Rényi div. $\alpha = 0.25$	80	100	5120	99
RCV-128	Rényi div. $\alpha = 2$	80	99	20480	66
Wiki-128	Itakura-Saito	20	99	20480	80
Wiki-128	KL-div	40	99	20480	99
Wiki-128	Rényi div. $\alpha = 0.25$	160	99	5120	99
Wiki-128	Rényi div. $\alpha = 2$	80	99	20480	87
RandHist-32	Itakura-Saito	5120	96	20480	99
RandHist-32	KL-div	160	100	2560	99
RandHist-32	Rényi div. $\alpha = 0.25$	20	100	1280	100
RandHist-32	Rényi div. $\alpha = 2$	2560	99	20480	100
Manner	BM25	1280	100	N/A	N/A

distance. Note that we do not learn a distance function for the Manner data set that contains extremely high dimensional sparse TF×IDF vectors.

In all cases, the distance is trained as a classifier that learns to distinguish between close and distant data points. More specifically, we create sets of positive and negative examples. A positive example set contains pairs of points that should be treated as similar, i.e., near points, while the negative example set contains pairs of points that should be treated as dissimilar ones. The underlying idea is to learn a distance that (1) pulls together points from the positive example set and (2) pushes points from the negative example set apart. More details are given in [5].

Symmetrization. Given a non-symmetric distance, there are two folklore approaches to make it symmetric, which use the value of the original distance $d(x, y)$ as well as the value of the distance function obtained by reversing arguments: $d_{\text{reverse}}(x, y) = d(y, x)$. Informally, we call the latter an *argument-reversed* distance. In the case of an average-based symmetrization, we compute the symmetrized distance as an average of the original and argument-reversed distances:

$$d_{\text{sym}} = \frac{d(x,y) + d_{\text{reverse}}(x,y)}{2} = \frac{d(x,y) + d(y,x)}{2} \tag{2}$$

In the case of a min-based symmetrization, we use their minimum:

$$d_{\text{sym}} = \min\left(d(x,y), d_{\text{reverse}}(x,y)\right) = \min\left(d(x,y), d(y,x)\right) \tag{3}$$

Symmetrization techniques given by Eqs. (2) and (3) are suboptimal in the sense that a *single* computation of the symmetrized distance entails *two* computations of the original distance. We can be more efficient when a distance function permits a more *natural* symmetrization, in particular, in the case of BM25 (see Eq. 1) we can compute the query term frequency using the same formula as the document term frequency. Furthermore, we can "share" a value of IDF_i between the query and the document vectors by "assigning" each vector the value $\sqrt{\text{IDF}_i}$. Although the resulting function is symmetric, it is not equivalent to the original BM25. More formally, in this "shared" setting a query vector is represented by the values $\text{TF}(x_i) \cdot \sqrt{\text{IDF}(x_i)}$, whereas a document vector is represented by the values $\text{TF}(y_i) \cdot \sqrt{\text{IDF}(y_i)}$. The pseudo-BM25 similarity is computed as the inner product between query and document vectors in the following way:

$$d(x,y) = -\sum_{x_i = y_i} \left(\text{TF}(x_i)\sqrt{\text{IDF}(x_i)}\right) \cdot \left(\text{TF}(y_i)\sqrt{\text{IDF}(y_i)}\right) \tag{4}$$

Discussion of Results. All the code in this section is implemented in Python. Thus, for efficiency reason, we limit the number of data points to 200K in the symmetrization experiment and to 500K in the distance learning experiment. Experimental results for $k = 10$ are presented in Table 3, where we measure how many candidates k_c we need to achieve a nearly perfect recall with respect to the original distance (we test all $k_c = k \cdot 2^i$, $i \leq 7$). We employ several symmetrization and distance learning methods: Yet, in the table, we show only the best recall for a given k_c. More specifically, we post the first k_c for which recall reaches 99%. If we cannot reach 99%, we post the maximum recall reached. We omit most low-dimensional results, because they are similar to Wiki-8 results (again, see [5] for a more detailed report).

From Table 3 we can immediately see that distance learning results in a much worse approximation of the original distance than symmetrization. For high-dimensional data, it is not always possible to achieve the recall of 99% for 10-NN search. When it is possible we need to retrieve from one thousand to 20 thousand candidate entries! Even for the low-dimensional Wiki-8 data set, achieving such high recall requires at least 640 candidate entries. We conclude that using distance learning is not a promising direction, because retrieving that many candidate entries accurately is hardly possible without resorting to the brute force search with the proxy distance (which is, in turn, not efficient).

In contrast, in the case of symmetrization, the number of required candidate entries is reasonably small except for Manner and RandHist-32 data sets. We, therefore, explore various symmetrization approaches in more details in the following section. Also note that KL-divergence can be symmetrized with little loss in accuracy, i.e., on the histogram-like data KL-divergence is only mildly non-symmetric. There is prior work on non-metric k-NN search that demonstrated good results specifically for KL-divergence [24,25] for Wiki-d and RCV-d

data sets. However, as our experiments clearly show, this work does not use a substantially non-symmetric distance.

Experiments with Index- and Query-Time Symmetrization for SW-Graph. In this section, we evaluate the effect of the distance symmetrization in two scenarios (for 10-NN search):

- A symmetrized distance is used for *both* indexing and retrieval. We call this a full symmetrization scenario. The search procedure is carried out using an SW-graph index [22] (see Sect. 2.1). This search generates a list of k_c candidates. Then, candidates are compared exhaustively with the query. This filter-and-refine experiment is analogous to the previous-subsection experiments except here we use approximate instead of the exact brute-force search.
- The second scenario relies on a partial, i.e., index-time only, symmetrization. Specifically, the symmetrized distance is used only to construct a proximity/neighborhood graph via SW-graph. Then, the search procedure uses the *original, non-symmetrized* distance to "guide" the search through the proximity graph.

Overall, we have 31 combinations of data sets and distances, but in this paper we present the results for most interesting cases (again see [5] for a complete set of plots). We randomly split data three times into queries and indexable data set points. For all distances except Rényi divergence we use 1 K queries for each split, i.e., the total number of queries is 3K. Because Rényi divergence is slow to compute, we use only 200 queries per split (i.e., the overall number of queries is 600).

Experiments are carried out using a `nmslib4a_bigger_reruns` branch[3] of NMSLIB [6]. We did not modify the standard NMSLIB code for SW-graph: Instead, we created a new implementation (file `small_world_rand_symm.cc`).

In the second scenario, we experiment with index- and query-time symmetrization in an actual indexing algorithm SW-graph rather than relying on the brute-force search. This approach generates a *final* list of k nearest neighbors rather than k_c candidates. *No* further re-ranking is necessary. We use two actual symmetrization strategies (the minimum- and the average-based symmetrization) as well as two types of *quasi*-symmetrization. For the first quasi-symmetrization type, we build the proximity graph using the Euclidean distance between vectors. The second quasi-symmetrization consists in building the proximity graph using the argument-reversed distance (see p. 6).

We verified that *none* of these quasi-symmetrization approaches would produce a better list of candidates in the filter-and-refine scenario (where the brute-force search is used to produce a candidate list). For example, for Wiki-128 and KL-divergence, it takes $k_c = 40$ candidates to exceed a 99% recall in a 10-NN search for the minimum-based symmetrization. For the L_2-based symmetrization, it takes as many as $k_c = 320$ candidates. The results are even worse for the

---

[3] https://github.com/nmslib/nmslib/tree/nmslib4a_bigger_reruns.

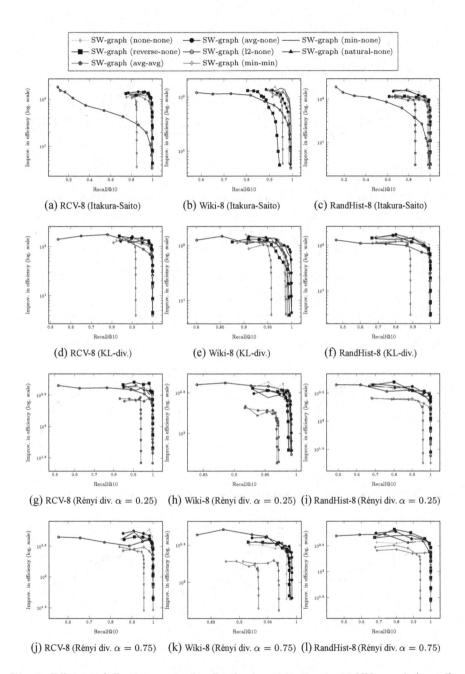

Fig. 1. Efficiency/effectiveness trade-offs of symmetrization in 10-NN search (part I). The number of data points is at most 500K. Best viewed in color.

138   L. Boytsov and E. Nyberg

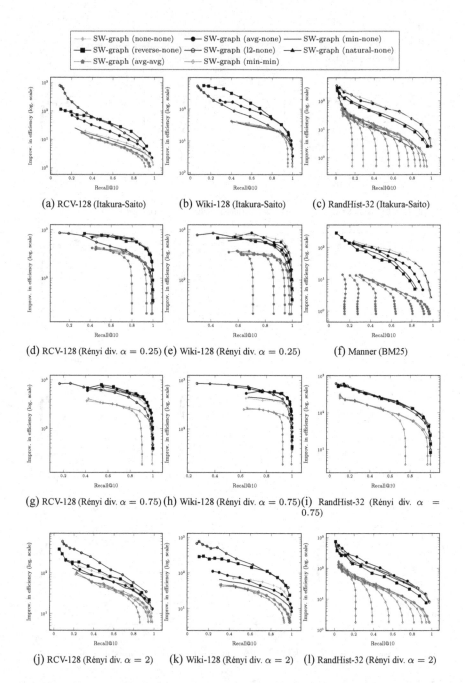

Fig. 2. Efficiency/effectiveness trade-offs of symmetrization in 10-NN search (part II). The number of data points is at most 500K. Best viewed in color.

filtering based on the argument-reversed distance: By using as many as $k_c = 1280$ candidates we obtain a recall of only 95.6%. It clearly does not make sense to evaluate these quasi-symmetrization methods in the complete filter-and-refine scenario. Yet, we need to check if it is beneficial to build the graph using a distance *different* from the original one.

Discussion of Results. Experiments were run on a laptop (i7-4700MQ @ 2.40 GHz with 16 GB of memory). Results are presented in Fig. 1 (low-dimensional data) and Fig. 2 (high-dimensional data). These are efficiency-effectiveness plots: Recall@10 is shown on the x-axis, improvement in efficiency—i.e., the speed up over the brute-force search—is shown on the y-axis. Higher and to the right is better. We test several modifications of SW-graph each of which has an additional marker in the form: **a–b**, where **a** denotes a type of index-time symmetrization and **b** denotes a type of query-time symmetrization. Red plots represent the original SW-graph, which is labeled as SW-graph (none-none).

Black plots represent modifications, where symmetrization is used only during indexing: SW-graph (avg-none), SW-graph (min-none), SW-graph (l2-none), SW-graph (reverse-none), and SW-graph (natural-none). The first two types of symmetrization are average- and minimum-based. SW-graph (l2-none) is a quasi-symmetrization approach that builds the graph using L_2, but searches using the original distance. SW-graph (reverse-none) builds the graph using the reversed-argument distance, but searches using the original distance. SW-graph (natural-none) is a natural symmetrization of BM25 described by Eq. (4), which is used only for *Manner*.

Blue plots represent the case of full (both query- and index-time) symmetrization. The index is used to carry out a k_c-NN search, which produces a list of k_c candidates for further verification. Depending on which symmetrization approach was more effective in the first series experiments (with brute-force search), we use either SW-graph (min-min) or SW-graph (avg-avg), which stand for full minimum- or average-based symmetrization. Because we do not know an optimum number of candidate records, we experiment with $k_c = k \cdot 2^i$ for successive integer values i. The larger is i, the more accurate is the filtering step and the less efficient is retrieval. However, it does not make sense to increase i beyond the point where the filtering accuracy reaches 99%. For this reason, the minimum value of k_c is k and the largest value of k_c is taken from Table 3.

For the remaining parameters of SW-graph we choose values that are known to perform well in other experiments: NN=15, efConstruction=100, and efSearch $= 2^j$ for $0 \le j \le 12$. Analogous to the first scenario (with brute-force search), we use 31 combination of data sets and distances. In each test, we randomly split data (into queries and indexable data) three times and average results over three splits.

From Figs. 1 and 2, we can see that in some cases there is little difference among best runs with the fully symmetrized distance (a method SW-graph (min-min) or SW-graph (avg-avg)) the runs produced by methods with true index-time symmetrization (SW-graph (min-none), SW-graph (avg-none)), and the original unmodified search algorithm (SW-graph (none-none)). Furthermore, we can see that there is often no difference between SW-graph (min-none),

SW-graph (avg-none), and SW-graph (none-none). However, sometimes all fully-symmetrized runs (for all values of k_c) are noticeably less efficient (see, e.g., Panels 1h and k). This difference is more pronounced in the case of high-dimensional data. Here, full symmetrization leads to a substantial (up to an order of magnitude) loss in performance in most cases.

Effectiveness of index-time symmetrization varies from case to case and there is no definitive winner. First, we note that in four cases index-time symmetrization is beneficial (Panels 2a, b, j, k). In particular, in Panels 2a, b, k, there is an up to 10× speedup. Note that it can sometimes be achieved by using an argument-reversed distance (Panels 2a, b) or L_2 (2k). This a *surprising* finding given that these quasi-symmetrization approaches do not perform well in the re-ranking–filter-and-refine—experiments. In particular, for L_2 and Wiki-128 reaching a 99% recall requires $k_c = 640$ compared to $k_c = 80$ for min-based symmetrization. For the Itakura-Saito distance and data sets RCV-128 and Wiki-128, it takes $k_c \leq 80$ to get a 99% recall. However, using the argument-reversed distance, we do not even reach the recall of 60% despite using a large $k_c = 1280$. It is worth noting, however, that in several cases using argument-reversed distance at index time leads to substantial degradation in performance (see, e.g., Panels 1b and f).

To conclude the section, we emphasize that in all cases the best performance is achieved using either the unmodified SW-graph or the SW-graph with an index-time proxy distance. However, there is not a single case where performance is improved by using the fully symmetrized distance (at both indexing and querying steps). Furthermore, in three especially challenging cases: Itakura-Saito distance with RandHist-32, Rényi divergence with RandHist-32, and BM25 with Manner, SW-graph has excellent performance. In all three cases (see Panels 2c, l, f), there is more than a 10× speed up at 90% recall compared to the brute-force search. Note that in these three cases data is substantially non-symmetric: Depending on the case, to accurately retrieve 10 nearest neighbors with respect to the original metric, it requires to obtain 1–5K nearest neighbors using its symmetrized variant (see Table 3). Thus, in these challenging cases, a brute-force filter-and-refine symmetrization solution would be particularly ineffective or inefficient whereas SW-graph has strong performance.

4   Conclusion

We systematically evaluate effects of distance metrization, symmetrization and quasi-symmetrization on performance of brute-force and index-based k-NN search (with a graph-based retrieval method SW-graph). Unlike previous work [24, 25] we experiment with substantially non-symmetric distances. Coercion of the non-metric distance to a metric space leads to a substantial performance degradation. Distance symmetrization causes a lesser performance loss. However, in all the cases a full filter-and-refine symmetrization is always inferior to either applying the graph-based retrieval method directly to a non-symmetric distance or to building an index (which is a neighborhood graph) with a modified, e.g. symmetrized, distance. Quite surprisingly, sometimes the best performing index-time distance is neither the original distance nor its symmetrization. This observation motivates a new line of research of designing index-specific graph-construction distance functions.

Acknowledgments. This work was done while Leonid Boytsov was a PhD student at CMU. Authors gratefully acknowledge the support by the NSF grant #1618159.

References

1. Arya, S., Mount, D.M.: Approximate nearest neighbor queries in fixed dimensions. In: SODA, vol. 93, pp. 271–280 (1993)
2. Aumüller, M., Bernhardsson, E., Faithfull, A.: ANN-benchmarks: a benchmarking tool for approximate nearest neighbor algorithms. In: Beecks, C., Borutta, F., Kröger, P., Seidl, T. (eds.) SISAP 2017. LNCS, vol. 10609, pp. 34–49. Springer, Cham (2017). https://doi.org/10.1007/978-3-319-68474-1_3
3. Bellet, A., Habrard, A., Sebban, M.: A survey on metric learning for feature vectors and structured data. CoRR abs/1306.6709 (2013)
4. Blei, D.M., Ng, A.Y., Jordan, M.I.: Latent dirichlet allocation. J. Mach. Learn. Res. **3**, 993–1022 (2003)
5. Boytsov, L.: Efficient and accurate non-metric k-NN search with applications to text matching. Ph.D. thesis, Carnegie Mellon University (2017)
6. Boytsov, L., Naidan, B.: Engineering efficient and effective non-metric space library. In: Brisaboa, N., Pedreira, O., Zezula, P. (eds.) SISAP 2013. LNCS, vol. 8199, pp. 280–293. Springer, Heidelberg (2013). https://doi.org/10.1007/978-3-642-41062-8_28
7. Cayton, L.: Fast nearest neighbor retrieval for bregman divergences. In: Proceedings of the 25th International Conference on Machine Learning, pp. 112–119. ACM (2008)
8. Chávez, E., Navarro, G., Baeza-Yates, R.A., Marroquín, J.L.: Searching in metric spaces. ACM Comput. Surv. **33**(3), 273–321 (2001)
9. Chechik, G., Sharma, V., Shalit, U., Bengio, S.: Large scale online learning of image similarity through ranking. J. Mach. Learn. Res. **11**, 1109–1135 (2010)
10. Davis, J.V., Kulis, B., Jain, P., Sra, S., Dhillon, I.S.: Information-theoretic metric learning. In: Proceedings of ICML 2007, pp. 209–216. ACM (2007)
11. Dong, W., Moses, C., Li, K.: Efficient k-nearest neighbor graph construction for generic similarity measures. In: Proceedings of WWW 2011, pp. 577–586. ACM (2011)
12. Hajebi, K., Abbasi-Yadkori, Y., Shahbazi, H., Zhang, H.: Fast approximate nearest-neighbor search with k-nearest neighbor graph. In: IJCAI/AAAI 2011, pp. 1312–1317 (2011)
13. Harwood, B., Drummond, T.: FANNG: fast approximate nearest neighbour graphs. In: Proceedings of CVPR, pp. 5713–5722 (2016)
14. Hjaltason, G.R., Samet, H.: Properties of embedding methods for similarity searching in metric spaces. IEEE Trans. Pattern Anal. Mach. Intell. **25**(5), 530–549 (2003)
15. Houle, M.E., Sakuma, J.: Fast approximate similarity search in extremely high-dimensional data sets. In: ICDE 2005, pp. 619–630 (2005)
16. Itakura, F., Saito, S.: Analysis synthesis telephony based on the maximum likelihood method. In: Proceedings of the 6th International Congress on Acoustics, pp. C17–C20 (1968)

17. Jacobs, D.W., Weinshall, D., Gdalyahu, Y.: Classification with nonmetric distances: image retrieval and class representation. IEEE Trans. Pattern Anal. Mach. Intell. **22**(6), 583–600 (2000)
18. Kullback, S., Leibler, R.A.: On information and sufficiency. Ann. Math. Statist. **22**(1), 79–86 (1951)
19. Lewis, D.D., Yang, Y., Rose, T.G., Li, F.: RCV1: a new benchmark collection for text categorization research. J. Mach. Learn. Res. **5**, 361–397 (2004)
20. Li, W., Zhang, Y., Sun, Y., Wang, W., Zhang, W., Lin, X.: Approximate nearest neighbor search on high dimensional data - experiments, analyses, and improvement (v1.0). CoRR abs/1610.02455 (2016)
21. Liu, E.Y., Guo, Z., Zhang, X., Jojic, V., Wang, W.: Metric learning from relative comparisons by minimizing squared residual. In: IEEE ICDM 2012, pp. 978–983. IEEE (2012)
22. Malkov, Y., Ponomarenko, A., Logvinov, A., Krylov, V.: Approximate nearest neighbor algorithm based on navigable small world graphs. Inf. Syst. **45**, 61–68 (2014)
23. Malkov, Y.A., Yashunin, D.A.: Efficient and robust approximate nearest neighbor search using hierarchical navigable small world graphs. CoRR abs/1603.09320 (2016)
24. Naidan, B., Boytsov, L., Nyberg, E.: Permutation search methods are efficient, yet faster search is possible. PVLDB **8**(12), 1618–1629 (2015)
25. Ponomarenko, A., Avrelin, N., Naidan, B., Boytsov, L.: Comparative analysis of data structures for approximate nearest neighbor search. In: DATA ANALYTICS 2014, The Third International Conference on Data Analytics, pp. 125–130 (2014)
26. Qi, G., Tang, J., Zha, Z., Chua, T., Zhang, H.: An efficient sparse metric learning in high-dimensional space via l_1-penalized log-determinant regularization. In: ICML 2009, pp. 841–848. ACM (2009)
27. Rényi, A.: On measures of entropy and information. In: Proceedings of the Fourth Berkeley Symposium on Mathematical Statistics and Probability, vol. 1, pp. 547–561 (1961)
28. Robertson, S.: Understanding inverse document frequency: on theoretical arguments for IDF. J. Doc. **60**(5), 503–520 (2004)
29. Samet, H.: Foundations of Multidimensional and Metric Data Structures. Morgan Kaufmann Publishers Inc., San Francisco (2005)
30. Skopal, T., Bustos, B.: On nonmetric similarity search problems in complex domains. ACM Comput. Surv. **43**(4), 34 (2011)
31. Sutherland, D.J.: Scalable, flexible and active learning on distributions. Ph.D. thesis, Carnegie Mellon University (2016)
32. Tellez, E.S., Ruiz, G., Chávez, E., Graff, M.: Local search methods for fast near neighbor search. CoRR abs/1705.10351 (2017). http://arxiv.org/abs/1705.10351
33. Toussaint, G.T.: The relative neighbourhood graph of a finite planar set. Pattern Recogn. **12**(4), 261–268 (1980)
34. Wang, J., Shen, H.T., Song, J., Ji, J.: Hashing for similarity search: a survey. CoRR abs/1408.2927 (2014)
35. Weber, R., Schek, H.J., Blott, S.: A quantitative analysis and performance study for similarity-search methods in high-dimensional spaces. In: VLDB, vol. 98, pp. 194–205 (1998)
36. Weinberger, K.Q., Blitzer, J., Saul, L.K.: Distance metric learning for large margin nearest neighbor classification. In: NIPS 2005, pp. 1473–1480 (2005)
37. Xiong, C., Johnson, D.M., Xu, R., Corso, J.J.: Random forests for metric learning with implicit pairwise position dependence. In: KDD 2012, pp. 958–966. ACM (2012)

Indexability-Based Dataset Partitioning

Angello Hoyos[1], Ubaldo Ruiz[1], Stephane Marchand-Maillet[2],
and Edgar Chávez[1(✉)]

[1] Centro de Investigación Científica y de Educación Superior de Ensenada
(CICESE), Ensenada, Mexico
{uruiz,elchavez}@cicese.mx, ahoyos@cicese.edu.mx
[2] Viper Group, Department of Computer Science, CUI - University of Geneva,
Geneva, Switzerland
Stephane.Marchand-Maillet@unige.ch

Abstract. Indexing exploits assumptions on the inner structures of a
dataset to make the nearest neighbor queries cheaper to resolve. Datasets
are generally indexed at once into a unique index for similarity search.
By indexing a given dataset as a whole, one faces the parameters of its
global structure, which may be adverse. A typical well-studied example
is a high global dimensionality of the dataset, making any indexing strategy inefficient due to the curse of dimensionality.

We conjecture that a dataset may be partitioned into subsets of variable indexability. The strategy is, therefore, to define a procedure to
extract parts of the dataset with predictable indexability and to adapt
the index structure to this parameter.

In this paper, we define and discuss indexability related to the curse
of dimensionality and propose a related heuristic to partition the dataset
into low-dimensional parts. Each data object is ranked according to its
degree centrality, under a connected sparse graph, the Half-Space Proximal Graph (HSP). We postulate centrality measures are good predictors
of dimensionality and indexability.

In view of validation, we conducted an experiment using the degree
centrality of the HSP graph as unique dimensionality/indexability measure. We ranked the data objects by their respective centrality degree
under the HSP graph, then extracted the lower dimensional subsets,
recomputed the HSP and repeated. Subsets were then indexed with an
exact method in increasing, decreasing and random order. We measured
the complexity of a fixed set of queries for each of the three arrange-
ments. For each set we used a fixed dataset with 250 queries.

The above single experiment demonstrated that the heuristic can
extract low dimensional subsets, and also that those subsets are easier to index.

This initial results demonstrate the validity of our conjecture and
motivate the need for exploring further the notion of indexability and
related dataset partitioning strategies.

Keywords: Indexability · Dataset partitioning · Spanning graph ·
Centrality measure · Curse of dimensionality

© Springer Nature Switzerland AG 2019
G. Amato et al. (Eds.): SISAP 2019, LNCS 11807, pp. 143–150, 2019.
https://doi.org/10.1007/978-3-030-32047-8_13

1   Introduction

The nearest neighbor search in a dataset is at the core of data analysis because it is via neighborhoods that the data makes sense, as opposed to being a set of arbitrary unrelated items. Resolving effectively range queries or the k-nearest neighbor problem has countless applications in machine learning, data mining and many other fields of data processing. It is therefore critical to make this step both effective and accurate. It is well-known that the effectiveness of indexing structures is reduced as a function of the dimensionality of the dataset. In this paper we present a study that takes an alternative approach to the general index structure improvement proposed in most of the literature. Under the assumption that effective index structures exist for "well-behaved" datasets, we propose to attack the dataset rather than the index structure and make it suitable to be indexed by state-of-the-art structures (e.g. [14]).

In Sect. 2, we briefly review related work and introduce the notion *dataset indexability*, that will be our criterion for adapting the dataset to index structures. In Sect. 3, we present our strategy to boost indexability, resulting into our main conjecture that is initially tested in Sect. 4 and discussed in Sect. 5.

2   Indexability

Measuring the *performance of an index structure* generally means evaluating the performance of an indexing strategy over standard benchmarks (datasets, queries and measures). Measuring the *indexability of a dataset* takes the problem upside down and looks at whether or not a given dataset can benefit from an index structure to answer nearest neighbor queries. Intuitively, a dataset is said to be indexable if one can build an exact index able to answer reasonably selective queries in time that is not proportional to the size of the dataset. A trivial example of an indexable dataset is a set of points on a line, the plane or with "small" dimension in general. In this example case, a classical data structure like the kd-tree [3], can handle the indexing task.

There are several dimensions to index fitness. Any index computes an index distance that approximates the true metric while being cheaper to compute. The effectiveness of the index measures how good the index distance bounds the original distance. The efficiency measures how fast the index distance can be computed for the entire dataset. These two measures are complemented with the memory usage and the speed at which the index can be constructed. It is usually the case that the effectiveness of an index can be boosted at the expense of its efficiency and memory usage or construction speed.

2.1   The Importance of Local Dimensionality

Indexability is therefore related to the deep foundations of distance-based indexing, essentially related to distance computation. From this perspective, indexability has been studied in relation to the curse of dimensionality and much has been discussed around this concept. Essentially, the main result of [4] and subsequent papers (e.g., [13,17,19]) is that, as defined in [20] (Definition 2.2), a *workload* $W = (S, F, n, d)$ consisting of a dataset S of n objects drawn iid from a distribution F and measured via distance function $d(.,.)$ can be made into a series W_i which will be said to have *vanishing variance* if there exists $\alpha > 0$ such that

$$\lim_{m \to \infty} \text{var} \left(\frac{D_m^\alpha}{\mathbb{E}[D_m^\alpha]} \right) = 0,$$

where D_m is the distance distribution of W_m (i.e. the distribution of distances between points in S_m). In that case, ([4], Theorem 1), for every $\varepsilon > 0$

$$\lim_{m \to \infty} P[D_m^{\max} \leq (1 + \varepsilon)D_m^{\min}] = 1.$$

Simply said, all distance values become indistinguishable as m increases. This is even more true in a fixed precision environment. A typical example of such a workload is a dataset with coordinates *iid* distributed in all m dimensions. As a result, the use of sum-based distance functions (such as Minkowski metrics) for high-dimensional datasets impedes their indexability.

Directly considering the global dimensionality of the dataset therefore appears as a crude approximation for indexability. Rather, provided one knows how to exploit local structures from within the data, the effective indexability should be boosted. The workload may have high representational dimension but an intrinsic low dimension, and be indexable using a classic metric indexing method like the BK-tree [6]. For other cases of intrinsic high dimension, the dataset would not be indexable, even in the approximate sense, as stated in a recent theoretical result on the conditional hardness of nearest neighbor search using polynomial preprocessing time [19]. In that paper the authors prove that computing a $(1 + \epsilon)$-approximation to the nearest neighbor requires $\Omega(N - \delta)$ time, with N the size of the dataset.

It is therefore critical to obtain a proper understanding of what dimensionality means locally. There are several proposals to measure local intrinsic dimension. An excellent review is provided by Michael Houle [10], who also proposed the *expansion dimension* for that purpose. The idea is to measure locally how many points are contained in a ball as its radius increases. Since in Euclidean spaces the volume of a ball of radius r is about r^m with m the number of dimensions, fitting the increase of the number of points contained in a ball of increasing radius allows for the estimation of the local intrinsic dimension. In this line of work, authors [2,11] advocate for feature selection for removing "spurious" dimensions while preserving original distances. The aim is to provide an equivalent but better indexable dataset. Alternatively, ranks may also be used as a robust replacement to distance values [7,12].

2.2   Dataset Shattering

An interesting alternative avenue for investigation is that of the VC dimension [22]. The relationship between the VC dimension and indexing has already been put forward by Pestov in [18]. Although the VC dimension is related to measuring the complexity of a class of functions, the notion of *shattering* is easily related to that of indexing. If a dataset is shattered, any of its elements can be particularized as a result of such shattering. Indexing has a similar objective. For example, the capabilities of permutation-based indexing schemes to shatter a dataset are explored in [1,15].

3   Boosting Dataset Indexability

In this exploratory work, we propose an alternative approach to combat the curse of dimensionality. Rather than considering the dataset as an integral entity, we seek a decomposition that will extract parts with higher indexability than the whole. Indexes can then be built over these parts individually and a query sent to the multiple index structures and recomposed globally.

3.1   Dataset Layering

We assume we are given a non-indexable dataset. Our aim is to decompose it into easily indexable parts. From the above discussion, non-indexability allows us to model the dataset as a blob of high dimension, which we will partition into fragments of low dimension.

Hence, we construct a partition by iteratively *peeling* the dataset (blob) into layers corresponding to surfaces of points equidistant from the blob center. We therefore inherit from the notion of centrality measures to define the layers which will be indexable. Centrality is classically defined in relation to a spanning graph. Various definitions of centrality exist [5,9], from the simplest based on node degree, to those exploiting a spectral decomposition of the graph (such as PageRank and others [21]).

We initially base our study on a degree-based centrality measure applied over the Half Space Proximal (HSP) graph constructed over the dataset, as detailed next.

3.2   The Half Space Proximal Graph

The Half Space Proximal is a local test for building a directed graph, which is a bounded dilation spanner over a set of objects in a metric space. Without needing synchronization, each node can compute its neighbors using the simple rule described below.

Let S a finite subset of a metric space. Let $u \in S$, we take its nearest element $v \in S$ and add an edge from u to v. We remove all the elements that are closer to v than to u. The region of objects closer to v than to u is called the *forbidden region* from the point u with respect to v. From the remaining points we take the nearest point to u and repeat until we have removed all points in S. We do this process for every point in S. In the end, we will have a directed graph with vertex set S and the edges found with the previous mechanism. The HSP, presented in [8], has maximum out-degree of six for points in the plane.

We conjecture that the out-degree of each node in the HSP depends only on the local intrinsic dimension of the node. Hence, in particular, it can be used as an estimator of the indexability of a point collection. The rationale behind this conjecture is related to the test conducted at each step of the construction. Every edge from the node is associated to an hyperplane, and the out-degree will be related to the number of hyperplanes needed to isolate the node.

Please notice that the HSP test in each node requires searching for the nearest neighbor of the node, then splitting the set into two parts and repeat until the set is empty. A careful implementation will require a quadratic number of distance computations. This imposes a severe limitation in practice, because interesting datasets are quite large.

4   Layered Indexing with the HSP

As an empirical validation of the above stated conjecture, we conducted an experiment using a set of 100'000 deep feature vectors of 4'096 real values. For this set we computed the HSP and ranked the nodes according to their degree. After this, we removed the 1'000 nodes with the smallest degree in the graph, recomputed the HSP in the remaining objects, and repeated. Note that the nodes linked to the removed objects are the most likely to have its out-degree modified. We only recomputed the edges of those nodes in the next iteration. Figure 1 shows the evolution of the average degree centrality when adding different layers in the dataset.

In this experiment, we noted that the number of changes in the out-degree of the touched nodes was slowly decreasing, and after some 40% of the dataset, stopped changing. This supports the existence of a hard kernel in the dataset.

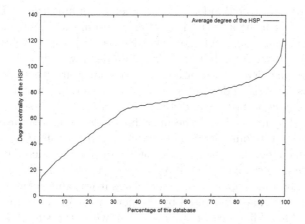

Fig. 1. Average degree centrality of different layers of dataset objects.

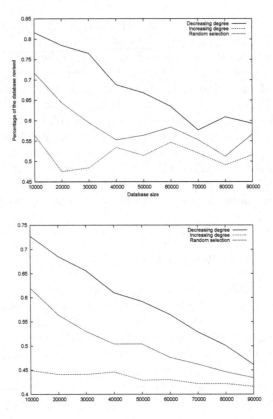

Fig. 2. Layered indexing of a dataset of 100,000 deep feature vectors of dimension 4096 with two exact indices, with the SAT (top) and with the VP-tree (bottom). Note the large difference for the low and large degree nodes. See text for more details.

The result of this layered indexing experiment is summarized in Fig. 2. In the plot, chunks of increasing size (horizontal axis) are indexed independently with an exact indexing method (SAT [16] and VP-Tree [23] respectively). Increasing the size of the dataset from 10'000 to 90'000 objects was first done by adding nodes of decreasing degree. According to our conjecture, this corresponds to going from least to most indexable subsets (least favorable indexing setup). We compare to the case of increasing degree, again varying the size from 10'000 to 90'000 objects, hoping to create the most favorable setup. We also included a control plot with a random selection of the dataset of the same size. For the rest, we kept the same index and the same set of 250 queries not included at indexing time. The results plotted correspond to the average over 250 queries in the index. The differences become apparent, in some places it was almost twice the number of distance computations (indicated by the percentage of the dataset visited on vertical axis). This difference becomes smaller when the subset is almost the entire dataset.

Notice that the difference in indexability persists across different indexes. The SAT is more sensitive the centrality of the collection, while the VP-tree is almost not affected when the dataset excludes its 10% least indexable part.

5 Discussion

The preliminary experimental results discussed in this communication are encouraging. They are an empirical corroboration of the intuition that indexability, local intrinsic dimensionality and centrality are related. This paper certainly does not propose a new indexing method, mainly because of the large cost of computing the degree centrality of the HSP graph. It rather motivates the quest for a faster-to-compute latent graph of the dataset and gives some hope in dealing with the curse of dimensionality.

Some open questions remain. What type of guarantees is it possible to give in a layered index? In other words, assume each part, from the most to least indexable, is indexed independently using a mixture of exact and approximate methods, and then queried at once, there will be an answer from each one of the indexes, some from the exact and some from the approximate methods. Even if the nearest neighbor belongs to an exact index, it is not sure that it is the true global nearest neighbor. What is then a good heuristic to assign a probability to the global answer?

It is also interesting to explore additional properties of the most or least indexable parts of the dataset. In the case of such deep features of images, what are the most representative objects of a class? Is it the most central, i.e. the least indexable? Or is it the opposite? Is it possible to build a classifier based only on the centrality of the objects in a class?

References

1. Amato, G., Falchi, F., Rabitti, F., Vadicamo, L.: Some theoretical and experimental observations on permutation spaces and similarity search. In: Traina, A.J.M., Traina, C., Cordeiro, R.L.F. (eds.) SISAP 2014. LNCS, vol. 8821, pp. 37–49. Springer, Cham (2014). https://doi.org/10.1007/978-3-319-11988-5_4
2. Amsaleg, L., et al.: Extreme-value-theoretic estimation of local intrinsic dimensionality. Data Min. Knowl. Disc. **32**(6), 1768–1805 (2018)
3. Bentley, J.L.: Multidimensional binary search trees used for associative searching. Commun. ACM **18**(9), 509–517 (1975)
4. Beyer, K., Goldstein, J., Ramakrishnan, R., Shaft, U.: When is "nearest neighbor" meaningful? In: Beeri, C., Buneman, P. (eds.) ICDT 1999. LNCS, vol. 1540, pp. 217–235. Springer, Heidelberg (1999). https://doi.org/10.1007/3-540-49257-7_15
5. Boldi, P., Vigna, S.: Axioms for centrality. Internet Math. **10**(3–4), 222–262 (2014)
6. Burkhard, W.A., Keller, R.M.: Some approaches to best-match file searching. Commun. ACM **16**(4), 230–236 (1973)
7. Chavez, E., Figueroa, K., Navarro, G.: Effective proximity retrieval by ordering permutations. IEEE Trans. Pattern Anal. Mach. Intell. **30**(9), 1647–1658 (2008)

8. Chavez, E., et al.: Half-space proximal: a new local test for extracting a bounded dilation spanner of a unit disk graph. In: Anderson, J.H., Prencipe, G., Wattenhofer, R. (eds.) OPODIS 2005. LNCS, vol. 3974, pp. 235–245. Springer, Heidelberg (2006). https://doi.org/10.1007/11795490_19
9. Grando, F., Granville, L.Z., Lamb, L.C.: Machine learning in network centrality measures: tutorial and outlook. ACM Comput. Surv. **51**(5), 102:1–102:32 (2018)
10. Houle, M.E.: Local intrinsic dimensionality I: an extreme-value-theoretic foundation for similarity applications. In: Proceedings of the 10th International Conference Similarity Search and Applications, SISAP 2017, Munich, Germany, 4–6 October 2017, pp. 64–79 (2017). https://doi.org/10.1007/978-3-319-68474-1_5
11. Houle, M.E., Ma, X., Oria, V., Sun, J.: Efficient algorithms for similarity search in axis-aligned subspaces. In: Traina, A.J.M., Traina, C., Cordeiro, R.L.F. (eds.) SISAP 2014. LNCS, vol. 8821, pp. 1–12. Springer, Cham (2014). https://doi.org/10.1007/978-3-319-11988-5_1
12. Houle, M.E., Nett, M.: Rank-based similarity search: reducing the dimensional dependence. IEEE Trans. Pattern Anal. Mach. Intell. **37**(1), 136–150 (2015)
13. Hsu, C.M., Chen, M.S.: On the necessary and sufficient conditions of a meaningful distance function for high dimensional data space, pp. 12–23 (2006)
14. Johnson, J., Douze, M., Jégou, H.: Billion-scale similarity search with GPUs. CoRR abs/1702.08734 (2017)
15. Marchand-Maillet, S., Roman-Rangel, E., Mohamed, H., Nielsen, F.: Quantifying the invariance and robustness of permutation-based indexing schemes. In: Amsaleg, L., Houle, M.E., Schubert, E. (eds.) SISAP 2016. LNCS, vol. 9939, pp. 79–92. Springer, Cham (2016). https://doi.org/10.1007/978-3-319-46759-7_6
16. Navarro, G.: Searching in metric spaces by spatial approximation. VLDB J. **11**(1), 28–46 (2002)
17. Pestov, V.: On the geometry of similarity search: dimensionality curse and concentration of measure. Inf. Process. Lett. **73**, 1–2 (2000)
18. Pestov, V.: Indexability, concentration, and VC theory. J. Discret. Algorithms **13**, 2–18 (2012)
19. Rubinstein, A.: Hardness of approximate nearest neighbor search. In: Proceedings of the 50th Annual ACM SIGACT Symposium on Theory of Computing, STOC 2018, pp. 1260–1268. ACM, New York (2018)
20. Shaft, U., Ramakrishnan, R.: Theory of nearest neighbors indexability. ACM Trans. Database Syst. **31**(3), 814–838 (2006)
21. Sun, K., Morrison, D., Bruno, E., Marchand-Maillet, S.: Learning representative nodes in social networks. In: Pei, J., Tseng, V.S., Cao, L., Motoda, H., Xu, G. (eds.) PAKDD 2013. LNCS (LNAI), vol. 7819, pp. 25–36. Springer, Heidelberg (2013). https://doi.org/10.1007/978-3-642-37456-2_3
22. Vapnik, V.N.: The Nature of Statistical Learning Theory, 2nd edn. Springer, New York (2000). https://doi.org/10.1007/978-1-4757-3264-1
23. Yianilos, P.N.: Data structures and algorithms for nearest neighbor search in general metric spaces. In: SODA, vol. 93, pp. 311–321 (1993)

Permutation's Signatures for Proximity Searching in Metric Spaces

Karina Figueroa[1](✉) ⓘ and Nora Reyes[2] ⓘ

[1] Facultad de Ciencias Físico-matemáticas Morelia,
Universidad Michoacana de San Nicolás de Hidalgo, Michoacán, Mexico
karina@fismat.umich.mx
[2] Universidad Nacional de San Luis, San Luis, Argentina
nreyes@unsl.edu.ar

Abstract. In a multimedia database, similarity searching is the only significant way to retrieve the most similar objects to a given query. The usual approach to efficiently solve this kind of search is building an index, which allows reducing the response time of online queries. Recently, the permutation-based algorithms (PBA) were presented, and from then on, this technique has been very successful. A PBA index consists of storing the permutation of any database element with respect to a set of permutants. If the cardinality of the set of permutants is β, any permutation needs storing a sequence of β small integers, whose values are between 1 to β. However, if we have space restrictions over the index, the only way of reducing its size is by considering fewer permutants. Hence, the index performance could be severely affected.

We present in this paper a novel way to reduce the index size of PBA, without removing any permutant, by storing instead of the permutation of each element its signature regarding pairs of permutants from the set. Furthermore, our proposal achieves a good search performance, regarding both time and quality of solving a query. We can reduce almost 50% of the space needed for the index. Moreover, according to our experimental evaluation, we can reduce the original technique costs while preserving its exceptional answer quality.

Keywords: Similarity searching · Nearest neighbor ·
Permutation-based algorithm

1 Introduction

Similarity searching is one of the most important problems to solve in multimedia databases. In this context, elements should be comparable, hopefully with a distance function between them to calculate how similar are. The simplest way

Universidad Michoacana de San Nicolás de Hidalgo.

to solve these queries is by using brute-force algorithms (sequential scan), which in huge databases it is unthinkable. Hence, considering that there is a distance function, the problem can be modeled as a metric space.

A metric space is a pair $(\mathbb{X}, d)$, where $\mathbb{X}$ is the universe of valid of objects and d is a distance function which defines how similar objects are; that is, $d : \mathbb{X} \times \mathbb{X} \rightarrow \mathbb{R}^+$. Usually, d is expensive to compute. The distance function must satisfy the following properties, which make d a metric; that is, let $x, y, z \in \mathbb{X}$, d: symmetry $d(x, y) = d(y, x)$, reflexivity $d(x, x) = 0$, strict positiveness $d(x, y) > 0 \Leftrightarrow x \neq y$, and triangle inequality $d(x, y) \leq d(x, z) + d(z, y)$. The database is a finite subset of valid objects $\mathbb{U} \subseteq \mathbb{X}$, $n = |\mathbb{U}|$.

The kind of queries can be solved in a metric space are basically two: *range query* consists of retrieving the elements from $\mathbb{U}$ within a given radius to the query q; that is, $R(q, r) = \{d(u, q) \leq r, \forall u \in \mathbb{U}\}$; and *k-nearest neighbors query* retrieves k elements of $\mathbb{U}$ that are closest to q, $|NN_k(q)| = k$, and $\forall v \in NN_k(q), u \in \mathbb{U} - NN_k(q), d(v, q) \leq d(u, q)$. In case of ties we choose any k-element set that satisfies the query.

In order to solve the queries more efficiently, several interesting algorithms have been proposed. Some of these algorithms are unquestionably effective in low dimensional spaces; however, their performance worsens as the intrinsic dimension increases (this problem is known as *curse of dimensionality* [4]). On the other hand, there are also algorithms designed for high dimension, but their performance tends to do a sequential scan [5].

Basically, there are few effective algorithms to face searches on high - dimensional metric spaces without making sequential scan. Usually, these proposals identify *quickly* some relevant elements within the correct answer, however the answer set returned could contains some irrelevant objects. In this context, one of the techniques which works in high dimension with a remarkable performance are the *Permutation-Based Algorithms* (PBA) [1,3,8].

In this paper, we present a novel way to use the distances between the database objects and the set of permutants in order to improve the performance of the permutant-based algorithms. With this proposal, we obtain a good search performance, both in search costs and in answer quality, while reduce significantly the space needed to store the index.

The rest of this paper is organized as follows. Section 2 reviews the PBA's basic concepts and some related work. We introduce our proposal in Sect. 3, and show its experimental evaluation in Sect. 4. Finally, we give some conclusion remarks and mention some lines of future work in Sect. 5.

2   Related Works

There are several algorithms for similarity searching in metric spaces [5]. Most of them, if the size of its index is reduced, the algorithm's performance gets worse. Basically, there are three families of indexes: pivot-based, partition-based, and permutation-based algorithms. For lack of space, we just describe the third one.

Permutation-based algorithms (PBA) [3] selects a set of elements $\mathbb{P} = \{p_1, p_2, \ldots, p_\beta\}$ (called *permutants*). Each element of the space $u \in \mathbb{U}$

defines a permutation Π_u, by ordering the permutants according to the distances to them. That is, for all $1 \leq i < \beta$, it satisfies $d(\Pi_u(i), u) \leq d(\Pi_u(i+1), u)$. Where $\Pi_u(i)$ is the permutant in the i-th postion. During the query, the comparison is between query's permutation and objects's permutations using Spearman Footrule distance. The idea is that the permutations closest to the query's permutation, considering Spearman Footrule distance [6], are probably the closest elements to q according to d. This method provides an approximate answer to the query, and it has proved to be unbeatable in high dimensional spaces.

In the literature of PBA, there are some proposals that uses auxiliary data structures to avoid the examination of the whole dataset. For example, in [1] the authors propose to use an inverted index to avoid the sequential scan between permutations. Unfortunately, they work with a big set of permutants, in order to increase its recall. In [7], the authors suggest using a suffix tree to find the most similar permutations improving the search costs.

There have been some proposals to reduce the size of the index on PBA. In [9] authors proposes a bit-encoding of the permutation; that is, let be Π_u a permutation of $u \in \mathbb{U}$ and Π_u^{-1} its inverse, and α a threshold, for each permutant $p_i \in \mathbb{P}$ if $|i - \Pi_u^{-1}(i)| > \alpha$ they used 1 to code the information of this permutant for u, otherwise, they used 0 (the details of α can be observed at [9]). Hence, each element codes its permutation by a sequence of bits. Then, they used Hamming distance to classify permutations and to answer the query. Authors reported good performance however they had to use a lot of permutants (at least 512).

3 Proposed Algorithm

In this article, we propose a new way to represent the information obtained from the distances between $u \in \mathbb{U}$ and $\mathbb{P}$. It allows reducing significantly the needed space of the index while maintaining a good search performance, both in search costs and answer quality.

As the usually methods proceed, our algorithm also considers two stages: building an index and solving queries. The first stage can be made off-line, while the second stage is performed on-line.

3.1 Building the Index

Let $\mathbb{U}$ the database, and $\mathbb{P} = \{p_1, p_2, \ldots, p_\beta\} \subseteq \mathbb{U}$ a random selected permutants, $\beta = |\mathbb{P}|$. Each $u \in \mathbb{U}$ computes all distances to the set of permutants $\mathbb{P}$. Let be $Q_u = \{d(u, p_1), \ldots, d(u, p_\beta)\}$. We set a parameter r, which is considered a tolerable difference of distances between an element and two permutants. With this information we define a *signature* of u. Formally, the signature S_u of an element u is a concatenation of some $B_u(p_i, p_j)$ pair of bits, where $B_u(p_i, p_j)$ is defined as follows:

Let be $p_i, p_j \in \mathbb{P}$ and an r the parameter, there are 3 cases:

$$B_u(p_i, p_j) = \begin{cases} 1\,1 & \text{if } |d(p_i, u) - d(p_j, u)| <= r \\ 1\,0 & \text{if } d(p_i, u) < d(p_j, u) \\ 0\,1 & \text{if } d(p_i, u) > d(p_j, u) \end{cases} \tag{1}$$

As it can be noticed, the value of each possible $B_u(p_i, p_j)$ is one element of the set $\{11, 10, 01\}$, which represents: 11 means that u is at the almost as nearby (r-near) to p_i as p_j; 10 represents that u is closest to p_i than to p_j, and 01 in other case. Of course, it is known that there are $n \times (n-1)$ possible pairs of permutants (i.e. pairs (p_i, p_j) is the same for (p_j, p_i)) and therefore, there would be calculated all of them only knowing the distances of the set Q_u. If we use all the possible $B_u(p_i, p_j)$, we do not reduce so much the space needed to store the signature with respect to the space needed to store the permutation. Consequently, we propose to use only the following set of permutants pairs $\mathbb{R} = \{(p_1, p_2), (p_2, p_3), \ldots, (p_{\beta-1}, p_\beta)\}$, and $\mathbb{R}' = \{(p_1, p_3), (p_2, p_4), \ldots, (p_{\beta-2}, p_\beta)\}$ to form the signature of an element. With this restriction, the signature is formally defined as:

$$S_u = B_u(p_1, p_2)|B_u(p_2, p_3)|\ldots|B_u(p_{\beta-1}, p_\beta)|B_u(p_1, p_3)|B_u(p_2, p_4)|\ldots|B_u(p_{\beta-2}, p_\beta)$$

Therefore, all the $B_u(p_i, p_j)$ are using only 2 bits each, and the concatenation of them forms the signature of u. It can be regarded that for values of $\beta > 3$, each signature will need more than one byte of storage. In general, the size in bytes of the signature S_u of any $u \in \mathbb{U}$ is $\gamma = \lceil \frac{(\beta * 2 - 3) * 2}{8} \rceil$. Hence, the size of the whole index will be $n \times \gamma$ bytes.

The main idea of this signature is to keep the information represented in permutations. Two equal elements will have the same order by each pair; while two similar objects could keep almost the same orders for all proposed pairs.

As an example, let be consider the set of permutants $\mathbb{P} = \{p_1, p_2, p_3, p_4, p_5, p_6\}$, the parameter $r = 2$, and the set of distances $Q_u = \{3, 2, 5, 7, 2, 4\}$. The resulting signature S_u will be the following sequence of bits and it occupies $\gamma = 3$ bytes of space:

$$S_u = \underbrace{1\,1}_{B_u(p_1,p_2)} \underbrace{1\,0}_{B_u(p_2,p_3)} \underbrace{1\,1}_{B_u(p_3,p_4)} \underbrace{0\,1}_{B_u(p_4,p_5)} \underbrace{1\,1}_{B_u(p_5,p_6)} \underbrace{1\,0}_{B_u(p_1,p_3)} \underbrace{1\,0}_{B_u(p_2,p_4)} \underbrace{0\,1}_{B_u(p_3,p_5)} \underbrace{0\,1}_{B_u(p_4,p_6)}$$

$$\underbrace{}_{\text{first byte}} \quad \underbrace{}_{\text{second byte}} \quad \underbrace{}_{\text{third byte}}$$

3.2 Searching

For a given query q, let be Q_q the set of distances between q and all $p_i \in \mathbb{P}$. The signature S_q of the query q is also computed by using $B_q(p_i, p_j)$, with the pair of permutants in $\mathbb{R} \cup \mathbb{R}'$. With this information (S_q) and the signatures of the elements stored into the index, we hopefully could retrieve the most similar elements to q.

In order to select just the relevant objects from the database, the signature of the query S_q is compared with all signatures S_u, $u \in \mathbb{U}$. If an element u is very similar to q, possibly the signature S_u will be similar to S_q. In order to determine this similarity, the main idea is counting how many pairs of bits are in the same order. We consider as candidate those signatures with at least two equal pairs per byte. The Algorithm 1 exposes this idea.

Algorithm 1. Compare2signatures (S_q, S_u)

---

let be S_q and S_u the signatures of a query q and an element u, respectively
let be *sizeSig* the size of the signature
let be masks= {0xC0,0x30,0x0C,0x03}
let be *count* = 0
for $i = 0$ to *sizeSig* **do**
  for m in masks **do**
    if $((S_q[i]\&m) == (S_u[i]\&m))$ **then**
      count + +
    end if
  end for
end for
Return *count*

As an example, we consider the same set of permutants $\mathbb{P} = \{p_1, p_2, p_3, p_4, p_5, p_6\}$, the parameter $r = 2$, and a query q whose set of distances from $\mathbb{P}$ is $Q_q = \{3, 3, 5, 7, 2, 3\}$. The signature S_q will be:

$$S_q = \underbrace{1\ 1}_{B_q(p_1,p_2)}\ \underbrace{1\ 1}_{B_q(p_2,p_3)}\ \underbrace{1\ 1}_{B_q(p_3,p_4)}\ \underbrace{0\ 1}_{B_q(p_4,p_5)}\ \underbrace{1\ 1}_{B_q(p_5,p_6)}\ \underbrace{1\ 0}_{B_q(p_1,p_3)}\ \underbrace{1\ 0}_{B_q(p_2,p_4)}\ \underbrace{0\ 1}_{B_q(p_3,p_5)}\ \underbrace{0\ 1}_{B_q(p_4,p_6)}$$

Therefore, the invocation of Compare2signatures (S_q, S_u) will return the count of 8, because $B_q(p_i, p_j) \approx B_u(p_i, p_j)$ eight times. The only pair of permutants where these signatures do not coincide is: (p_2, p_3). For all the other eight pairs of permutants considered they have the same value.

Then by using the Algorithm 1, Compare2signatures (S_q, S_u) is computed and order the elements considering an decreasing order of the returned value. Hence, to answer any kind of query we compare a fraction of the first elements (in the ordered list) directly by using the distance d from q, and return the relevant elements seen.

In order to solve ties between the values returned for Compare2signatures with two different elements and q, we use the Hamming distance HD between the signatures. That is, if for two different elements $u, v \in X$, Compare2signatures (S_q, S_u) = Compare2signatures (S_q, S_v), then we use $HD(S_q, S_u)$ and $HD(S_q, S_v)$ to solve the tie, or any order will be valid. For this purpose, we define HD as:

$$HD(S_q, S_u) = \sum_{i=1}^{2(2\beta-3)} (S_{q_i} \oplus S_{u_i})$$

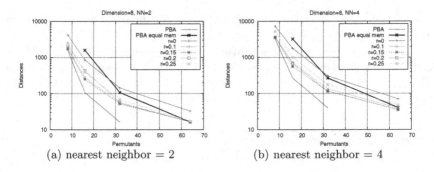

Fig. 1. Performance of the indexes for 2 and 4 NN, using the space in dimension 8.

where S_{q_i} and S_{u_i} are the i-th bit of S_q and S_u respectively, and $\oplus$ is the exclusive or operator between bits.

4   Experimental Results

In order to evaluate the performance of the proposed algorithm, it was tested with synthetic databases in several dimensions (8, 32). This kind of dataset was randomly generated in the unitary hypercube with a uniformly distribution. All these spaces have 100,000 vectors. Although they are vector spaces, they are treated as any other metric space, disregarding the information of the coordinates of each vector. These collections allow us to control the exact dimensionality of the space we are working with. Euclidean distance was used for all these spaces. For each dataset there are 500 objects as queries (synthetic also).

In Fig. 1(a) and (b) the dataset in dimension 8 was used and the queries were for 2−NN, and 4-NN (left and right). In the axis x, the number of permutants is changing. The axis y shows the number of computed distances to retrieve the answer. In order to make a comparison of the proposed index performance, it is also shown the performance of PBA algorithm. However, the tiny line (serie marked as PBA) is an unfair comparison, because our proposal uses less than 50% of PBA space. Hence, to make a fair comparison, we need to use the same amount of memory for the index; the serie labeled as *PBA equal mem* is the performance of the PBA that utilizes the same size of our index. As it was expected, as r is higher more candidates have to be reviewed (more calculations of the distance function are needed to get the correct answer). In both figures, at Fig. 1, we show that with the radii r less than 0.2 we can improve the performance, otherwise only with 64 permutants *PBA equal mem* overcomes our technique. It can be noticed that with $r = 0$ and with 32 or 64 permutants *PBA equal mem* can beat us. Moreover, the Fig. 2 illustrate the same type of experiments considering the dataset of vectors in dimension 32. As it can be seen, the behavior of our technique stays also better than *PBA equal mem* in most of the values of r, showing that it also resists the increasing of the space dimension.

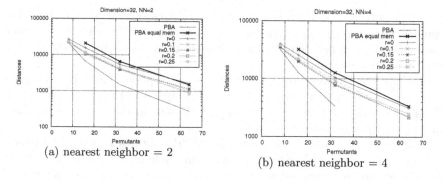

(a) nearest neighbor = 2

(b) nearest neighbor = 4

Fig. 2. Performance of the indexes for 2 and 4 NN, using the space in dimension 32.

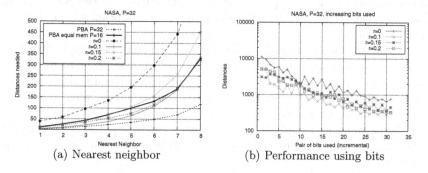

(a) Nearest neighbor

(b) Performance using bits

Fig. 3. Performance of the indexes for retrieving 8 nearest neighbors of the database.

4.1 Real Databases

In order to also test the performance of our proposal on real databases, we consider a database consisting of 40,150 vectors, from NASA images http://www.dimacs.rutgers.edu/Challenges/Sixth/software.html. Each element is a feature vectors in $\mathbb{R}^{20}$. The Euclidean distance is used to compare the objects.

At Fig. 3(a) the number of distances needed to retrieve some nearest neighborh is shown. The novel proposal has a competitive performance as oracle of which elements are similar. In some experiments, the proposal is better up to 66% (5NN). On the other hand, at Fig. 3(b) the performances of using more pairs is shown. At axis $x = 1$ is using only the 1st pair, $x = 2$ is using the 1st and 2st pairs, and so on. As it is expected, using more pairs improve the performance.

5   Conclusions and Future Works

As we have mentioned, the *Permutation Based Algorithm* (PBA) is a novel technique to solve approximate similarity searching.

In order to use more efficiently the space, we propose a new alternative for PBA that allows reducing the index space and obtains a competitive performance. For this purpose, we introduce the signature of each element regarding the set of permutants. Each signature concatenates several pairs of bits that represent how the element sees some pairs of permutants. By this way, we need to store fewer bytes per element than the space needed to maintain the relation between some permutants.

Furthermore, our index do not degrade its search performance due this space reduction. The basic idea is to represent in a simple way the information given for some pairs of permutants, based on that the similar objects could see these pairs of permutants in a similar way.

As it can be shown, this new alternative of PBA obtains a good performance, both in search costs and answer quality, while allows us to save more than the 50% of the space needed for the original PBA with the same number of permutants. Moreover, we also save CPU time at evaluating the dissimilarity between signatures, with respect to evaluate distances between permutations. Hence, we consider that the impact of our proposal is significant for the similarity search community.

As future works, we plan to evaluate the performance of using different sets of pairs to conform the signatures. We also consider studying how to combine permutations and signatures, in order to take advantage when we have available storage space. Moreover, we can consider how to affect another ways to select the set of permutants the search performance of our proposal. For example, we can select the permutants by using the technique of Sparse Spatial Selection (SSS) [2]. Also, the proposal in this article can be used instead Spearman Footrule for other indexes.

References

1. Amato, G., Savino, P.: Approximate similarity search in metric spaces using inverted files. In: 3rd International ICST Conference on Scalable Information Systems, INFOSCALE 2008, Vico Equense, Italy, 4–6 June 2008, p. 28 (2008)
2. Brisaboa, N.R., Fariña, A., Pedreira, O., Reyes, N.: Similarity search using sparse pivots for efficient multimedia information retrieval. In: Proceedings of the 8th IEEE International Symposium on Multimedia (ISM 2006), San Diego, USA, pp. 881–888 (2006)
3. Chávez, E., Figueroa, K., Navarro, G.: Effective proximity retrieval by ordering permutations. IEEE Trans. Pattern Anal. Mach. Intell. (TPAMI) 30(9), 1647–1658 (2009)
4. Chávez, E., Navarro, G.: A probabilistic spell for the curse of dimensionality. In: Buchsbaum, A.L., Snoeyink, J. (eds.) ALENEX 2001. LNCS, vol. 2153, pp. 147–160. Springer, Heidelberg (2001). https://doi.org/10.1007/3-540-44808-X_12
5. Chávez, E., Navarro, G., Baeza-Yates, R., Marroquín, J.: Proximity searching in metric spaces. ACM Comput. Surv. 33(3), 273–321 (2001)
6. Diaconis, P., Graham, R.L.: Spearman's footrule as a measure of disarray. J. Roy. Stat. Soc.: Ser. B (Methodol.) 39(2), 262–268 (1977)
7. Esuli, A.: Use of permutation prefixes for efficient and scalable approximate similarity search. Inf. Process. Manag. 48(5), 889–902 (2012)

8. Naidan, B., Boytsov, L., Nyberg, E.: Permutation search methods are efficient, yet faster search is possible. VLDB **8**(12), 1618–1629 (2015)
9. Téllez, E.S., Chávez, E., Camarena-Ibarrola, A.: A brief index for proximity searching. In: Bayro-Corrochano, E., Eklundh, J.-O. (eds.) CIARP 2009. LNCS, vol. 5856, pp. 529–536. Springer, Heidelberg (2009). https://doi.org/10.1007/978-3-642-10268-4_62

A k–Skyband Approach for Feature Selection

Marcos Bedo[1(✉)], Paolo Ciaccia[2], Davide Martinenghi[3],
and Daniel de Oliveira[4]

[1] INFES, Fluminense Federal University, Santo Antônio de Pádua, Brazil
marcosbedo@id.uff.br
[2] DISI, Università di Bologna, Bologna, Italy
paolo.ciaccia@unibo.it
[3] DEIB, Politecnico di Milano, Milan, Italy
davide.martinenghi@polimi.it
[4] IC, Fluminense Federal University, Niterói, Brazil
danielcmo@ic.uff.br

Abstract. Distance concentration is a phantom menace for the labeling of high dimensional data by distance-based classifiers. Filter methods reduce data dimensionality, but they also add their ranking bias indirectly into the classification procedure. In this study, we examine the filtering problem from another perspective, in which multiple filters are aggregated according to classifiers' constraints. Our approach, named S-Filter, is designed as a top-k skyline (k-skyband) search over multiple rankings by relying on the concept of $\mathcal{F}$–dominance for weighted and monotone linear functions. Unlike existing approaches, S-Filter provides a deterministic strategy for joining multiple filters and avoids the semantic problem of breaking top-k ties. S-Filter's first stage uses labeling-driven measures, *e.g.*, F1-Score, for assessing the quality of each filter with regards to a particular classifier, whereas range-tolerance intervals around the initial quality measures define the partial search weights. Next, S-Filter applies the FSA instance-optimal algorithm for selecting all the dimensions that can be among the top-k for a weight within the range-tolerance intervals. Experiments on high dimensional datasets show that S-Filter outperforms state-of-the-art filters in two scenarios: *(i)* exploratory analysis on varying k and range-tolerance intervals, and *(ii)* data reduction to its intrinsic dimensionality.

Keywords: Feature selection · Filters · Skyline queries · Classification

1 Introduction

Distance concentration is a counter-intuitive phenomenon that affects the behavior of several *metric* distance functions under certain conditions [10].

The authors thank the National Council for Scientific and Technological Development and Faperj (G. E-26/203.215/2016 and I. Sed. 2018) for their financial support.

G. Amato et al. (Eds.): SISAP 2019, LNCS 11807, pp. 160–168, 2019.
https://doi.org/10.1007/978-3-030-32047-8_15

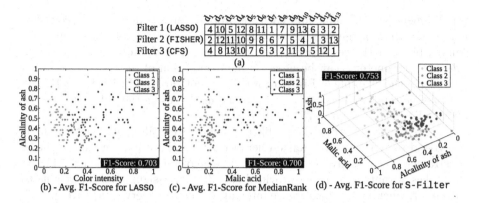

Fig. 1. (a) Scores of WINE dimensions ranked by 3 filters (higher values are better). (b) Plot of top-2 dimensions from Filter 1 (LASSO). (c) Plot of top-2 dimensions by MedianRank. (d) Plot of dimensions selected by S-Filter.

Consequently, distance-based classifiers may struggle in the handling of high dimensional datasets [7]. Supervised filter methods enable choosing the most relevant dimensions[1] from the high dimensional set according to several statistical criteria (scores), which softens concentration as a classification pre-processing step [1]. While several studies discuss whether a filter criterion outperforms others [11], a more comprehensive approach is the joining of multiple filters into a single ranking [4]. Apart from data-driven approaches, *e.g.*, ensembles, the finding of the dimensions with best aggregate scores from a list of filters resemble the optimization problem in [5], in which the result set is constructed by a *monotone* scoring function applied upon two or more rankings produced by distinct filters.

Figure 1 presents an example for the 13-dimensional UCI[2] WINE dataset and three different filters that assign a ranking r_{ij} to dimension d_i according to criterion j, *e.g.*, $r_{23} = 8$. On a reference classifier C_1, the top-2 dimensions selected by the LASSO filter (Fig. 1(b)), i.e., d_{10} and d_4, correspond to an average F1-Score of 0.703, whereas if method MedianRank [6] is applied for aggregating the rankings (Fig. 1(c)), then the top-2 dimensions are d_3 and d_4, with a comparable F1-Score of 0.700. MedianRank assumes rankings are equally relevant, which, in some sense, is similar in spirit to using a linear aggregation function f in which all weights are equal, i.e., $f(d_i) = \sum_{j=1}^{3} w_j \cdot r_{ij}$, with $w_1 = w_2 = w_3 = 1/3$. Uneven weights can be drawn from labeling-driven measures whenever classifiers are used for assessing the quality of dimensions within each ranking. For instance, suppose weights $w_1 = 0.31, w_2 = 0.41, w_3 = 0.28$ are assigned to the rankings in Fig. 1(a) through classifier C_1, then the top-2 dimensions are d_4 and d_2. However, different weights could result by using a different classifier, say, C_2, e.g., $w_1 = 0.25, w_2 = 0.35, w_3 = 0.4$, thereby determining d_4 and d_3 as the

---

[1] The most relevant *dimensions* for a particular set of points are the most prominent data *features*. Accordingly, we use the terms dimensions and features alternately.

[2] archive.ics.uci.edu/ml/datasets.

top-2 dimensions. A viable approach we propose for handling such uncertainty is the *relaxation* of the weights through a set of *constraints* that determine the compatibility of the result with the maximization of f. Accordingly, a *range-tolerance* parameter ξ can be used for the exploration of results around initial weights. In the example of Fig. 1(a), if C_1's weights are used with $\xi = 0.01$ then the *possible* top-2 dimensions are d_4, d_2, and d_3. By using these dimensions, the average F1-Score rises to 0.753 (Fig. 1(d)).

In this paper, we study the problem of aggregating filters as a deterministic k-skyband search whenever range-tolerance intervals are imposed as weighting constraints. Accordingly, we design an embedded approach, named S-Filter, that takes advantage of a user-provided classifier for assigning initial weights to each filter, whereas range-tolerance intervals define the query setup. We investigate S-Filter with varying k values and range-tolerance intervals, and results indicate our method outperforms state-of-the-art filters.

2   Preliminaries and Related Work

Dimensionality Reduction by Filtering. Consider a dataset $\mathcal{S} = \{s_1, \ldots, s_n\}$ where each element s_i is described in a d-dimensional normalized space, *i.e.*, $s_i \in [0,1]^d$ and associated with a categorical label. A supervised filter ranks labeled features d_j of the elements s_i in $\mathcal{S}$ by assigning a real value to them so that top-valued dimensions can be seen as the best *data features* according to the filter perspective. Additionally, a classifier can be employed for the evaluation of ranked features, which results in an *embedded approach* that uses both filter and classifier bias for finding the most suitable subset of dimensions [1,4].

Although we take advantage of such an embedded pipeline, our goal is rather determining the *dominant* dimensions (in the skyline sense [2]) within multiple rankings. Accordingly, we build upon representative filters, such as FISHER, M-INFFS, CFS, RELIEFF, LASSO, ECFS, UDFS, INFFS, and ILFS [1,11].

Choosing the Number of Relevant Dimensions. The *intrinsic dimension* is a common measure employed for estimating the number of relevant features within a dataset. As the asymptotic approximation of statistical learning theory indicates, the generalization of distance-based classifiers depends on the intrinsic representation of the data [7,9]. Two groups of estimators stand out regarding their computational efficiency and quality of predictions: *(i)* methods of power-law approximation, and *(ii)* methods of concentration of measure [8]. A power-law approximation is based on the notion that the number of possible elements under the same distance range grows exponentially with dimensions. An effective strategy for quantifying such behavior is the use of a *Distance Plot* that depicts the joint pairwise distance distribution in the logarithm scale. The trimmed points in the plot are interpolated by a line whose linear slope is the *distance exponent* $\mathcal{D}$ that approximates the intrinsic dimension [8]. On the other hand, an efficient approach for quantifying the concentration of measure is the ρ-score of Pestov [10], calculated as $\rho = \mu_T^2/2 \cdot \sigma_T^2$, where functions μ_T and σ_T are the mean and standard deviation of the distance distribution, respectively.

Multiple Ranking Aggregation. If m filters are applied upon a labeled dataset $\mathcal{S}$ then a list of rankings $\mathcal{R} = \{r_{ij}\}$, $i = 1, \ldots, d$, $j = 1, \ldots, m$ is obtained. Higher scores indicate better features so that an order of relevancy can be inferred for each filter j. A scoring function f assigns a score to each dimension d_i by aggregating the m rankings of d_i, i.e., $f(d_i) = f(r_{i1}, \ldots, r_{im})$. Function f is monotone if $f(d_i) \geq f(d_i')$ for pairs of dimensions d_i, d_i' in which $r_{ij} \geq r_{ij}'$, $j = 1, \ldots, m$. The number of monotone functions is infinite, and the whole set of such functions is noted $\mathcal{M}$. A top-k query on $\mathcal{R}$ returns the k dimensions with the highest scores according to a function $f \in \mathcal{M}$. A dimension d_i *dominates* another dimension d_i', denoted $d_i \succ d_i'$, whenever $r_{ij} \geq r_{ij}'$ holds for *all* filters $j = 1, \ldots, m$ and at least one of the inequalities is strict. Dominance is at the core of *skyline* queries, since the skyline of $\mathcal{R}$ is the set of non-dominated dimensions in $\mathcal{R}$, i.e., $Sky(\mathcal{R}) = \{d_i \in \mathcal{R} \mid \nexists d_i' \in \mathcal{R}, \ d_i' \succ d_i\}$. Notice $Sky(\mathcal{R})$ equals the set of dimensions that can be the top-1 result according to any monotone function [2], i.e., $d_i \in Sky(\mathcal{R}) \Leftrightarrow \exists f \in \mathcal{M}. \ \forall d_i' \in \mathcal{R} \setminus \{d_i\}. \ f(d_i) > f(d_i')$. Dominance can be generalized to $\mathcal{F}$-dominance by considering an arbitrary set of monotone functions $\mathcal{F}$, which generalizes the concept of skyline as well [2].

Definition 1 ($\mathcal{F}$-dominance). *Given a set of monotone functions $\mathcal{F}$, a dimension d_i $\mathcal{F}$-dominates dimension d_i', denoted by $d_i \succ_\mathcal{F} d_i'$, iff $\forall f \in \mathcal{F}. \ f(d_i) \geq f(d_i')$ and $\exists f \in \mathcal{F}. \ f(d_i) > f(d_i')$.*

Definition 2 (Non-$\mathcal{F}$-dominated skyline). *Given a set of monotone functions $\mathcal{F}$, the non-$\mathcal{F}$-dominated skyline of $\mathcal{R}$, denoted $ND(\mathcal{R}, \mathcal{F})$ is the set of non-$\mathcal{F}$-dominated dimensions, i.e., $ND(\mathcal{R}, \mathcal{F}) = \{d_i \in \mathcal{R} \mid \nexists d_i' \in \mathcal{R}, \ d_i' \succ_\mathcal{F} d_i\}$.*

Note that, since $\mathcal{F} \subseteq \mathcal{M}$, then $ND(\mathcal{R}, \mathcal{F}) \subseteq Sky(\mathcal{R})$. In particular, $ND(\mathcal{R}, \mathcal{M}) = Sky(\mathcal{R})$ and $ND(\mathcal{R}, \{f\})$ is the set of possible top-1 dimensions according to f [2]. The k-skyband of $\mathcal{R}$, $Sky_k(\mathcal{R})$, is the set of dimensions in $\mathcal{R}$ that are dominated by less than k dimensions and, therefore, includes all possible top-k dimensions. The generalization of $ND(\mathcal{R}, \mathcal{F})$ is provided by Definition 3.

Definition 3. *Given a set of monotone functions $\mathcal{F}$, the (non-$\mathcal{F}$-dominated) k-skyband, $ND_k(\mathcal{R}, \mathcal{F})$, contains all the dimensions that are $\mathcal{F}$-dominated by less than k dimensions.*

The authors in [3] show that $ND_k(\mathcal{R}, \mathcal{M}) = Sky_k(\mathcal{R})$, whereas $ND_k(\mathcal{R}, \mathcal{F}) \subseteq Sky_k(\mathcal{R})$ whenever $\mathcal{F} \subset \mathcal{M}$. If $\mathcal{F}$ contains a single function f then $ND_k(\mathcal{R}, \{f\})$ is the set of *all* possible top-k dimensions according to f. Clearly, if a dimension $d_i \notin ND_k(\mathcal{R}, \mathcal{F})$, then no scoring function $f \in \mathcal{F}$ can make d_i a top-k dimension. From a computational viewpoint, the cost of computing $ND_k(\mathcal{R}, \mathcal{F})$ using the instance-optimal FSA algorithm [3] is comparable to that of top-k queries.

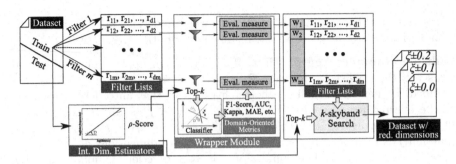

Fig. 2. S-Filter pipeline. Features are ranked by filtering criteria and evaluated by a wrapped classifier defining the initial weights for the $\mathrm{ND}_k(\mathcal{R}, \mathcal{F})$ query.

3   A Two-Worlds Skyline: Embedding Filters and Classifiers

Our approach evaluates multiple filter rankings from a user-provided classifier perspective, in which the most relevant features are the non-$\mathcal{F}$-dominated ones. Figure 2 summarizes all steps related to the execution of the embedded method, which was named S-Filter. The main idea behind S-Filter is the aggregation of multiple filters according to a non-$\mathcal{F}$-dominated k-skyband query, in which weights depend on labeling-driven measures indicated by a wrapped classifier and user-provided range-tolerance intervals. Such a parameterization enables users to tune two important classification aspects, namely (i) choose the levels and importance of classification "hits" and "misses" by changing labeling-driven measures, e.g., F1-Score and Kappa Coefficient, according to characteristics of the problem at hand, and (ii) delimit the uncertainty of the measures through range-tolerance intervals. S-Filter applies the user-provided classifier for the calculation of labeling-driven measures by relying on a *holdout* data split. These values are then employed as the initial search weights for the $\mathrm{ND}_k(\mathcal{R}, \mathcal{F})$ query.

A cross-validation execution of the classifier runs for the ranking list produced by each filter, in which the number of employed features is defined by the data intrinsic dimension, i.e., the top-k features of each ranking are selected by setting k as the round value of either $\mathcal{D}$ or ρ. An average labeling-driven value a_j is assigned to each filter $j = 1, \ldots, m$ at the end of the classification stage, and S-Filter estimates the initial search weights proportional to such measures, i.e., $w_j = a_j / \sum_{j=1}^{m} a_j$. S-Filter's last step is the execution of a $\mathrm{ND}_k(\mathcal{R}, \mathcal{F})$ query on the rankings, where $\mathcal{F}$ is determined by (i) the set of initial weights w_j, and (ii) range-tolerance intervals ξ, as in Eq. 1.

$$\mathcal{F} = \{f(d_i) = 1/d \cdot \sum_j w'_j \cdot r_{ij}\}, \quad \text{where } (1 - \xi) \cdot w_j \leq w'_j \leq (1 + \xi) \cdot w_j. \quad (1)$$

The set of all non-$\mathcal{F}$-dominated dimensions is returned by the FSA algorithm.

Table 1. Datasets employed in the experiments.

Name	$|\mathcal{S}|$	d	L_p	$\mathcal{D}$	ρ	Labels	Description
DAILY	9,120	5,625	L_1	3	7	480/480	*Sensor measures for daily activities*
GISETTE	13,500	5,000	L_1	9	21	6,750/6,750	*Handwritten digits "4" and "9"*
MFEAT	2,000	649	L_1	2	13	200/200	*Features from handwritten numerals*
SMART	7,767	561	L_2	4	4	23/1,223	*Smartphone sensors for human activity*
UJIN	19,935	520	L_2	5	19	1,102/5,048	*Characters in a UNIPEN-like format*

4 Experiments

This section compares the performance of S-Filter and state-of-the-art filters FISHER, M-INFFS, CFS, RELIEFF, LASSO, ECFS, UDFS, INFFS and ILFS regarding: *(i) parameter exploration*, which aims at investigating the impact of varying top-k and range-tolerance values, and *(ii) labeling quality impact*, whose goal is assessing the gains of S-Filter when data are reduced to its intrinsic dimension.

Datasets. We used five UCI datasets (see Table 1) in association with different distance functions. In all experiments, F1-Scores of top-k features were used as labeling-driven measures to assess the quality of individual rankings, and the intrinsic dimension was calculated by the ρ-score.

Classifiers. We evaluated three classifiers that are impacted in different ways by high-dimensional datasets. First, we employed the baseline Instance-based Learning (IBL) classifier, which depends on the distance function for defining its search space [1]. We also evaluated the Discriminant Analysis (DA) classifier since the error of its internal regression is affected by the number of dimensions [7]. Finally, we experimented with pruned Decision-Trees (DT), whose number of levels and construction are impacted by data dimensionality [7].

Parameter Exploration. We evaluated the impact of S-Filter parameters in data classification by performing 3-fold cross-validations on the test portion of the datasets. We varied parameter k exponentially from top-2 features to top-256 features, while range-tolerance intervals were set from 0 to 50% of uncertainty, *i.e.*, $\xi \pm \{0.0, \ldots, 0.5\}$. Figure 3 reports the average classification ratio as blue-to-yellow heatmaps of F1-Score (also employed for S-Filter training). Results show that S-Filter not only outperformed the competitors in most cases for increasing values of range-tolerance but also reached a *stable* (absolute variation $\leq 1\%$) ratio with fewer dimensions in comparison to other methods. In particular, S-Filter outperformed all filters on set DAILY ($\mathcal{D} = 3$) for $k = \{2, 4\}$ and $\xi \pm \{0.0, 0.1\}$ in up to 13.4% with regards to the closest competitor (CFS). Intrinsic high-dimensional datasets GISETTE and SMART showed that S-Filter required a larger uncertainty ratio $\xi \geq 0.3$ for surpassing the RELIEFF filter. Such findings indicate that S-Filter demanded fewer expected dimensions to reach better quality, whereas range-tolerance restrained the weight estimation error.

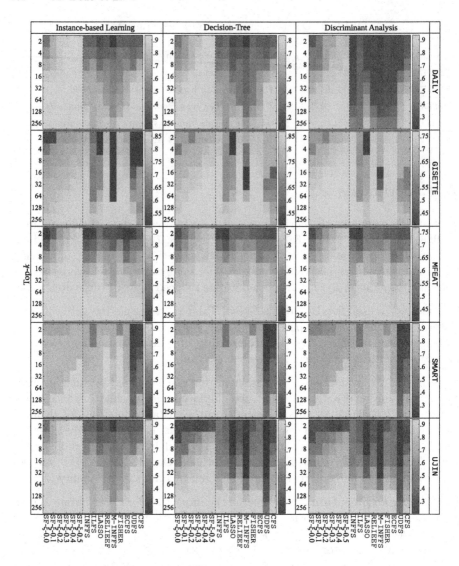

Fig. 3. F1-Scores for varying k and range-tolerance intervals.

Reduction to Data Intrinsic Dimension. We examined the filter-based competitors in the practical case where the number of features is set to the intrinsic dimension. Figure 4 reports the average and standard deviation F1-Scores obtained by the classifiers IbL, DT, and DA and the nine compared filters as a single heatmap. Results indicate S-Filter outperformed the competitors for $\xi \pm 0.0$ considering most of the scenarios, whereas S-Filter with $\xi \pm 0.3$ either tied or outperformed isolated filters. In particular, S-Filter without range-tolerance achieved 15.3%, 17.6%, and 23.4% higher F1–Scores, on average, regarding IbL,

	S-Filter ξ=0.0	S-Filter ξ=0.1	S-Filter ξ=0.2	S-Filter ξ=0.3	S-Filter ξ=0.4	S-Filter ξ=0.5	INFFS	ILFS	LASSO	RELI-EFF	MUTI-INFFS	FISHER	ECFS	UDFS	CFS	
DAILY	0.679	0.798	0.866	0.913	0.904	0.912	0.530	0.639	0.394	0.528	0.495	0.530	0.566	0.467	0.714	
	±.002	±.002	±.003	±.003	±.003	±.003	±.002	±.003	±.002	±.001	±.002	±.003	±.002	±.002	±.003	
GISETTE	0.648	0.653	0.685	0.764	0.717	0.739	0.747	0.552	0.581	0.690	0.387	0.748	0.744	0.508	0.615	
	±.003	±.003	±.003	±.003	±.002	±.003	±.002	±.002	±.003	±.002	±.002	±.003	±.004	±.002	±.002	
MFEAT	0.718	0.853	0.910	0.931	0.955	0.955	0.570	0.651	0.848	0.707	0.615	0.798	0.534	0.597	0.676	IBL
	±.004	±.003	±.003	±.003	±.003	±.004	±.001	±.001	±.002	±.001	±.002	±.001	±.002	±.002	±.001	
SMART	0.653	0.577	0.572	0.628	0.629	0.666	0.605	0.379	0.614	0.593	0.612	0.627	0.605	0.182	0.206	
	±.040	±.042	±.051	±.050	±.051	±.050	±.030	±.042	±.035	±.037	±.041	±.038	±.039	±.041	±.043	
UJIN	0.344	0.483	0.549	0.602	0.623	0.636	0.328	0.394	0.110	0.336	0.092	0.388	0.327	0.096	0.338	
	±.063	±.061	±.062	±.063	±.061	±.065	±.094	±.101	±.095	±.098	±.105	±.087	±.084	±.095	±.098	
DAILY	0.592	0.671	0.747	0.844	0.863	0.859	0.475	0.550	0.353	0.478	0.380	0.484	0.493	0.434	0.658	
	±.003	±.003	±.003	±.003	±.003	±.003	±.002	±.004	±.004	±.005	±.002	±.003	±.003	±.003	±.002	
GISETTE	0.654	0.665	0.705	0.757	0.733	0.743	0.733	0.535	0.568	0.702	0.306	0.741	0.743	0.470	0.397	
	±.004	±.004	±.005	±.004	±.004	±.003	±.003	±.003	±.002	±.003	±.003	±.002	±.001	±.003	±.002	
MFEAT	0.693	0.764	0.819	0.812	0.871	0.868	0.658	0.686	0.667	0.733	0.711	0.813	0.659	0.583	0.649	DEC. TREE
	±.006	±.007	±.006	±.005	±.005	±.006	±.002	±.002	±.003	±.002	±.003	±.002	±.005	±.004	±.003	
SMART	0.675	0.668	0.670	0.686	0.693	0.692	0.623	0.409	0.653	0.660	0.608	0.634	0.623	0.175	0.220	
	±.043	±.044	±.048	±.047	±.045	±.043	±.043	±.044	±.045	±.044	±.046	±.047	±.045	±.046	±.045	
UJIN	0.399	0.523	0.581	0.628	0.635	0.656	0.362	0.374	0.140	0.406	0.022	0.384	0.351	0.029	0.392	
	±.068	±.066	±.067	±.067	±.065	±.067	±.073	±.074	±.077	±.077	±.076	±.078	±.078	±.078	±.079	
DAILY	0.292	0.367	0.477	0.621	0.613	0.652	0.119	0.214	0.128	0.120	0.106	0.140	0.123	0.129	0.290	
	±.001	±.002	±.002	±.002	±.002	±.002	±.002	±.003	±.003	±.002	±.002	±.003	±.003	±.002	±.002	
GISETTE	0.675	0.688	0.703	0.730	0.735	0.734	0.729	0.560	0.586	0.716	0.386	0.718	0.724	0.514	0.497	
	±.004	±.003	±.003	±.004	±.003	±.003	±.004	±.003	±.002	±.003	±.007	±.002	±.001	±.008	±.005	
MFEAT	0.727	0.824	0.902	0.910	0.935	0.936	0.600	0.732	0.765	0.685	0.694	0.820	0.602	0.543	0.750	DISC. AN
	±.003	±.002	±.003	±.003	±.003	±.002	±.002	±.002	±.001	±.002	±.001	±.002	±.003	±.003	±.002	
SMART	0.612	0.623	0.635	0.644	0.649	0.663	0.649	0.402	0.623	0.633	0.647	0.647	0.642	0.102	0.137	
	±.098	±.097	±.092	±.093	±.095	±.094	±.069	±.072	±.077	±.079	±.081	±.072	±.077	±.081	±.082	
UJIN	0.306	0.353	0.362	0.415	0.409	0.415	0.235	0.285	0.106	0.322	0.022	0.272	0.243	0.026	0.305	
	±.062	±.063	±.062	±.063	±.061	±.063	±.104	±.112	±.120	±.101	±.121	±.114	±.117	±.118	±.106	

Fig. 4. F1-Scores for data reduction to the intrinsic dimension.

DT, and DA, respectively. Such outcomes reinforce the findings of the previous evaluation, where S-Filter provided better F1-Scores than the competitors.

Conclusions. Our solution, S-Filter, is an embedded and deterministic k-skyband-based approach that aggregates multiple filters with classifier-estimated weights, mitigating estimation uncertainty via tolerance intervals. Experiments show that S-Filter attains a higher classification ratio than existing filters.

References

1. Aggarwal, C.C.: Data Mining: The Textbook. Springer, Cham (2015). https://doi.org/10.1007/978-3-319-14142-8
2. Ciaccia, P., Martinenghi, D.: Reconciling skyline and ranking queries. PVLDB **10**(11), 1454–1465 (2017)
3. Ciaccia, P., Martinenghi, D.: $FA + TA < FSA$: flexible score aggregation. In: CIKM, pp. 57–66. ACM (2018)
4. Drotár, P., Gazda, M., Vokorokos, L.: Ensemble feature selection using election methods and ranker clustering. Inf. Sci. **480**, 365–380 (2019)
5. Fagin, R.: Combining fuzzy information from multiple systems. In: PODS, pp. 216–226 (1996)
6. Fagin, R., Kumar, R., Sivakumar, D.: Efficient similarity search and classification via rank aggregation. In: SIGMOD, pp. 301–312. ACM (2003)

7. James, G., Witten, D., Hastie, T., Tibshirani, R.: An Introduction to Statistical Learning, vol. 112. Springer, New York (2013). https://doi.org/10.1007/978-1-4614-7138-7
8. Navarro, G., Paredes, R., Reyes, N., Bustos, C.: An empirical evaluation of intrinsic dimension estimators. Inf. Syst. **64**, 206–218 (2017)
9. Pestov, V.: An axiomatic approach to intrinsic dimension of a dataset. Neural Netw. **21**(2–3), 204–213 (2008)
10. Pestov, V.: Lower bounds on performance of metric tree indexing schemes for exact similarity search in high dimensions. Algorithmica **66**(2), 310–328 (2013)
11. Roffo, G., Melzi, S., Castellani, U., Vinciarelli, A.: Infinite latent feature selection: a probabilistic latent graph-based ranking approach. In: CVPR (2017)

Clustering and Outlier Detection

Faster k-Medoids Clustering: Improving the PAM, CLARA, and CLARANS Algorithms

Erich Schubert[1($\boxtimes$)] (iD) and Peter J. Rousseeuw[2] (iD)

[1] Technische Universität Dortmund, Dortmund, Germany
erich.schubert@tu-dortmund.de
[2] Department of Mathematics, KU Leuven, Leuven, Belgium
peter@rousseeuw.net

Abstract. Clustering non-Euclidean data is difficult, and one of the most used algorithms besides hierarchical clustering is the popular algorithm Partitioning Around Medoids (PAM), also simply referred to as k-medoids.

In Euclidean geometry the mean—as used in k-means—is a good estimator for the cluster center, but this does not exist for arbitrary dissimilarities. PAM uses the medoid instead, the object with the smallest dissimilarity to all others in the cluster. This notion of centrality can be used with any (dis-)similarity, and thus is of high relevance to many domains and applications.

A key issue with PAM is its high run time cost. We propose modifications to the PAM algorithm that achieve an $O(k)$-fold speedup in the second ("SWAP") phase of the algorithm, but will still find the same results as the original PAM algorithm. If we slightly relax the choice of swaps performed (while retaining comparable quality), we can further accelerate the algorithm by performing up to k swaps in each iteration. With the substantially faster SWAP, we can now explore faster intialization strategies. We also show how the CLARA and CLARANS algorithms benefit from the proposed modifications.

Keywords: Cluster analysis · k-Medoids · PAM · CLARA · CLARANS

1 Introduction

Clustering is a common unsupervised machine learning task, in which the data set has to be automatically partitioned into "clusters", such that objects within the same cluster are more similar, while objects in different clusters are more different. There is not (and likely never will be) a generally accepted definition of a cluster, because "clusters are, in large part, in the eye of the beholder" [7], meaning that every user may have different enough needs and intentions to want a different algorithm and notion of cluster. And therefore, over many years of

© Springer Nature Switzerland AG 2019
G. Amato et al. (Eds.): SISAP 2019, LNCS 11807, pp. 171–187, 2019.
https://doi.org/10.1007/978-3-030-32047-8_16

research, hundreds of clustering algorithms and evaluation measures have been proposed, each with their merits and drawbacks. Nevertheless, a few seminal methods such as hierarchical clustering, k-means, PAM [9], and DBSCAN [6] have received repeated and widespread use. One may be tempted to think that these *classic* methods have all been well researched and understood, but there are still many scientific publications trying to explain these algorithms better (e.g., [19]), trying to parallelize and scale them to larger data sets, trying to better understand similarities and relationships among the published methods (e.g., [18]), or proposing further improvements – and so does this paper for the widely used PAM algorithm.

A classic method taught in textbooks is k-means (for an overview of the complicated history of k-means, refer to [3]), where the data is modeled using k cluster means, that are iteratively refined by assigning all objects to the nearest mean, then recomputing the mean of each cluster. This converges to a local optimum because the mean is the least squares estimator of location, and both steps reduce the same quantity, a measure known as sum-of-squared errors:

$$SSQ := \sum_{i=1}^{k} \sum_{x_j \in C_i} ||x_j - \mu_i||_2^2. \tag{1}$$

In k-medoids, the data is modeled similarly, using k representative objects m_i called medoids (chosen from the data set; defined below) that serve as "prototypes" for the clusters in order to allow using arbitrary other dissimilarities and arbitrary input domains, using the absolute error criterion ("total deviation", TD) as objective:

$$TD := \sum_{i=1}^{k} \sum_{x_j \in C_i} d(x_j, m_i), \tag{2}$$

which is the sum of dissimilarities of each point $x_j \in C_i$ to the medoid m_i of its cluster. If we use squared Euclidean as distance function (i.e., $d(x, m) = ||x - m||_2^2$), we almost obtain the usual SSQ objective used by k-means, except that k-means is free to choose any $\mu_i \in \mathbb{R}^d$, whereas in k-medoids $m_i \in C_i$ must be one of the original data points. For *squared* Euclidean distances and Bregman divergences, the arithmetic mean is the optimal choice for μ. For L_1 distance (i.e, $\sum |x_i - y_i|$), also called Manhattan distance, the component-wise median is a better choice in $\mathbb{R}^d$ [4]. For unsquared Euclidean distances, we get the much harder Weber problem [14], which has no closed-form solution [4]. For other distance functions, finding a closed form to compute the best m_i would require a separate mathematical analysis. Furthermore, our input domain is not necessarily a $\mathbb{R}^d$ vector space. In k-medoids clustering, we therefore constrain m_i to be one of our data samples. The medoid of a set C is defined as the object with the smallest sum of dissimilarities (or, equivalently, smallest average) to all other objects in the set:

$$\text{medoid}(C) := \arg\min_{x_i \in C} \sum_{x_j \in C} d(x_i, x_j).$$

This definition does not require the dissimilarity to be a metric, and by using arg max it can also be applied to similarities. The algorithms discussed below all can trivially be modified to maximize similarities rather than minimizing distances, and none assumes the triangular inequality. Partitioning Around Medoids (PAM, [9]) is the most widely known algorithm to find a good partitioning using medoids, with respect to TD (Eq. 2).

2 Partitioning Around Medoids (PAM) and Its Variants

The "Program PAM" [9] consists of two algorithms, BUILD to choose an initial clustering, and SWAP to improve the clustering towards a local optimum (finding the global optimum of the k-medoids problem is, unfortunately, NP-hard). The algorithms require a dissimilarity matrix, which requires $O(n^2)$ memory and typically $O(n^2 d)$ time to compute (but much more for expensive distances such as earth movers distance).

In order to find a good initial clustering, BUILD chooses k times the point which yields the smallest distance sum TD (this means first choosing the point with the smallest distance to all others; afterwards always adding the point that reduces TD most). The motivation here was to find a good starting point, in order to require fewer iterations of the refinement procedure. The second part, SWAP, improves the clustering by considering all possible simple changes to the set of k medoids, which effectively means replacing (swapping) some medoid with some non-medoid, which gives $k(n-k)$ candidate swaps. If it reduces TD, the best such change is then applied, in the spirit of a greedy steepest-descent method, and this process is repeated until no further improvements are found.

The algorithm CLARA [10]) repeatedly applies PAM on a subsample with $n' \ll n$ objects, with the suggested value $n' = 40 + 2k$. Afterwards, the remaining objects are assigned to their closest medoid. The run with the least TD (on the entire data) is returned. If the sample size is chosen $n' \in O(k)$ as suggested, the run time reduces to $O(k^3)$, which explains why the approach is typically used only with small k [12].

Lucasius et al. [12] propose a genetic algorithm for k-medoids, by performing a randomized exploration based on "mutation" of the best solutions found so far. The algorithm CLARANS [13] interprets the search space as a high-dimensional hypergraph, where each edge corresponds to swapping a medoid and non-medoid. On this graph it performs a randomized greedy exploration, where the first edge that reduces the loss TD is followed until no edge can be found with $p = 1.25\% \cdot k(n - k)$ attempts. Other proposals include optimizations for Euclidean space and tabu search heuristics [8].

Reynolds et al. [16] discuss an interesting trick to speed up PAM. They show how to decompose the change in the loss function into two components, where the first depends only on the medoid removed, the second part only on the new point. This decomposition forms the base for our approach, and we will thus discuss it in Sect. 3 in more detail.

Park and Jun [15] propose a "k-means like" algorithm for k-medoids (actually already considered by [16] before), where in each iteration the medoid is chosen to be the object with the smallest distance sum to other members of the cluster, then each point is assigned to the nearest medoid until TD no longer decreases. This is, unfortunately, not very effective at improving the clustering: new medoids are only chosen from within the cluster, and *have* to cover the entire current cluster. This misses many improvements where cluster members can be reassigned to *other* clusters with little cost; such improvements are considered by SWAP. Furthermore, the discrete nature of medoids makes this much more likely to get stuck in a local optimum. In our experiments this approach produced much worse results than PAM, as previously observed by [16].

3   Finding the Best Swap

The algorithm SWAP evaluates every swap of each medoid m_i with any non-medoid x_j. Recomputing the resulting TD using Eq. 2 every time requires finding the nearest medoid for every point, which causes many redundant computations. Instead, PAM only computes the *change* in TD for each object x_o if we swap m_i with x_j:

$$\Delta TD = \sum_{x_o} \Delta(x_o, m_i, x_j) \tag{3}$$

In the function $\Delta(x_o, m_i, x_j)$ we can often detect when a point remains assigned to its current medoid (if $c_k \neq c_i$, and this distance is also smaller than the distance to x_j), and then immediately return 0. Because of space restrictions, we do not repeat the original "if" statements used in [9], but instead condense them into the equation:

$$\Delta(x_o, m_i, x_j) = \begin{cases} \min\{d(x_o, x_j), d_s(o)\} - d_n(o) & \text{if } i = nearest(o) \\ \min\{d(x_o, x_j) - d_n(o), 0\} & \text{otherwise} \end{cases}, \tag{4}$$

where $d_n(o)$ is the distance to the nearest medoid of o, and $d_s(o)$ is the distance to the second nearest medoid. Computing them on the fly increases the runtime by a factor of $O(k)$, but we can cache these values, and only update them when performing a swap.

Reynolds et al. [16] note that we can decompose ΔTD into: (i) the loss of removing medoid m_i, and assigning all of its members to the next best alternative, which can be computed as $\sum_{o \in C_i} d_s(o) - d_n(o)$ (ii) the (negative) loss of adding the replacement medoid x_j, and reassigning all objects closest to this new medoid. Since (i) does not depend on the choice of x_j, we can make the loop over all medoids m_i outermost, reassign all its points to the second nearest medoid (cache the distance to the now nearest neighbor), and compute the resulting loss. We then iterate over all non-medoids and compute the benefit of using them as the missing medoid instead. In the Δ function, we no longer have to consider the second nearest now (we virtually removed the old medoid already). The authors observed roughly a two-fold speedup using this approach.

Our approach is based on a similar idea of exploiting redundancy in these computations (by caching shared computations), but we instead move the loops over the medoids m_i into the *innermost* for loop. The reason for this is to further remove redundant computations. This becomes apparent when we realize that in Eq. 4, the second case does not depend on the current medoid i. If we transform the second case into an if statement, we can often avoid to iterate over all k medoids.

3.1   Making PAM SWAP Faster: FastPAM1

Algorithm 1 shows the improved SWAP algorithm. In lines 4–5 we compute the benefit of making x_j a medoid. As we do *not yet* decide which medoid to remove, we use an array of ΔTDs for each possible medoid to replace. We can now for each point compute the benefit when removing its current medoid (line 9), or the benefit if the new medoid is closer than the current medoid (line 10), which corresponds to the two cases in Eq. 4. Because the second case does not depend on i, we can replace the min statement with an if conditional *outside* of the loop in lines 10–12. After iterating over all points, we choose the best medoid, and remember the overall best swap. If we always prefer the smaller index i on ties, we choose *exactly the same* swap as the original PAM algorithm.

Assuming that the new medoid is closest in $O(1/k)$ cases on average, we can compute the change for all k medoids with $O(k \cdot 1/k) = O(1)$ effort, by saving on the innermost loop. Therefore, we expect a typical speedup on the order of $O(k)$ compared to the original PAM SWAP (but it may be hard to guarantee this for any *useful* assumption on the data distribution; the worst case supposedly remains unaffected) at the slight cost of storing one ΔTD for each medoid m_i (compared to the cost of the distance matrix and the distances to the nearest and second nearest medoids, the cost of this is negligible).

Algorithm 1. FastPAM1: Improved SWAP algorithm

```
1  repeat
2  │   (ΔTD*, m*, x*) ← (0, null, null) ;            // Empty best candidate storage
3  │   foreach x_j ∉ {m_1, ..., m_k} do
4  │   │   d_j ← d_nearest(x_j) ;                      // Distance to current medoid
5  │   │   ΔTD ← (−d_j, −d_j, ..., −d_j) ;             // Change if making j a medoid
6  │   │   foreach x_o ≠ x_j do
7  │   │   │   d_oj ← d(x_o, x_j) ;                      // Distance to new medoid
8  │   │   │   (n, d_n, ds) ← (nearest(o), d_nearest(o), d_second(o)) ;  // Cached values
9  │   │   │   ΔTD_n ← ΔTD_n + min{d_oj, d_s} − d_n ;    // Loss change
10 │   │   │   if d_oj < d_n then                        // Reassignment check
11 │   │   │   │   foreach m_i ∈ {m_1, ..., m_k} \ m_n do
12 │   │   │   │   │   ΔTD_i ← ΔTD_i + d_oj − d_n;        // Update loss change
13 │   │   i ← arg min ΔTD_i ;                          // Choose best medoid i
14 │   │   if ΔTD_i < ΔTD* then (ΔTD*, m*, x*) ← (ΔTD_i, m_i, x_j) ;  // Store
15 │   break loop if ΔTD* ≥ 0;
16 │   swap roles of medoid m* and non-medoid x* ;
17 │   TD ← TD + ΔTD* ;
18 return TD, M, C;
```

3.2   Swapping Multiple Medoids: FastPAM2

A second technique to make SWAP faster is based on the following observation: PAM will always identify the *single* best swap, then restart search; whereas the classic k-means updates all means in each iteration. Choosing the best swap has the benefit that this makes the algorithm independent of the data order [9] as long as there are no ties, and it means we need to execute fewer swaps than if we would greedily perform any swap that yields an improvement (where we may end up replacing the same medoid several times). But on the other hand, in particular for large k, we can assume that many clusters will be independent, and we could therefore update the medoids of these clusters in the same iteration, hence reduce the number of iterations by up to a factor of k.

Based on this observation, we propose to consider the best swap for *each* medoid, i.e., perform up to k swaps. This is a fairly simple modification shown in Algorithm 2, as we can use an array of swap candidates $(\Delta TD^*_i, x^*_i)$ in line 3, storing the best candidate for each current medoid m_i, and update these in line 15. After evaluating all possible swaps, we find the best swap within these up to k candidates (if we did not find a candidate, the algorithm has converged). We perform the best of these swaps in line 18. Then we recompute in line 22 for each remaining swap candidate if it still improves the clustering, otherwise this additional swap is not performed.

Algorithm 2. FastPAM2: SWAP with multiple candidates

1 **repeat**
2   **foreach** x_o **do** compute nearest(o), $d_{\text{nearest}}(o)$, $d_{\text{second}}(o)$; ;
3      $\Delta TD^*, x^* \leftarrow [0, \ldots, 0], [\text{null}, \ldots, \text{null}]$;   // Empty best candidates array
4   **foreach** $x_j \notin \{m_1, \ldots, m_k\}$ **do**
5          $d_j \leftarrow d_{\text{nearest}}(x_j)$;   // Distance to current medoid
6          $\Delta TD \leftarrow (-d_j, -d_j, \ldots, -d_j)$;   // Change for making j a medoid
7   **foreach** $x_o \neq x_j$ **do**
8              $d_{oj} \leftarrow d(x_o, x_j)$;   // Distance to new medoid
9              $(n, d_n, ds) \leftarrow (\text{nearest}(o), d_{\text{nearest}}(o), d_{\text{second}}(o))$;   // Cached
10             $\Delta TD_n \leftarrow \Delta TD_n + \min\{d_{oj}, ds\} - d_n$;   // Loss change for x_o
11   **if** $d_{oj} < d_n$ **then**   // Reassignment check
12   **foreach** $m_i \in \{m_1, \ldots, m_k\} \setminus m_n$ **do**
13                     $\Delta TD_i \leftarrow \Delta TD_i + d_{oj} - d_n$;   // Update loss change
14   **foreach** i where $\Delta TD_i < \Delta TD^*_i$ **do**
15             $(\Delta TD^*_i, x^*_i) \leftarrow (\Delta TD_i, x_j)$;   // Remember the best swap for i
16   **break loop if** $\min \Delta TD^* \geq 0$;   // Stop if no improvements were found
17   **while** $i \leftarrow \arg\min \Delta TD^*$ and $\Delta TD^*_i < 0$ **do**   // Execute all improvements
18   swap roles of medoid m_i and non-medoid x^*_i ;
19         $TD \leftarrow TD + \Delta TD^*_i$;
20         $\Delta TD^*_i \leftarrow 0$;   // Disable the swap just performed
21   **foreach** j where $\Delta TD^*_j < 0$ **do**   // For remaining swap candidates
22             $\Delta TD^*_j \leftarrow \sum_{x_o \notin \{m_1, \ldots, m_k\} \setminus m_j} \Delta(x_o, m_j, x^*_j)$;   // Recompute TD
23 **return** TD, M, C;

The improvements of this strategy are, unsurprisingly, much smaller than those of FastPAM1. In early iterations we see multiple swaps being executed, but in the later iterations it is common that only few medoids change. Nevertheless, this simple modification yields another measurable performance improvement. However—in contrast to the first improvement—this no longer guarantees to yield the same result. From a theoretical point of view, both the original PAM, and FastPAM2 perform a steepest descent optimization strategy; where PAM only permits descends consisting of a single swap, whereas FastPAM2 can perform multiple swaps at once as long as they use different medoids. Therefore, both are able to find results of equivalent quality. In our experiments, FastPAM2 would often find *marginally better* results than PAM, and faster.

3.3   Faster Initialization with Linear Approximative BUILD (LAB): FastPAM

With these optimizations to SWAP, reducing the time from $O(k(n-k)^2)$ to $O((n-k)^2)$, the bottleneck of PAM becomes the BUILD phase. In our experiments with large k, PAM would spend 99% of the run time in SWAP. With above optimizations this reduces to about 15%. About 16% is the time to compute the distance matrix, and 69% of the time is spent in BUILD. The complexity of BUILD is in $O(kn^2)$, so for large k this is expected to happen. Because we made SWAP much faster, we can afford to begin with slightly worse starting conditions, even if we need more iterations of SWAP afterwards.

Algorithm 3. FastPAM LAB: Linear Approximate BUILD initialization.

```
1  (TD, m₁) ← (∞, null);
2  S ← subsample of size 10 + ⌈√n⌉ from X ;                          // Subsample
3  foreach xⱼ ∈ S do                                                 // First medoid
4  │   TDⱼ ← 0 ;
5  │   foreach xₒ ∈ S ∧ xₒ ≠ xⱼ do  TDⱼ ← TDⱼ + d(xₒ, xⱼ);
6  │   if TDⱼ < TD then (TD, m₁) ← (TDⱼ, xⱼ);      // Smallest distance sum
7  for i = 1 ... k − 1 do                                            // Other medoids
8  │   (ΔTD*, x*) ← (∞, null);
9  │   S ← subsample of size 10 + ⌈√n⌉ from X \ {m₁, ..., mᵢ} ;      // Subsample
10 │   foreach xⱼ ∈ S do
11 │   │   ΔTD ← 0 ;
12 │   │   foreach xₒ ∈ S ∧ xₒ ≠ xⱼ do
13 │   │   │   δ ← d(xₒ, xⱼ) − min₍ₒ∈m₁,...,mᵢ₎ d(xₒ, o);
14 │   │   │   if δ < 0 then ΔTD ← ΔTD + δ;
15 │   │   if ΔTD < ΔTD* then (ΔTD*, x*) ← (ΔTD, xⱼ);   // best reduction
16 │   (TD, mᵢ₊₁) ← (TD + ΔTD*, x*);
17 return TD, {m₁, ..., mₖ};
```

An elegant way of initializing k-means is k-means++ [2]. The beautiful idea of this approach is to choose seeds with the probability proportional to their squared distance to the nearest seed (the first seed is picked uniformly). If we assume there is a cluster of points and no seed nearby, the probability mass of this cluster is substantial, and we are likely to place a seed there; afterwards the probability mass of this cluster reduces. Furthermore, this initialization is (in expectation) $O(\log k)$ competitive to the optimal solution, so it will theoretically generate good starting conditions. But as seen in our experiments, this guarantee is pretty loose; and BUILD empirically produces much better starting conditions than k-means++ (we are not aware of a detailed theoretical analysis). The reason is that k-means++ picks *random* points (usually) from different clusters, but makes no effort to find good centers of the clusters (which is not that important for k-means, where the mean is in between of the data points). Therefore, with k-means++-style initialization we need around k additional swaps to pick the medoid of each cluster (and hence, k iterations of original PAM SWAP, although much fewer with FastPAM2). Because a single iteration of swap used to take as much time as BUILD, the k-means++ initialization only begins to shine if we use FastPAM1 to reduce the cost of iterating together with the FastPAM2 strategy of doing as many swaps as possible.

We experimented with k-means++, but eventually settled for a different strategy we call LAB (Linear Approximative BUILD). What we title "FastPAM" then is the combination of LAB with the optimizations of FastPAM2. As the name indicates, LAB is a linear approximation of the original PAM BUILD. In order to achieve linear runtime in n, we simply subsample the data set. Before choosing each medoid, we sample $10 + \lceil \sqrt{n} \rceil$ points from all non-medoid points. From this subsample we choose the one with the largest decrease ΔTD with respect to the current subsample only. Results were slightly better with sampling k times, and not just once; since each object has k chances of being in the sample, and if we draw a bad sample it only affects a single medoid. A pseudocode of LAB is given as Algorithm 3. Clearly, the complexity is down to $O(kn)$.

3.4   Integration: FastCLARA and FastCLARANS

Since CLARA [10] uses PAM as a subroutine, we can trivially use our improved FastPAM with CLARA. In the experiments we will denote this variant as FastCLARA.

CLARANS [13] uses a randomized search instead of considering all possible swaps. For this, it chooses a random pair of a non-medoid object and a medoid, computes whether this improves the current loss, and then greedily performs this swap. Adapting the idea from FastPAM1 to the random exploration approach of CLARANS, we pick only the non-medoid object at random, but can consider all medoids at a similar cost to looking at a single medoid. This means we can either explore k times as many edges of the graph, or we can reduce the number of samples to draw by a factor of k. In our experiments we opted for the second choice, to make the results comparable to the original CLARANS in the number of edges considered; but as the edges chosen involve the same non-medoids, we

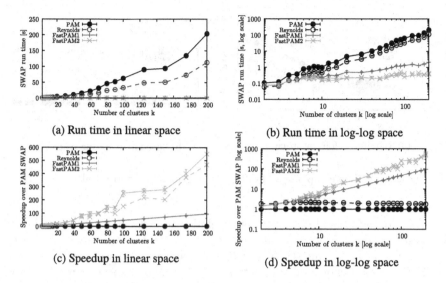

(a) Run time in linear space

(b) Run time in log-log space

(c) Speedup in linear space

(d) Speedup in log-log space

Fig. 1. Run time of PAM SWAP (SWAP only, without DAISY, without BUILD)

expect a slight loss in quality that should be easily countered by increasing the subsampling rate of non-medoids. By varying the subsampling rate, the user can control the tradeoff between computation time and exploration.

4   Experiments

Theoretical considerations show that we must expect an $O(k)$ speedup of Fast-PAM1 over the original PAM algorithm, so our experiments primarily need to verify that there is no trivial error (in contrast to much work published in recent years, the speedup is not just empirical). Nevertheless constant factors and implementation details can make a big difference [11], and we want to ensure that we do not pay big overheads for theoretical gains that would only manifest for infinite data.[1] For FastPAM2 the speedup is expected to be only a small factor due to the reduction in iterations. In contrast to FastPAM1, it does not guarantee the exact same results; therefore we also want to verify that they are of the expected equivalent quality. The worse starting conditions of LAB should not affect the final result, but will require additional iterations of SWAP. We observed increased runtimes when using k-means++ for PAM initialization, therefore it needs to be verified experimentally that LAB does not require excessive additional iterations.

We showcase results from the "one-hundred plant species leaves" data set (texture features only) from the well-known UCI repository [5], but we verified

---

[1] Clearly, our $O(k)$ fold speedup must be immediately measurable, not just asymptotically, because the constant overhead for maintaining the fixed array cache is small.

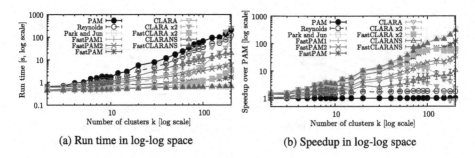

(a) Run time in log-log space   (b) Speedup in log-log space

Fig. 2. Run time comparison of different variations and derived algorithms.

the results on additional data sets (not included because of space restrictions). We chose this data set because it has 100 classes, and 1600 instances, a fairly small size that PAM can still easily handle. Naively, one would expect that $k = 100$ is a good choice on this data set, but some leaf species are likely not distinguishable by unsupervised learning. We used the ELKI open-source data mining toolkit [20] in Java to develop our version. For comparison, we also ported FastPAM2 to the R `cluster` package, which is based on the original PAM source code and written in C. Experiments were run on an Intel i7-7700 at 3.6 GHz with turbo boost disabled. We perform 25 runs, and plot the average, minimum and maximum. Both implementations and all data sets show similar behavior, so we are confident that the results are not just due to implementation differences [11].

4.1   Run Time Speedup

In Fig. 1, we vary k from 2 to 200, and plot the run time of the PAM SWAP phase *only* (the cost of computing the distance matrix and the BUILD phase is not included), using the original PAM, the Reynolds version, and the proposed improvements. Figure 1a shows the run time in linear space, to visualize the drastic run time differences observed. Reynolds' was quite consistently two times faster than the original PAM; but our proposed methods were faster by a factor that grows approximately linearly with the number of clusters k. In log-log-space, Fig. 1b, we can differentiate the three variants studied.

In Fig. 1c we plot the speedup over PAM. Reynolds' SWAP clearly was about twice as fast as the original PAM. The FastPAM1 improvement gives an empirical speedup factor of about $\frac{1}{2}k$, while the second improvement contributed an additional speedup of about 2–2.5× by reducing the number of iterations. Because of the multiplicative effect of these savings, the linear plot in Fig. 1c gives the false impression that this second contribution yields the larger benefit. The logspace plot in Fig. 1d more accurately reflects the contribution of the two factors, resulting in a speedup of over 250 times at $k = 150$; while at $k = 2$ and $k = 3$ the speedup was just 1.4× resp. 1.75×, and less than our implementation of Reynolds (in R, our method is as fast as Reynolds for $k = 2$). In the most extreme case tested, a speedup of about 1000× at $k = 200$ is measured – but because the

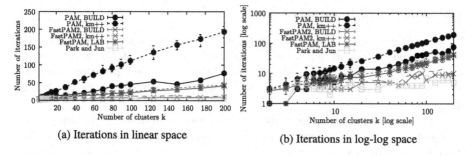

(a) Iterations in linear space

(b) Iterations in log-log space

Fig. 3. Number of iterations for PAM vs. FastPAM2 and BUILD vs. LAB initialization

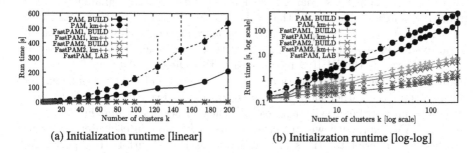

(a) Initialization runtime [linear]

(b) Initialization runtime [log-log]

Fig. 4. Runtime impact of k-means++ and LAB initialization

speedup depends on $O(k)$, the exact values are meaningless, furthermore, we excluded the distance matrix computation and initialization in this experiment.

In Fig. 2, we study the run time of approximations to PAM (including the distance matrix computation and initialization time now). We only present the log-log space plots, because of the extreme differences. The run time of CLARA, as k increases, approaches the run time of PAM. This is expected, because the subsample size for CLARA is chosen as $40 + 2k$, and necessary because the subsample size needs to be sufficiently larger than k. For CLARA x2 we also evaluate doubling this value to $80 + 4k$, and we also double the number of restarts from 5 to 10. CLARA x2 is thus expected to take 8 times longer, but should give better results. FastCLARA is CLARA using our FastPAM approach, and performs much better, but for large k also eventually becomes slower than Fast-PAM. The run time of CLARANS on this data set (see later for CLARANS problems) is in between the original PAM and CLARA, and with our optimizations FastCLARANS becomes the fastest method tested (at similar quality to CLARANS, and with the same problems). Park and Jun's [15] approach is similarly fast to FastCLARANS for large k, but its quality is quite poor, as we will see and discuss in Sect. 4.3.

4.2   Number of Iterations

We are not aware of theoretical results on the number of iterations needed for PAM. Based on results for k-means, we must assume that the worst case is superpolynomial like k-means [1], albeit in practice a "few" iterations are usually enough. Because of this, we are also interested in studying the number of iterations.

Figure 3 shows the number of iterations needed with different methods, both in linear space and log space. In line with previous empirical results, only few iterations are necessary. Because PAM only performs the best swap in each iteration, a linear dependency on k is to be assumed; interestingly enough we usually observed much less than k iterations, so many medoids remain unchanged from their initial values (note that this may be due to the rather small data set size, too). The k-means++ initialization required roughly 2–4× as many iterations for PAM; with the original algorithm where each iteration would cost about as much as the BUILD initialization, this choice is detrimental even for small k. With the improvements of this paper, these additional iterations are cheaper than the rather slow BUILD initialization by a factor of $O(k)$ now, hence we can now begin with a worse but cheaper starting point. Furthermore, the FastPAM2 approach which performs up to k swaps in each iteration does reduce the number of iterations substantially. FastPAM2 with BUILD performed the second-lowest numbers of iterations. Our proposed LAB initialization of FastPAM saves a few extra iterations compared to the k-means++ strategy, at better initial quality, and hence is measurably faster in the end. Park and Jun [15] at first seems to perform very well in this figure, with slightly fewer iterations than FastPAM2 with BUILD. Unfortunately, this is because the "k-means style" algorithm misses many improvements to the clustering, and hence produces much worse results as we will observe next.

In Fig. 4 we revisit the runtime experiment, and focus on initialization. As we can see, the increased number of iterations hurts runtime with the original PAM algorithm as well as its Reynolds variant substantially (the reasons for this are explained in Sect. 3.3); for FastPAM1, the use of k-means++ only comes at a small performance penalty (while it still needs as many iterations as the original PAM, these have become $O(k)$ times faster, and the initialization cost begins to matter much more), and with FastPAM2's ability to perform multiple swaps per iteration, a linear-time initialization such as the proposed LAB clearly becomes the preferred initialization method, in particular for large k.

4.3   Quality

Any algorithmic change and optimization comes at the risk of breaking some things, or negatively affecting numerics (see, e.g., [17] on how common numerical issues are even with basic statistics such as variance in SQL databases). In order to check for such issues, we made sure that our implementations pass the same unit tests as the other algorithms in both ELKI and R. We do not expect numerical problems, and Reynolds' variant and FastPAM1 are supposed

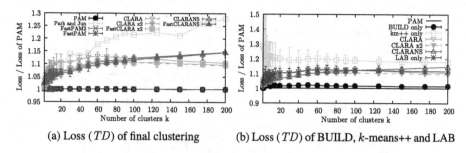

(a) Loss (TD) of final clustering   (b) Loss (TD) of BUILD, k-means++ and LAB

Fig. 5. Loss (TD) compared to PAM

to give the same result (and do so in the experiments, so we exclude them from the plot). The FastPAM2 algorithm is greedy in performing swaps, and may therefore converge to a different solution, but of the same quality.

In Fig. 5a we visualize the loss, i.e., the objective TD, of different approximations compared to the solution found by the original PAM approach (which is not necessarily the global minimum). For large k, the solution found by the approach of Park and Jun [15] is over 25% worse here, for the reasons discussed before. Our strategy FastPAM2 gives results comparable to PAM as expected (sometimes slightly better, sometimes slightly worse). The cheaper LAB initialization (full FastPAM) does not cause a noticeable loss in quality either, but further improves the total run time. CLARA (which only uses a subsample of the data) finds considerably worse results. By doubling the subsample size to $80 + 4k$ and doing twice as many restarts (CLARA x2) the results only improve slightly for large k (but much more for small k). CLARA x2 is until about $k = 70$ as good as CLARANS here, but faster; for larger k it becomes even better than CLARANS, but also slower. FastCLARA has the same quality as CLARA x2 (we use the x2 parameters, too), but it was much faster. FastCLARANS is slightly better than CLARANS, and was considerably faster. All the CLARANS results degrade with increasing k, so it may become necessary to increase the subsample size there, which will increase the run time (it is up to the user to choose his preferences, quality or runtime). In conclusion, all our "Fast" approaches perform as well as their older counterparts, but are $O(k)$ times faster.

In Fig. 5b, we evaluate the quality of LAB, k-means++, and BUILD initialization compared to the converged PAM result. As seen in the previous experiments, all three initializations will yield similar results after PAM, but we can compare the quality of the initial medoids to the full PAM result. As we can see, the BUILD approach produces the best initial results (and as noted by [9], the BUILD result may be usable without further refinement). While k-means++ offers some theoretical advantages (c.f., Sect. 3.3), the initial result is quite bad as this strategy only attempts to pick a random point from each cluster, and not the medoids. Our proposed LAB initialization is in between k-means++ and BUILD, and by itself performs similar to CLARA. As it only considers a subset of the data, its medoids will be worse than BUILD; but because it chooses the

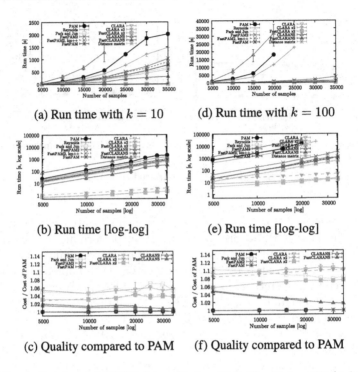

(a) Run time with $k = 10$

(d) Run time with $k = 100$

(b) Run time [log-log]

(e) Run time [log-log]

(c) Quality compared to PAM

(f) Quality compared to PAM

Fig. 6. Results on MNIST data with $k = 10$ (left) and $k = 100$ (right)

best medoid of the sample it performs better than k-means++. As it reduces the runtime for $O(n^2 k)$ to $O(nk)$ it is the preferred choice for FastPAM nevertheless.

4.4 Scalability Experiments

Just as PAM, our method also requires the entire distance matrix to be pre-computed. This will require $O(n^2)$ time and memory, making the method as-is unsuitable for big data (for real big data problems, it will however often be enough to cluster a subsample that fits into memory). Our improvements focus on reducing the dependency on k, but we nevertheless experimented with scal-ability in n, too (and we already included FastCLARA and FastCLARANS in the previous experiments). The behavior of the PAM variants is as expected $O(n^2)$, but we see nevertheless quite big differences between PAM, FastPAM, and sampling-based approaches. In this experiment, we use the well-known MNIST data set from the UCI repository [5], which has 784 variables (each corresponding to a pixel in a 28×28 grid) and 60000 instances. We used the first $n = 5000, 10000, \ldots, 35000$ instances with a time limit of 6 h and compare $k = 10$ and $k = 100$. The high number of variables makes this data set expensive for CLARANS.

The problem of quadratic runtime is best seen in the linear scale plots Fig. 6a and d. As a reference, we give the time needed for computing the distance matrix as dotted line, which is also quadratic. Except for CLARA and FastCLARANS, the runtime is dominated by computing the distance matrix (and hence CLARA, which only uses a constant-size sample, shines for large n). The original CLARANS suffers from excessive distance re-computations. The authors assumed that distances are cheap to compute, and noted that it may be necessary to cache the distances. FastCLARANS reduces the number of distance computations of CLARANS by a factor of $O(k)$, and is still cheaper than the full distance matrix here. For more expensive distances such as dynamic time warping, FastPAM will outperform FastCLARANS, and it will almost always give better results. For $k = 10$, only CLARANS, PAM and Reynolds' variant are problematic at this data size, but at $k = 100$ the benefits of our improvements become very noticeable. The CLARA methods are squeezed to the axis in the linear plot, and hence we also provide log-log plots in Fig. 6b and e. For $k = 10$, the lines of CLARA and FastCLARA x2 almost coincide by chance (note that FastCLARA x2 produces a result comparable to the slower CLARA x2 method; expected to be 8 times slower), but at $k = 100$ it is faster than CLARA demonstrating that our improvements also accelerate CLARA.

While the scalability in n is quadratic, we observe that if you can afford to compute the pairwise distance matrix, then you will *now* also be able to run FastPAM. For $k = 10$, the additional runtime of FastPAM was about 30% the runtime of computing the distance matrix computation, and at $k = 100$ FastPAM took about as much time as the distance matrix. Hence, if you can compute the distance matrix, you can also run FastPAM for reasonable values of $k \ll n$, and the main scalability problem is the memory consumption of the distance matrix.

If computing the distance matrix is prohibitive, it may still be possible to use FastCLARA, which is $O(k)$ times faster than CLARA, and will scale linearly in n. But as seen in Fig. 6c and f, CLARA will usually give worse results (about 10% in our experiments). For many users this difference will be acceptable, as a clustering result is never "perfect". For large data sets, FastCLARANS will usually give better results, unless the sample size of CLARA is increased considerably. But on the other hand, FastCLARANS is only advisable for inexpensive distance functions such as (low-dimensional) Euclidean distance, and requires a non-trivial distance cache otherwise.

5   Conclusions

In this article we proposed a modification of the popular PAM algorithm that typically yields an $O(k)$ fold speedup, by clever caching of partial results in order to avoid recomputation. This caching was enabled by changing the nesting order of the loops in the algorithm, showing once more how much seemingly minor looking implementation details can matter [11]. As a second improvement, we propose to find the best swap for each medoid, and execute as many as possible in each iteration, which reduces the number of iterations needed for convergence without loss of quality.

The major speedups obtained enable the use of this classic clustering method on much larger data, in particular with large k. With the faster refinement procedure, it now pays off to use cheaper initialization methods with PAM. For this, we propose LAB initialization, a linear-time approximation of the original PAM BUILD algorithm.

Methods based on PAM, such as CLARA, CLARANS, and the many parallel and distributed variants of these algorithms for big data, all benefit from this improvement, as they either use PAM as a subroutine (CLARA), or employ a similar swapping method (CLARANS) that can be modified accordingly as seen in Sect. 3.4.

The proposed methods are included in the open-source framework ELKI 0.7.5 [20], and FastPAM2 (but not yet LAB, FastCLARA, nor FastCLARANS) is included in the R cluster package 2.0.9, to make it easy for others to benefit from these improvements. With the availability in two major clustering tools, we hope that many users will find using PAM possible on much larger data sets with higher k than before.

References

1. Arthur, D., Vassilvitskii, S.: How slow is the k-means method? In: ACM Symposium on Computational Geometry (2006). https://doi.org/10.1145/1137856.1137880
2. Arthur, D., Vassilvitskii, S.: k-means++: the advantages of careful seeding. In: ACM-SIAM SODA (2007)
3. Bock, H.: Clustering methods: a history of k-means algorithms. In: Brito, P., Cucumel, G., Bertrand, P., de Carvalho, F. (eds.) Selected Contributions in Data Analysis and Classification. Springer, Heidelberg (2007). https://doi.org/10.1007/978-3-540-73560-1_15
4. Bradley, P.S., Mangasarian, O.L., Street, W.N.: Clustering via concave minimization. In: NIPS (1996)
5. Dheeru, D., Karra Taniskidou, E.: UCI machine learning repository (2017). http://archive.ics.uci.edu/ml
6. Ester, M., Kriegel, H., Sander, J., Xu, X.: A density-based algorithm for discovering clusters in large spatial databases with noise. In: KDD (1996)
7. Estivill-Castro, V.: Why so many clustering algorithms: a position paper. SIGKDD Explor. 4(1), 65–75 (2002). https://doi.org/10.1145/568574.568575
8. Estivill-Castro, V., Houle, M.E.: Robust distance-based clustering with applications to spatial data mining. Algorithmica 30(2), 216–242 (2001). https://doi.org/10.1007/s00453-001-0010-1

9. Kaufman, L., Rousseeuw, P.J.: Clustering by means of medoids. In: Dodge, Y. (ed.) Statistical Data Analysis Based on the L_1 Norm and Related Methods (1987). ISBN 0444702733

10. Kaufman, L., Rousseeuw, P.J.: Clustering large data sets. In: Pattern Recognition in Practice. Elsevier (1986). https://doi.org/10.1016/b978-0-444-87877-9.50039-x

11. Kriegel, H., Schubert, E., Zimek, A.: The (black) art of runtime evaluation: are we comparing algorithms or implementations? Knowl. Inf. Syst. **52**(2), 341–378 (2017). https://doi.org/10.1007/s10115-016-1004-2

12. Lucasius, C., Dane, A., Kateman, G.: On k-medoid clustering of large data sets with the aid of a genetic algorithm. Anal. Chim. Acta **282**(3), 647–669 (1993). https://doi.org/10.1016/0003-2670(93)80130-D

13. Ng, R.T., Han, J.: CLARANS: a method for clustering objects for spatial data mining. IEEE TKDE **14**(5), 1003–1016 (2002). https://doi.org/10.1109/TKDE. 2002.1033770

14. Overton, M.L.: A quadratically convergent method for minimizing a sum of euclidean norms. Math. Program. **27**(1), 34–63 (1983). https://doi.org/10.1007/ BF02591963

15. Park, H., Jun, C.: A simple and fast algorithm for k-medoids clustering. Expert Syst. Appl. **36**(2), 3336–3341 (2009). https://doi.org/10.1016/j.eswa.2008.01.039

16. Reynolds, A.P., Richards, G., de la Iglesia, B., Rayward-Smith, V.J.: Clustering rules: a comparison of partitioning and hierarchical clustering algorithms. J. Math. Model. Algorithms **5**(4), 475–504 (2006). https://doi.org/10.1007/s10852-005-9022-1

17. Schubert, E., Gertz, M.: Numerically stable parallel computation of (co-)variance. In: SSDBM (2018). https://doi.org/10.1145/3221269.3223036

18. Schubert, E., Hess, S., Morik, K.: The relationship of DBSCAN to matrix factorization and spectral clustering. In: LWDA. CEUR Workshop Proceedings, vol. 2191 (2018)

19. Schubert, E., Sander, J., Ester, M., Kriegel, H., Xu, X.: DBSCAN revisited, revisited: why and how you should (still) use DBSCAN. ACM Trans. Database Syst. **42**(3), 19 (2017). https://doi.org/10.1145/3068335

20. Schubert, E., Zimek, A.: ELKI: a large open-source library for data analysis - ELKI release 0.7.5 "Heidelberg". arXiv:1902.03616 (2019)

MORe++: k-Means Based Outlier Removal on High-Dimensional Data

Anna Beer$^{(\boxtimes)}$, Jennifer Lauterbach, and Thomas Seidl

Ludwig-Maximilians-Universität München, Munich, Germany
{beer,seidl}@dbs.ifi.lmu.de, j.lauterbach@campus.lmu.de

Abstract. MORe++ is a k-Means based **O**utlier **Re**moval method working on high dimensional data. It is simple, efficient and scalable. The core idea is to find local outliers by examining the points of different k-Means clusters separately. Like that, one-dimensional projections of the data become meaningful and allow to find one-dimensional outliers easily, which else would be hidden by points of other clusters. MORe++ does not need any additional input parameters than the number of clusters k used for k-Means, and delivers an intuitively accessible degree of outlierness. In extensive experiments it performed well compared to k-Means-- and ORC.

Keywords: Outlier detection · High-dimensional · Histogram-based · K-means

1 Introduction

As outlier detection in general delivers valuable results for fraud detection, medical problems, or finding errors in data, most techniques do not regard the plethora of attributes which is gathered for each data point in modern applications. An outlier, which is often defined as "an observation which deviates so much from the other observations as to arouse suspicions that it was generated by a different mechanism" [14], is more difficult to find in high-dimensional data than in low-dimensional, since the mechanisms generating data are difficult to identify in high-dimensional data due to the curse of dimensionality. Thus, most classic outlier detection algorithms are not applicable to high-dimensional data. Density based algorithms for example, are not meaningful in high-dimensional data, which usually is per se sparse. Also angular based outlier factors like, e.g., ABOD [18], are not interpretable anymore for high-dimensional data. Moreover, most of those algorithms do not scale with the number of dimensions.

Thus we introduce MORe++ (k-Means-based Outlier Removal using k-Means++), a fast method to score outliers in high-dimensional data. We achieve scalability w.r.t. the number of dimensions and retain explainability of the scores by regarding each dimension separately. In contrast to other methods we can even find clusters or outliers overlapping in some dimensions, since we do not

© Springer Nature Switzerland AG 2019
G. Amato et al. (Eds.): SISAP 2019, LNCS 11807, pp. 188–202, 2019.
https://doi.org/10.1007/978-3-030-32047-8_17

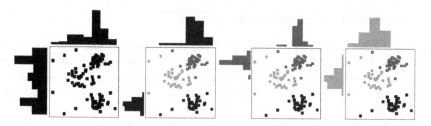

Fig. 1. Histogram of complete dataset in black, in contrast to histograms of points belonging to the same centroid according to k-means in green, yellow and purple. (Color figure online)

regard all points at once, but only those which are in one cluster according to k-Means. Using histograms, we accelerate our algorithm even further in regards to the number of points. Figure 1 shows how overlapping in a dimension prevents finding outliers if regarding the whole dataset at once, which is why we regard only a part of the datapoints at once.

Summarizing, our main contributions are:

1. We introduce a meaningful score for outliers in high-dimensional data
2. Our method is fast and scales linearly with the number of dimensions and points
3. It is based on k-Means and compatible to a lot of k-Means extensions
4. It is easy to implement
5. It is easily parallelizable and suitable for high-dimensional data, since it does not rely on distance measures operating on the full-dimensional space.

The remainder is structured as follows: in Sect. 2 we give an overview over other k-Means extensions and outlier detection methods using k-Means or a histogram-based approach. We also investigate diverse approaches of histogram segmentation. The complete algorithm is explained in detail in Sect. 3. In Sect. 4 we examine our algorithm regarding a plethora of aspects in overall 40 synthetic as well as real data experiments. Section 5 concludes this paper giving a short summary and prospect to promising future work.

2 Related Work

We first give the foundations looking at k-Means Clustering and existing recent extensions in Sect. 2.1. As there are already several methods combining k-Means and outlier removal, we give an overview over those in Sect. 2.2 and discuss the advantages of MORe++ in contrast to them. In Sect. 2.3 we look at histograms and outlier detection algorithms using them. We note that regarding only the projections of the complete dataset at once cannot lead to an effective outlier detection.

2.1 K-Means Clustering and Extensions

k-Means [19,20] is one of the most famous clustering algorithms and is still frequently used for diverse tasks. Given the number of clusters k, centers are randomly initialized in the original algorithm. All points are assigned to their closest center and the cluster centers are recomputed. Those two steps are repeated until no point changes its cluster membership any more and the algorithm converges against a local minimum of the mean distance from points to their cluster centers.

There exist several improvements of k-Means. For example, k-Means++ [2] optimizes the initial cluster centers by regarding the shortest distance to already chosen cluster centers. We will use this extension for our algorithm MORe++, as it usually improves the quality of clustering and reduces the variance of results. Another improvement is k-Median [5], which uses the median instead of the mean when calculating the new cluster centers to minimize the negative impact of outliers. Nevertheless, this comes at the cost of an increased runtime. On the other hand, kmeans|| [3] reduces runtime by parallelizing k-Means++. Instead of sampling single points for the initialization like k-Means++, $O(k)$ points are sampled $O(\log n)$ times. Also high-dimensional data can be clustered better with diverse variants of k-Means developed for subspace clustering, like NR-kmeans [22] or Sub-kmeans [21]. Where Sub-kmeans finds a "clustered" space containing all structural information and a noise space, NR-kMeans looks for an optimal arbitrarily oriented subspace for each partition. Those improvements, of which we apply only k-Means++ in this paper, are compatible to MORe++ and will be regarded in future work.

2.2 Outlier Detection and K-Means

There are several algorithms combining k-Means and outlier removal, of which we introduce the most common ones in the following. In Sect. 4 we will compare MORe++ to the first both introduced, k-means-- [7] and ORC [13].

k-means-- [7] combines outlier removal and k-Means by alternately removing outliers and performing k-Means iterations. In every step l points which are farthest away from their center are removed from the dataset for the next calculation of cluster centers, where l is given by the user. Even for $k = 1$, its running time is $O(n^{d^3})$, which is infeasible for high-dimensional data.

ORC (Outlier Removal Clustering) [13] assigns an outlyingness factor o_i to every point after a complete pass of k-Means. Points with o_i higher than a user given threshold T are removed from the set of points, and k-Means is performed again. o_i is, similar to k-means--, based on the distance to the nearest cluster center, and normalized by division by the highest distance between a point and its center. The algorithm is quite sensitive to the choice of T, and our experiments will show that ORC cannot handle high dimensional data well.

NEO-K-Means [28] considers outliers and overlapping clusters, for which it requires two parameters α and β. Using those parameters, it strives for a "trade-off between a clustering quality measure, overlap among the clusters, and non-exhaustiveness (the number of outliers not assigned to any group)" [28] in every k-Means step. Reaching this trade-off requires iterations until convergence, there seems to be neither an upper bound for the running time, nor do the authors perform a complexity analysis. The main criterion is again the distance between points and their closest center, which becomes more and more useless with increasing number of dimensions.

KMOR [10] uses an additional cluster for all outliers, needing two parameters to control the number of outliers. A point is considered an outlier if the distance to *all* cluster centers is at least $\gamma \times d_{avg}$, which forbids finding local outliers.

Other methods combining k-Means and outlier detection are, e.g., ODC (Outlier Detection and Clustering) [1], where outliers are points having a distance to their cluster centers larger than p times the average distance. CBOD [17] and [16] are two-stage algorithms, where [16] additionally creates a minimum spanning tree on which they work on. [29] is a three stage algorithm first finding local outliers, then global outliers, and lastly combining clusters with similar densities and overlapping clusters.

All mentioned algorithms have in common that they rely on distance measures in the full-dimensional space, usually the Euclidean distance. As with increasing number of dimensions, all distances become similar due to the curse of dimensionality [4], the results get distorted for high-dimensional data. In contrast, MORe++ does not need any distance measure working on high-dimensional space. Additionally, due to the separate consideration of each dimension, it is already faster than these methods plus it is easy parallelizable.

2.3   Outlier Detection with Histograms

As using histograms is an established way to simplify data, several construction possibilities regarding the bin-width and bin-quantity exist: One of the most common possibilities, Sturges' rule [15], tends to oversmooth histograms and does not work well with large datasets and not normally distributed data. Other common rules are Scott's rule [26] and Freedman and Diaconis's rule [9], which are both better for larger samples. MORe++ uses Scott's rule which suggests h as the number of bins: $h = \frac{3.45\hat{\sigma}}{n^{1/3}}$, where $\hat{\sigma}$ is the sample standard deviation.

There are methods using histograms to find outliers: HBOS [12] constructs a histogram for each dimension and calculates an anomaly score for each data instance using the inverse estimated densities and supposing feature independence. [11] finds sparse regions in the dataset using histograms and a nearest neighbour approach. Based on those regions, local outlier candidates are identified, which can be removed from the set of outliers in a later optional reconsideration phase.

Looking at higher dimensional datasets and subspace clustering, objects may belong to different clusters in different subspaces, thus they could be outliers in some subspaces, but not in others. OutRank [27] addresses this issue by introducing a "degree of outlierness", the outlier rank. With that, points which are only in a subset of attributes anomalies, can also be detected as outliers [23]. In contrast, many outlier detection algorithms, like for example, LOCI (local correlation integral) [24], look for outliers in the full dimensional space, or even micro-clusters. LOF [6], the local outlier factor, is another approach returning the degree of outlierness. It regards the isolation of a point with respect to the surrounding neighborhood, and was even extended for high-dimensional data [18,30].

3   MORe++

In the following we describe and analyze the outlier detection algorithm MORe++ in detail: Sect. 3.1 gives an overview, Sect. 3.2 explains how we find one-dimensional outliers in histograms, and Sect. 3.3 gives a complexity analysis.

3.1   Outline of MORe++

Based on the idea already shown in Fig. 1, MORe++ regards the points of every k-Means cluster separately. Like that, one-dimensional projections already enable the detection of outliers. Section 3.2 explains how to get one-dimensional outliers based on the according histogram. The higher the number of dimensions in which a point is considered an outlier, the higher is its outlier score. Thus, MORe++ finds a degree of outlierness. Algorithm 1 describes our approach in detail: on the basis of the clustering returned by k-Means++, we build a histogram for every dimension for every cluster. The method *calculate1dOutliers* returns one-dimensional outliers for each dimension and each cluster given the according histogram, as explained in Sect. 3.2. The outlier score is the relation between the number of dimensions in which the point is considered a (one-dimensional) outlier and the total number of dimensions. Users can now either use this degree of outlierness, or give a threshold *ost* (outlier score threshold), so that points with an outlier score higher than *ost* are outliers in the full dimensional space. This approach delivers several advantages:

1. Our distance measure does not get skewed with increasing number of dimensions, as we regard every dimension separately
2. We can easily parallelize the calculation of outliers as we regard all clusters and also all dimensions independently. Thus, MORe++ is suitable for many points as well as for high- dimensional data.
3. Users do not have to know the number of outliers beforehand
4. A degree of outlierness gives more information than a hard classification
5. The threshold users *can* give is quite intuitive, as it is simply the minimal percentage of dimensions in which a point should be a one-dimensional outlier. As experiments will show, MORe++ is quite robust w.r.t. *ost* (we use the same value $ost = 0.2$ for 35 out of 40 experiments in total), thus a hard classification using a fixed *ost* is also a promising idea for future work
6. MORe++ is very fast with only $O(nd)$, where, e.g., k-means-- is exponential.

Algorithm 1. Pseudo-Code of MORe++

Data: Data X, number of clusters k
Result: Clustering labels, outlierScore for data points
1 **foreach** $x \in X$ **do**
2 | $numberOfOutlierDims[x] \leftarrow 0$;
3 **end**
4 **foreach** $c \in clusters$ **do**
5 | **foreach** $d \in range(dimensions)$ **do**
6 | | Build histogram;
7 | | $1dOutliers \leftarrow calculate1dOutliers(histogram)$;
8 | | **foreach** $1dOutlier \in 1dOutliers$ **do**
9 | | | $numberOfOutlierDims[1dOutlier] + +$;
10 | | **end**
11 | **end**
12 **end**
13 **foreach** $x \in X$ **do**
14 | $outlierScore \leftarrow numberOfOutlierDims[x]/dimensions$;
15 **end**

3.2 Detecting One-Dimensional Outliers in Histograms

To detect one-dimensional outliers in histograms, it is important that we only look at points assigned to one cluster by k-Means. Else, points of other clusters would cover outliers in the one-dimensional projections, as Fig. 1 already showed: see for example the outlier in the bottom middle, which is later in the first, purple cluster. Using histograms of the complete data, it is covered in both dimensions by the purple resp. the yellow cluster. Looking at the histogram of only the points assigned to the first (purple) cluster for dimension 1 (horizontal), it can be detected quite easily using the following approach:

If there are empty bins in the histogram, as shown in Fig. 2 on the left, then we partition the data along these empty bins. If there are no empty bins, we split the data where the height of the bins changes most from one bin to the next, relatively to the higher bin of both, as can be seen in Fig. 2 on the right. If there are several changes $s_0, ..., s_j$ which are (relatively) equally high, then we perform the split in the middle at $s_{\lfloor j/2 \rfloor}$. After the dataset is partitioned, all points which are not in the partition containing the majority of the points are marked as outliers for this dimension.

3.3 Complexity Analysis

For n points of dimensionality d the complexity of k-Means itself is $O(nkdi)$ with i the number of iterations until convergence and k the number of clusters. We build a histogram for every cluster and every dimension, which sums up to $O(kdb)$ for histograms with $b < n$ bins. Using Scott's rule for the construction of histograms (see Sect. 2.3), $b \in O(n^{-\frac{1}{3}})$. One-dimensional outliers are calculated

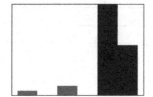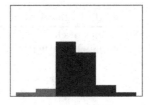

Fig. 2. Outliers (marked red) are detected using either empty bins or the highest relative difference between two adjacent bins. (Color figure online)

in $O(b+n) \subseteq O(n)$ and the calculation of all outlier scores can be done in $O(nd)$. Thus, MORe++ lies with $O(nd + kdn^{-\frac{1}{3}}) \subseteq O(nd)$ in a smaller complexity class than k-Means itself and is only linear in the number of dimensions as well as in the number of points. Furthermore, it is easily parallelizable.

In contrast, the running time of k-means– is with $O(n^{d3})$ much larger. ORC, which delivers clearly worse results than MORe++, needs to run j iterations of k-Means alternately with determining the outlyingness factor, which is in $O(nd)$, resulting in a total runtime of $O(j * (nkdi + nd)) \subset O(j * nkdi)$.

4   Experiments

We performed several experiments regarding the quality of outlier detection compared to ORC and k-Means-- (see Sect. 2), based on the ROC AUC (Area Under the Receiver Operating Characteristic Curve) value [8][1] and F1-measure; due to the lack of space and a high similarity of the results we only show the former here. All synthetic datasets were constructed using cluster centers drawn from a uniform distribution function and generating Gaussian distributed clusters around them. Outliers were added following a uniform distribution function.

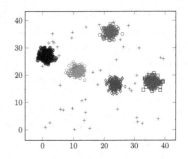

Fig. 3. 2d-projection of the base case experiment

In Sect. 4.1 we examine the influence of the following aspects onto the results of MORe++: size of dataset n, number of dimensions dim, percentage of outliers out, variance of clusters var, number of clusters k, and percentage of additional noise dimensions dim_n. For that, we created a base-case shown in Fig. 3 from which we kept all parameters but one at a time to investigate MORe++'s behaviour regarding that one aspect. In Sect. 4.2 we regard the behaviour of the algorithms on some real world datasets. They show, that even though MORe++ and k-Means-- deliver similar results

---

[1] Reminder: ROC AUC ranges from 0 to 1, where a perfect outlier prediction is 1. It regards the true positive rate vs. false positive rate. If ROC AUC is 0.5, the model has no class separation capacity.

in most of the previous test series, MORe++ clearly outperforms k-Means-- for real world data.

4.1   Influence of Various Aspects

To evaluate the behaviour of MORe++ regarding several aspects, we created a base case experiment as explained above with the following parameters: size of dataset $n = 1000$, number of dimensions $dim = 5$, percentage of outliers $out = 5\%$, variance of clusters $var = 1.2$, number of clusters $k = 5$, and percentage of additional noise dimensions $dim_n = 0$. Figure 3 shows two arbitrary dimensions of the base-case, where the ground truth outliers are marked by red crosses and clusters by colored shapes other than crosses. In each test series we changed exactly one parameter of that base case and compared the results to those of ORC and k-means--, where the base case is always marked with box brackets in the x-axis. To improve comparability, we also kept the (initially randomly chosen) cluster centers of the generated clusters the same, where possible.

As k-means-- is non-deterministic, we took the average of 100 executions. MORe++ and ORC are deterministic due to the use of cluster centers as in k-Means++ [2]. For each test series and each algorithm, we chose the parameter resulting in the best ROC AUC value after testing values from 0 to 1 in steps of 0.1 for the parameter ost in MORe++ and T in ORC, which resulted for both parameters and most of the test series in a value of 0.2, otherwise the best parameter settings are given in the according experiments. For k-Means-- the parameter l was taken from the ground truth (i.e. $l = 50$ for all experiments but the ones were the percentage of outliers or the size of the dataset was changed).

Experiments Regarding Number of Points. To examine MORe++'s behaviour with respect to the size of the dataset, we tested the base case with different values for the number of points $n = 500, 1000, 2500, 5000, 10000, 100000$. The results can be seen in Fig. 4 and show that MORe is better or comparable to ORC and k-Means-- in most cases. For 10000 points and only 5 dimensions, relatively many outliers are overlapping with the cluster itself, which explains the slight decline of results for a very high ratio between dimensions and points.

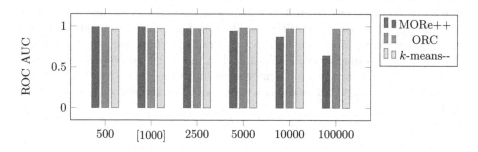

Fig. 4. ROC AUC Score for increasing number of points n

As MORe++ was developed for high-dimensional data, and especially for a high dimension to data ratio, this slight decline is predictable as well as manageable.

Experiments Regarding Number of Dimensions. We increased the number of dimensions up to 3000, which is a ratio of dimensions to data points of 3. For a lot of use cases, like data mining of textual data or image processing, the number of dimensions usually exceeds the number of data points. That often constitutes a problem for outlier detection algorithms as ORC: for a higher dimensionality than 30, the results of ORC[2], like it can be seen in Fig. 5 worsen a lot and from 1000 dimensions on, a constant ROC AUC of 0.6 is reached, which is only slightly better than guessing (which would be a ROC AUC of 0.5). k-Means-- on the other hand performs effectively as good as MORe++, and for both of them there is no decrease of quality subject to the dimensionality; they both perfom almost perfectly in high-dimensional space. For $dim \geq 5$, the ROC AUC for MORe++ is always at least 0.99, for k-Means-- the same holds for $dim \geq 30$. Note, that all clusters in this test series are in the full dimensional space, thus very similar results of MORe++ and k-Means-- were expectable.

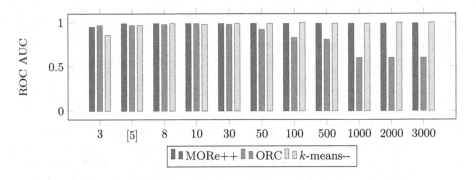

Fig. 5. ROC AUC Score for increasing number of dimensions dim

Experiments Regarding Percentage of Outliers. As Fig. 6 shows, MORe++ becomes less accurate for increasing number of outliers as well as ORC. A high percentage of outliers is difficult to handle well using our approach of finding one-dimensional outliers in histograms, as high amounts of outliers smooth the one-dimensional histograms. But those high percentages of outliers constitute rather noise than some interesting, outlying points, which MORe++ aims to find, thus, this is more a question of where "outlierness" ends and "noise" starts.

---

[2] Best values for T: 0.8 for $dim = \{50, 100\}$, 0.9 for $dim = \{100, 500\}$, else $T = 0.2$.

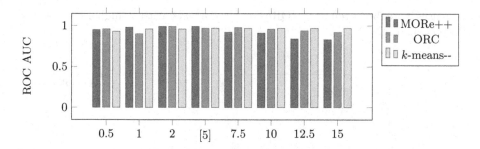

Fig. 6. ROC AUC Score for different outlier percentages

Experiments Regarding Variance of Clusters. For different variances $var = 0.7, 1.0, 1.2, 1.5, 2.0$ of the clusters MORe++ reached almost constant results, where ORC and k-Means-- get noteably worse with increasing variance, as Fig. 7 shows. That is, because with increasing variance the (full-dimensional) distances from points to their centers increase, too, thus they are more similar to the distances from outliers to cluster centers. As MORe++ does not rely on distinguishing high-dimensional distances of non-outliers and outliers to cluster centers, it is able to cope very well with diverse variances, in contrast to the comparative methods.

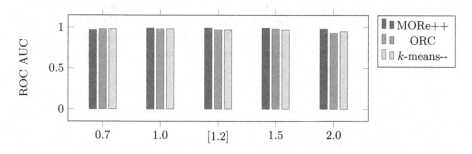

Fig. 7. ROC AUC Score for different cluster variances var

Experiments Regarding Number of Clusters. With increasing number of clusters, there are more overlapping clusters, thus k-Means and also all of the tested outlier detection algorithms[3] become worse, as can be seen in Fig. 8. But in contrast to k-means--, MORe++ gains an advantage, as the dataset is divided into more subsets (clusters found by k-Means) on which the outlier detection is performed separately. Thus, the outliers become more obvious in those smaller subsets, which counteracts the before mentioned negative effects. That results in a relative improvement to k-means--, although the quality of outlier detection

---

[3] Best values for T: 0.4 for $k = \{8, 10\}$, 0.6 for $k = \{25, 50\}$, else $T = 0.2$.

decreases notably even for MORe++ for datasets with a higher number of clusters than $k \geq 25$.

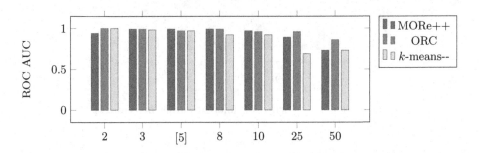

Fig. 8. ROC AUC Score for different numbers of clusters k

Experiments Regarding Noise Dimensions. Subspace clustering is based on the assumption, that with increasing number of dimensions more and more dimensions become "noise dimensions". According to that, we added noisy dimensions to our dataset, for which the results can be seen in Fig. 9[4]. With an increase of noise dimensions, distance measures in the full-dimensional space become more and more meaningless according to the curse of dimensionality, and the noise of some dimensions distorts the outlierness in other dimensions for algorithms using distances on the full-dimensional space. As MORe++ regards every dimension separately and counts the number of dimensions in which a point is an outlier, it is clearly more robust to additional noise dimension than its competitors.

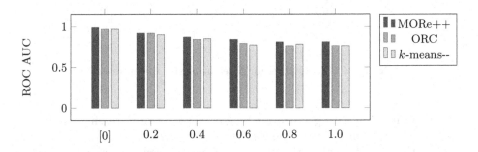

Fig. 9. ROC AUC Score for different number of noise dimensions

---

[4] Best ROC AUC values for ORC were achieved with $T = 0.5$ for $dim_n = 0.2$, $T = 0.7$ for $dim_n = 1.0$, and $T = 0.6$ else. For MORe++ $ost = 0.3$ delivered best results for $dim_n = \{0.8, 1.0\}$, else $ost = 0.2$.

Evaluation of Systematic Experiments. The biggest difference of results could be seen for high-dimensional data, where ORC was clearly outperformed by MORe++ and k-Means--. As many results seem to be similar or only slightly better than k-Means--, we want to emphasize the difference in runtime, where MORe++ only needs (including running the k-Means) $O(nkdi)$ and k-Means-- needs $O(n^{d^3})$. Further, MORe++ persuaded especially in the following points:

- For datasets of high dimensionality
- For datasets with clusters of higher variances
- For datasets with noise dimensions

4.2   Real World Datasets

We tested some real world datsets with different properties: sizes ranged between 148 and almost 95000, dimensionalities between 3 and 166, percentage of outliers between 0.4 and 7 and number of clusters between 1 and 5. Table 1 gives an overview over size, number (and percentage) of outliers, and number of clusters of the real world datasets which we used and which can be found in the ODDS library [25]. Table 2 gives the parameters chosen for all algorithms, where we used the parameter resulting in the best ROC AUC, resp. the ground truth value as number of outliers l for k-means--. For MORe++ and ORC we tested values between 0 and 1 in steps of 0.1. As in the previous section, we took the average of 100 executions of k-means-- due to its non-determinism. Figure 10 shows the results of the experiments: for Lympho, Shuttle, and Smtp MORe++ clearly outperforms both other algorithms. Glass is an interesting experiment, as ORC is the best performing algorithm here, followed by MORe++. For Musk, MORe++ achieved the best results, closely followed by k-means-- and ORC. So, even though MORe++ performed quite similar to k-means-- in most cases in the previous section, it seems to be more suitable for real world scenarios plus it is by far faster. ORC performed well on the Glass dataset, but cannot deal with very high number of dimensions as shown in Sect. 4.1. Thus, for outlier detection in high-dimensional datasets, MORe++ is preferable.

Table 1. Overview of real world datasets

Dataset	n	dim	outlier	k
Glass	214	9	9 (4.2%)	5
Lympho	148	18	6 (4.1%)	2
Musk	3062	166	97 (3.2%)	3
Shuttle	49097	9	3511 (7%)	1
Smtp	95156	3	2211 (0.4%)	1

Table 2. Chosen parameters for real world datasets

	MORe++	ORC	k-means--
	ost	T	l
Glass	0.1	0.3	9
Lympho	0.4	0.7	6
Musk	0.3	0.7	97
Shuttle	0.3	0.3	3511
Smtp	0.4	0.5	2211

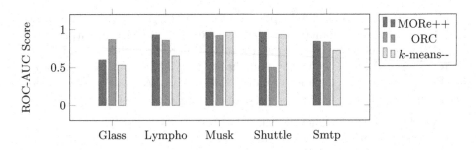

Fig. 10. ROC AUC for real world datasets

5   Conclusion

In conclusion we developed the outlier detection algorithm MORe++, which is based on histograms of one-dimensional projections of separately regarded k-Means clusters. By projecting onto single dimensions, we can circumvent some aspects of the curse of dimensionality: neither do we need a distance measure working in high-dimensional space nor is our runtime exponential in the number of dimensions. Users do not have to know the number of outliers beforehand and local outliers can easily be detected. The algorithm is easily parallelizable and easy to implement. A plethora of variations and improvements of k-Means could be used to further improve our already good results, and also using another algorithm than k-Means as foundation for the partitioning of the data could be tried.

Acknowledgement. This work has been funded by the German Federal Ministry of Education and Research (BMBF) under Grant No. 01IS18036A. The authors of this work take full responsibilities for its content.

References

1. Ahmed, M., Mahmood, A.N.: A novel approach for outlier detection and clustering improvement. In: 2013 IEEE 8th Conference on Industrial Electronics and Applications (ICIEA), pp. 577–582. IEEE (2013)
2. Arthur, D., Vassilvitskii, S.: k-means++: the advantages of careful seeding. In: Proceedings of the Eighteenth Annual ACM-SIAM Symposium on Discrete Algorithms, pp. 1027–1035. Society for Industrial and Applied Mathematics (2007)
3. Bahmani, B., Moseley, B., Vattani, A., Kumar, R., Vassilvitskii, S.: Scalable k-means++. Proc. VLDB Endow. **5**(7), 622–633 (2012)
4. Bellman, R.E.: Adaptive Control Processes: A Guided Tour, vol. 2045. Princeton University Press, Princeton (2015)
5. Bradley, P.S., Mangasarian, O.L., Street, W.N.: Clustering via concave minimization. In: Advances in Neural Information Processing Systems, pp. 368–374 (1997)
6. Breunig, M.M., Kriegel, H.P., Ng, R.T., Sander, J.: Lof: identifying density-based local outliers. In: ACM Sigmod Record, vol. 29, pp. 93–104. ACM (2000)

7. Chawla, S., Gionis, A.: k-means-: a unified approach to clustering and outlier detection. In: Proceedings of the 2013 SIAM International Conference on Data Mining, pp. 189–197. SIAM (2013)
8. Fawcett, T.: An introduction to ROC analysis. Pattern Recognit. Lett. **27**(8), 861–874 (2006)
9. Freedman, D., Diaconis, P.: On the histogram as a density estimator: L 2 theory. Probab. Theory Relat. Fields **57**(4), 453–476 (1981)
10. Gan, G., Ng, M.K.P.: K-means clustering with outlier removal. Pattern Recognit. Lett. **90**, 8–14 (2017)
11. Gebski, M., Wong, R.K.: An efficient histogram method for outlier detection. In: Kotagiri, R., Krishna, P.R., Mohania, M., Nantajeewarawat, E. (eds.) DASFAA 2007. LNCS, vol. 4443, pp. 176–187. Springer, Heidelberg (2007). https://doi.org/10.1007/978-3-540-71703-4_17
12. Goldstein, M., Dengel, A.: Histogram-based outlier score (HBOS): a fast unsupervised anomaly detection algorithm. In: KI-2012: Poster and Demo Track, pp. 59–63 (2012)
13. Hautamäki, V., Cherednichenko, S., Kärkkäinen, I., Kinnunen, T., Fränti, P.: Improving K-means by outlier removal. In: Kalviainen, H., Parkkinen, J., Kaarna, A. (eds.) SCIA 2005. LNCS, vol. 3540, pp. 978–987. Springer, Heidelberg (2005). https://doi.org/10.1007/11499145_99
14. Hawkins, D.M.: Identification of Outliers, vol. 11. Springer, Dordrecht (1980). https://doi.org/10.1007/978-94-015-3994-4
15. Hyndman, R.J.: The problem with Sturges rule for constructing histograms (1995)
16. Jiang, M.F., Tseng, S.S., Su, C.M.: Two-phase clustering process for outliers detection. Pattern Recognit. Lett. **22**(6–7), 691–700 (2001)
17. Jiang, S., An, Q.: Clustering-based outlier detection method. In: 2008 Fifth International Conference on Fuzzy Systems and Knowledge Discovery, vol. 2, pp. 429–433. IEEE (2008)
18. Kriegel, H.P., Zimek, A., et al.: Angle-based outlier detection in high-dimensional data. In: Proceedings of the 14th ACM SIGKDD International Conference on Knowledge Discovery and Data Mining, pp. 444–452. ACM (2008)
19. Lloyd, S.: Least squares quantization in PCM. IEEE Trans. Inf. Theory **28**(2), 129–137 (1982)
20. MacQueen, J., et al.: Some methods for classification and analysis of multivariate observations. In: Proceedings of the Fifth Berkeley Symposium on Mathematical Statistics and Probability, Oakland, CA, USA, vol. 1, pp. 281–297 (1967)
21. Mautz, D., Ye, W., Plant, C., Böhm, C.: Towards an optimal subspace for k-means. In: Proceedings of the 23rd ACM SIGKDD International Conference on Knowledge Discovery and Data Mining, pp. 365–373. ACM (2017)
22. Mautz, D., Ye, W., Plant, C., Böhm, C.: Discovering non-redundant k-means clusterings in optimal subspaces. In: Proceedings of the 24th ACM SIGKDD International Conference on Knowledge Discovery & Data Mining, pp. 1973–1982. ACM (2018)
23. Müller, E., Assent, I., Iglesias, P., Mülle, Y., Böhm, K.: Outlier ranking via subspace analysis in multiple views of the data. In: 2012 IEEE 12th International Conference on Data Mining, pp. 529–538. IEEE (2012)
24. Papadimitriou, S., Kitagawa, H., Gibbons, P.B., Faloutsos, C.: Loci: fast outlier detection using the local correlation integral. In: Proceedings 19th International Conference on Data Engineering (Cat. No. 03CH37405), pp. 315–326. IEEE (2003)
25. Rayana, S.: ODDS library (2016). http://odds.cs.stonybrook.edu

26. Scott, D.W.: On optimal and data-based histograms. Biometrika **66**(3), 605–610 (1979)
27. Seidl, T., Müller, E., Assent, I., Steinhausen, U.: Outlier detection and ranking based on subspace clustering. In: Dagstuhl Seminar Proceedings. Schloss Dagstuhl-Leibniz-Zentrum für Informatik (2009)
28. Whang, J.J., Dhillon, I.S., Gleich, D.F.: Non-exhaustive, overlapping k-means. In: Proceedings of the 2015 SIAM International Conference on Data Mining, pp. 936–944. SIAM (2015)
29. Zhou, Y., Yu, H., Cai, X.: A novel k-means algorithm for clustering and outlier detection. In: 2009 Second International Conference on Future Information Technology and Management Engineering, pp. 476–480. IEEE (2009)
30. Zimek, A., Schubert, E., Kriegel, H.P.: A survey on unsupervised outlier detection in high-dimensional numerical data. Stat. Anal. Data Min.: ASA Data Sci. J. **5**(5), 363–387 (2012)

A Generic Summary Structure for Arbitrarily Oriented Subspace Clustering in Data Streams

Felix Borutta[1(✉)], Peer Kröger[1(✉)], and Thomas Hubauer[2]

[1] Ludwig-Maximilians-Universität München, Munich, Germany
{borutta,kroeger}@dbs.ifi.lmu.de
[2] Siemens AG, Munich, Germany
thomas.hubauer@siemens.com

Abstract. Nowadays, as lots of data is gathered in large volumes and with high velocity, the development of algorithms capable of handling complex data streams in (near) real-time is a major challenge. In this work, we present the algorithm CORRSTREAM which tackles the problem of detecting arbitrarily oriented subspace clusters in high-dimensional data streams. The proposed method follows a two phase approach, where the continuous online phase aggregates data points within a proper microcluster structure that stores all necessary information to define a microcluster's subspace and is generic enough to cope with a variety of offline procedures. Given several such microclusters, the offline phase is able to build a final clustering model which reveals arbitrarily oriented subspaces in which the data tend to cluster. In our experimental evaluation, we show that CORRSTREAM not only has an acceptable throughput but also outperforms static counterpart algorithms by orders of magnitude when considering the runtime. At the same time, the loss of accuracy is quite small.

1 Introduction

A common approach to discover knowledge in high-dimensional data is *subspace clustering*. Among the subspace clustering algorithms, correlation clustering methods generally aim at detecting interesting dependencies between different features and thus reducing the feature space such that clusters can be clearly distinguished from other clusters by only considering the found combinations of features, i.e., so called arbitrarily oriented subspaces.

Several subspace clustering methods have been proposed in the past but most of them are limited to static databases and the few algorithms that can deal with streaming data basically concentrate on detecting subspace clusters in axis-parallel subspaces. However, in many applications it might happen that complex relationships in form of correlations between different features appear. Finding such correlations can be beneficial in many applications, e.g., having knowledge about hidden dependencies that occur during standard operation of a machine can be very useful in monitoring systems.

© Springer Nature Switzerland AG 2019
G. Amato et al. (Eds.): SISAP 2019, LNCS 11807, pp. 203–211, 2019.
https://doi.org/10.1007/978-3-030-32047-8_18

In this work, we present a generic framework for PCA-based correlation clustering called CORRSTREAM which is able to cope with data streams. Therefore, we propose a generic microcluster structure, that summarizes all the necessary statistical information of the incorporated data points during the online phase. In the offline phase, which can be initiated on demand or periodically, the information stored in the microclusters can be reused by any PCA-based clustering technique to generate a final correlation clustering model. An aging mechanism allows to "forget" the information contributed by stale data and hence keeps the microcluster model up-to-date.

2   Related Work

There has been a variety of work on subspace clustering algorithms that are applicable to high dimensional data. Some of the presented techniques [6,8] limit themselves to the detection of axis-parallel correlation clusters. One class of algorithms that are able to find arbitrarily oriented correlation clusters are PCA-based correlation clustering algorithms. These algorithms use the principal component analysis (PCA) to detect low dimensional subspaces defined by inter-attribute correlations. The $ORCLUS$ algorithm [7] is a k-means based clustering that is able to identify arbitrarily oriented subspace clusters in high dimensional data. The $4C$ algorithm [9], the $COPAC$ algorithm [3] and the $ERiC$ algorithm [2] are density-based approaches that tackle the problem of correlation clustering. The approaches presented in [1,14] follow a different and global approach and rely on hough transformation. Further approaches that consider correlation clustering but use slightly different concepts are presented in [4,10,11,17,19,20,22]. We refer to [15] for more detailed information on existing solutions for correlation clustering. However, none of the aforementioned methods can be applied to our problem because they all require multiple passes over the data, or costly learning procedures, which leads to infeasible costs when considering a streaming environment.

A very fundamental work on stream clustering can be found in [23], where the authors introduce the concept of *Microclusters*, i.e., data structures that encapsulate the information of a set of points such that it serves as a compact representative of these points. By maintaining the microclusters during an online phase, the $BIRCH$ algorithm finally refines the clustering model in an offline step to identify spherical clusters. This computing framework has been widely used in adopted forms for several stream clustering algorithms in the past, e.g. the k-means based *CluStream* algorithm [5], the density-based *DenStream* algorithm or the axis-parallel subspace methods *HDDStream* [18], or *PreDeConStream* [13]. There have been several other stream clustering algorithms proposed (see [21] for a survey). Although they all reveal interesting data processing schemes and summary structures, none of them is applicable when considering the problem of detecting arbitrarily oriented subspace clusters.

3   PCA-Based Correlation Clustering on Data Streams

As this work concentrates on streaming data, we define a data stream S as an ordered and possibly infinite sequence of data objects $x_1, x_2, ..., x_i, ...$ that must be accessed in the order they arrive and can be read only in one linear scan. To be able to deal with such streams appropriately, a proper algorithm requires a few design decision with respect to the data processing scheme, the data aggregation strategy and the data aging mechanism. Since our *PCA*-based correlation clustering algorithm shall be able to deliver up-to-date subspace clustering models at any time, CORRSTREAM processes each incoming data object individually rather than relying on batch processing. As it is infeasible to keep all data objects in memory we propose the following summary structure that captures the key statistics of the yet found clusters compactly.

Definition 1. *A microcluster CCMicro at time t for a set of d-dimensional points $C = \{p_1, p_2, ..., p_n\}$ arriving at different points in time is defined as CCMicro$(C, t) = (V(t), E(t), \mu(t), ts)$ with*

- $V(t)$ *being the eigenvector matrix of the covariance matrix of C at time t,*
- $E(t)$ *being the corresponding eigenvalues of the eigenvectors in $V(t)$,*
- $\mu(t)$ *being the mean of the data points contained in C at time t, and*
- *ts being the timestamp of the update of this microcluster.*

We design this data structure with the objective to compress correlation clusters. At the same time this data structure must be generic enough so that it can be used for any *PCA*-based correlation clustering technique. Another major criterion for such microcluster structures is the ability to be processed in an incremental manner since they may absorb further data objects or need to be merged eventually. We therefore borrow an *incremental principal component analysis* approach from [16]. As aging mechanism, we employ the *damped window model* on each microcluster structure. Since recent data is typically more important than old data objects, it is useful to "forget" stale data. We therefore employ the *exponential fading function* for data aging. This technique assigns a weight to each data object which decreases exponentially with time t by using the fading function $f(t) = 2^{-\lambda \cdot t}$. $\lambda > 0$ is the decay rate and determines the impact of stale data to the application. A high value of λ means low importance of old data and vice versa.

In general, the CORRSTREAM framework consists of two phases, i.e., an online phase in which microclusters are generated, maintained and/or discarded due to temporal expiration, and an offline phase to extract on demand clustering models of the current state of the stream. During the continuous online phase, the data stream is consumed and for each data object o a *rangeNN* query is performed to detect the closest microcluster. The *rangeNN* query retrieves the closest microcluster within a maximum distance of ϵ. If such a microcluster exists, it absorbs the current data object o, otherwise a new microcluster is created. Note that the *rangeNN* query uses two distance measures, i.e., the Euclidean distance and the correlation distance. This is due to the fact that we first collect

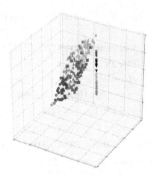

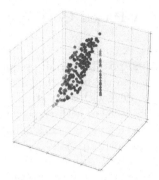

(a) Microcluster model.   (b) Macrocluster model.

Fig. 1. Micro- and macrocluster models on a toy dataset as retrieved by CORRSTREAM. Differently colored and shaped point sets describe different micro- resp. macroclusters. (Color figure online)

a bunch of spatially close (wrt. the Euclidean distance) data objects for each microcluster within a small buffer before we run an initial PCA to initialize the microcluster. This is necessary since the correlation distance relies on a microcluster's eigensystem and hence becomes only meaningful if eigenvectors and -values are determined with respect to a few points. Once the buffer is full, we initialize the microcluster and maintain its components that enable the usage of the correlation distance.

Definition 2. *Given a microcluster mc with mean point μ_{mc} and a data object o, the correlation distance between both is defined as*

$$distance_{corr}(mc, o) = \sqrt{(\mu_{mc} - o) \cdot \hat{M}_{mc} \cdot (o - \mu_{mc})}$$

with similarity matrix $\hat{M}_{mc} = V_{mc}\hat{E}_{mc}V_{mc}^T$, where $\hat{E}_{mc}$ is the adopted eigenvector matrix

$$\hat{E}_{mc}(i, i) = \begin{cases} 1 & if\ E_{mc}(i, i) \geq \alpha \\ \kappa & else. \end{cases}$$

Here, $\alpha \in [0; 1]$ is a threshold that defines the amount of variance that must be captured along the corresponding eigenvector for the eigenvector to be considered strong. For weak eigenvectors the value in $\hat{E}_{mc}$ is set to a constant value $\kappa \gg 1$.

The constant value κ specifies the allowed degree of deviation from the correlation subspace. Following [9], we set this value to $\kappa = 50$.

After determining the closest microcluster mc of the incoming data object o, the latter must be incorporated into mc properly. The proposed algorithm differs between adding the data object to the buffer of a microcluster that has not been initialized yet and adding the object to an already initialized microcluster. In the first case, the mean and the current timestamp of the microcluster are updated

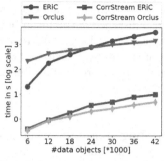
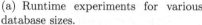

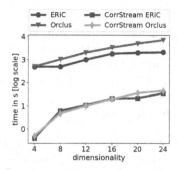

(a) Runtime experiments for various database sizes.

(b) Runtime experiments for varying dimensionalities.

Fig. 2. Runtime measurements for varying database sizes, resp. dimensionalities.

additionally, and in the second case the existing components of the microcluster are reused and the incremental PCA procedure is invoked to generate the new eigenvectors and -values additionally.

The offline phase of CORRSTREAM aims at constructing a high quality clustering model that describes correlation subspaces appearing in the data. Figure 1(a) exemplary depicts the outcoming microcluster model of the online phase for a small synthetic, 3-dimensional data set. As can be seen, some of the microclusters can be grouped so that finally two separated subspace clusters, i.e., a 1-dimensional cluster and a 2-dimensional one (cf. Fig. 1(b)), are formed. Finding such groupings of microclusters is the goal of the offline phase and the generic structure of the microclusters allows that a variety of static correlation clustering algorithms can be adopted to build the final clustering model.

Using the clustering scheme of the *ERiC* algorithm for the offline procedure for instance requires minor adaptations. After partitioning the set of microclusters into disjoint partitions according to their subspace dimensionality that can be derived from the eigenvalue matrices, the algorithm determines macroclusters within each partition. Therefore, we apply a *DBSCAN* [12] variant capable of dealing with the structure of the microclusters. The basic idea is to use the orientation of the microclusters given by the eigenvectors to group those microclusters whose eigenvectors span a similar subspace. According to [2], we use their correlation distance as the composition of the *approximate linear dependency* and the *affine distance* between two microclusters. Using this distance measure for *DBSCAN* within each partition finally yields a density-based correlation clustering model for each partition.

As for *ERiC*, the adaptations that have to be made for the *ORCLUS* algorithm are quite small. First, the parameter $k0$, i.e., the initial number of seeds that the algorithm starts with, has to be set to a fraction of the number of microclusters since the user usually do not know the exact number in advance. Further, as the algorithm has to work with the mean points of the microclusters as well as their eigenvectors, we propose to incorporate the orientations of both affected eigensystems when measuring the distance between two microclusters. If just assigning by measuring the projected pairwise distance of the mean points, as given in [7], it might happen that two microclusters that have a different subspace orientation are grouped because the mean of the one microcluster fits into the subspace of the other one although their subspace preferences are different.

4   Experiments

Since CORRSTREAM is the first method for detecting arbitrarily oriented subspace clusters in data streams, we compare it with the static equivalents, i.e., *ERiC* and *ORCLUS*, in terms of runtime and clustering quality. We further present the throughput of CORRSTREAM. We consider different database sizes, i.e., various numbers of objects delivered by the data stream, and various dimensionalities of the data points. All data points are distributed among 5 equi-sized clusters. For the experiments using variously sized data sets, each cluster has a random dimensionality between 1 and 2, and one full-dimensional noise cluster spanned over the entire normalized $\mathbb{R}^3$ space. For the experiments that investigate the performance under varying dimensionalities, the number of data points is fixed to 12'000 and except of one full-dimensional noise cluster, the correlation clusters have a random dimensionality which is below the full dimensionality. For all experiments, we report the results when using the best considered parameter setting. We consider different parameter settings by performing grid search over buff_size $\in \{10, 15, 20\}$; $\epsilon \in \{0.1, 0.15, 0.2, 0.3\}$; $minMcs \in \{1, 2, 3\}$. The parameters that were already introduced by previous methods are fixed.

Figure 2(a) shows the run times for varying numbers of data objects. The y-axis shows the measured runtime in log scale. While the static methods both show rather fast increasing runtimes, even for moderately sized databases, the CORRSTREAM variants are able to process 42'000 data objects within only a few seconds. When ranging the dimensionality of the feature space from 4 to 24 dimensions, the outcome is quite similar. As depicted in Fig. 2(b), the static algorithms need much more time compared to our method.

Next, we investigate the considered methods in terms of clustering quality. Therefore, we compare the resulting clusterings of each approach to the ground truth labels and use precision and recall values to examine the performance. Figure 3(a) and (b) compare the results of CORRSTREAM using the *ERiC* approach for the offline phase against the static *ERiC* algorithm. In both plots the *ERiC* algorithm shows higher precision and recall values than CORRSTREAM. This might be reasoned by aging as well as by treating microclusters as noise if their buffers have not been filled. Figure 3(c) and (d) show the corresponding results when using *ORCLUS* for the offline procedure. Interestingly, the CORRSTREAM results are better in the experiments when increasing the dimensionality. This might be due to the fact that during the first iteration of *ORCLUS*, all data points are assigned to the closest cluster center with respect to the Euclidean distance. This leads to a situation where clusters, or points that are treated as intermediate clusters during the iterations, are spanned across several actual clusters. Such intermediate clusters, that typically occur if differently

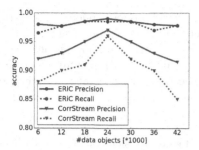

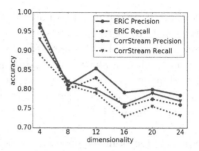

(a) Varying numbers of data objects.   (b) Varying dimensionalities.

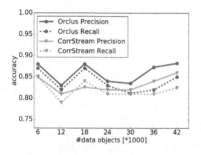

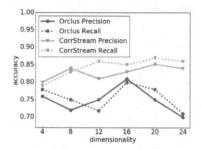

(c) Varying numbers of data objects.   (d) Varying dimensionalities.

Fig. 3. Precision and recall measurements.

oriented correlation clusters intersect, finally form a "false" subspace and thus might absorb data points that actually belong to another cluster. CORRSTREAM reduces this problem due to using the correlation distance for the assignment of a data point as soon as a microcluster is initialized.

Finally, as the throughput is one of the major criterions for streaming algorithms, we evaluate CORRSTREAM in terms of the number of data objects processed per millisecond. The plot in Fig. 4 shows the throughput of the online phase by using different dimensionalities for the feature space. As can be seen, the throughput of the algorithm decreases with increasing dimensionalities.

Nevertheless, the decline of the throughput decreases for higher dimensionalities and still is about 363 data objects per second for 24 dimensional feature spaces.

5   Conclusion

In this work we presented the first streaming algorithm capable of detecting arbitrarily oriented subspace clusters. We applied the established two-step approach by dividing the

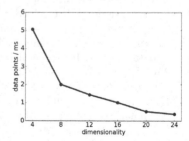

Fig. 4. The throughput of the online phase.

procedure in an online and an offline phase. A newly proposed microcluster structure is used to aggregate similar data objects and thus compressing the volume of data significantly while providing all necessary statistical information to compute a correlation clustering model during an offline procedure. Our experimental evaluation showed that CORRSTREAM outperforms its static counterparts clearly in terms of computational costs by just suffering a small loss concerning the clustering quality.

Acknowledgement. This work was partially funded by Siemens AG and has been developed in cooperation with the Munich Center for Machine Learning (MCML), funded by the German Federal Ministry of Education and Research (BMBF) under Grant No. 01IS18036A. The authors of this work take full responsibilities for its content.

References

1. Achtert, E., Böhm, C., David, J., Kröger, P., Zimek, A.: Global correlation clustering based on the hough transform. Stat. Anal. Data Min. **1**(3), 111–127 (2008)
2. Achtert, E., Böhm, C., Kriegel, H.P., Kröger, P., Zimek, A.: On exploring complex relationships of correlation clusters. In: Proceedings of SSBDM, p. 7 (2007)
3. Achtert, E., Böhm, C., Kriegel, H.P., Kröger, P., Zimek, A., et al.: Robust, complete, and efficient correlation clustering. In: SDM, pp. 413–418 (2007)
4. Achtert, E., Böhm, C., Kröger, P., Zimek, A.: Mining hierarchies of correlation clusters. In: Proceedings of SSBDM, pp. 119–128 (2006)
5. Aggarwal, C.C., Han, J., Wang, J., Yu, P.S.: A framework for clustering evolving data streams. In: Proceedings of VLDB, pp. 81–92 (2003)
6. Aggarwal, C.C., Wolf, J.L., Yu, P.S., Procopiuc, C., Park, J.S.: Fast algorithms for projected clustering. In: ACM SIGMoD Record, vol. 28, pp. 61–72 (1999)
7. Aggarwal, C.C., Yu, P.S.: Finding generalized projected clusters in high dimensional spaces, vol. 29 (2000)
8. Agrawal, R., Gehrke, J., Gunopulos, D., Raghavan, P.: Automatic subspace clustering of high dimensional data for data mining applications, vol. 27 (1998)
9. Böhm, C., Kailing, K., Kröger, P., Zimek, A.: Computing clusters of correlation connected objects. In: Proceedings of ACM SIGMOD, pp. 455–466 (2004)
10. Costeira, J.P., Kanade, T.: A multibody factorization method for independently moving objects. IJCV **29**(3), 159–179 (1998)
11. Elhamifar, E., Vidal, R.: Sparse subspace clustering: algorithm, theory, and applications. IEEE TPAMI **35**(11), 2765–2781 (2013)
12. Ester, M., Kriegel, H.P., Sander, J., Xu, X.: A density-based algorithm for discovering clusters in large spatial databases with noise. In: KDD, pp. 226–231 (1996)
13. Hassani, M., Spaus, P., Gaber, M.M., Seidl, T.: Density-based projected clustering of data streams. In: Hüllermeier, E., Link, S., Fober, T., Seeger, B. (eds.) SUM 2012. LNCS (LNAI), vol. 7520, pp. 311–324. Springer, Heidelberg (2012). https://doi.org/10.1007/978-3-642-33362-0_24
14. Kazempour, D., Mauder, M., Kröger, P., Seidl, T.: Detecting global hyper-paraboloid correlated clusters based on hough transform. In: Proceedings of SSDBM, p. 31. ACM (2017)
15. Kriegel, H.P., Kröger, P., Zimek, A.: Clustering high-dimensional data: a survey on subspace clustering, pattern-based clustering, and correlation clustering. ACM TKDD **3**(1), 1 (2009)

16. Li, Y.: On incremental and robust subspace learning. Pattern Recognit. **37**(7), 1509–1518 (2004)
17. Ng, A.Y., Jordan, M.I., Weiss, Y.: On spectral clustering: analysis and an algorithm. In: Proceedings of NIPS, pp. 849–856 (2002)
18. Ntoutsi, I., Zimek, A., Palpanas, T., Kröger, P., Kriegel, H.P.: Density-based projected clustering over high dimensional data streams. In: SDM, pp. 987–998 (2012)
19. Patel, V.M., Van Nguyen, H., Vidal, R.: Latent space sparse subspace clustering. In: Proceedings of ICCV, pp. 225–232 (2013)
20. Peng, X., Xiao, S., Feng, J., Yau, W.Y., Yi, Z.: Deep subspace clustering with sparsity prior. In: IJCAI, pp. 1925–1931 (2016)
21. Silva, J.A., Faria, E.R., Barros, R.C., Hruschka, E.R., de Carvalho, A.C., Gama, J.: Data stream clustering: a survey. ACM CSUR **46**(1), 13:1–13:31 (2013)
22. Tung, A.K., Xu, X., Ooi, B.C.: Curler: finding and visualizing nonlinear correlation clusters. In: Proceedings of SIGMOD, pp. 467–478 (2005)
23. Zhang, T., Ramakrishnan, R., Livny, M.: Birch: an efficient data clustering method for very large databases. In: ACM Sigmod Record, vol. 25, pp. 103–114 (1996)

Similarity Grouping in Big Data Systems

Yasin N. Silva[1(✉)], Manuel Sandoval[1], Diana Prado[1],
Xavier Wallace[1], and Chuitian Rong[2]

[1] Arizona State University, Glendale, USA
{ysilva,mosandov,dapradol,xgwallac}@asu.edu
[2] Tianjin Polytechnic University, Tianjin, China
chuitian@tjpu.edu.cn

Abstract. Distributed computing technologies have opened the door for a wide range of organizations to analyze massive amounts of data. Grouping (fast but based on exact semantics) and clustering (relatively slow but based on similarity-aware semantics) are among the most useful data analysis operations. Previous work introduced the Similarity Grouping (SG) operator, which aims to integrate the best features of grouping and clustering, i.e., fast execution times and similarity-aware grouping semantics. The SG operators, however, were proposed for single node relational database systems. This paper introduces the Distributed Similarity Grouping (DSG) operator, a highly parallel operator for identifying similarity groups in big datasets. DSG enables the identification of groups where all the elements are within a given threshold from each other. This paper presents DSG's design details, implementation guidelines on Spark and Hadoop (two important Big Data systems), and extensive performance and scalability evaluation.

Keywords: Similarity grouping · Big data systems · Performance evaluation · MapReduce · Spark · Hadoop · Clustering

1 Introduction and Related Work

In many business and scientific scenarios, organizations accumulate large amounts of data. While organizations could gain many insights from integrating and analyzing these big datasets, in many cases their dimensions prevent their efficient processing on single-node systems. Big data systems such as Hadoop [1], Spark [2] and Bigtable [4] represent an answer to the requirement of highly distributed and parallelized data analysis of big datasets. Many of these systems are now supported on cloud-based platforms. Hadoop and Spark are two popular Big Data systems. Hadoop [1] and its programming framework, MapReduce [3], support two key operations: *map* and *reduce*. Multiple map tasks process input chunks in parallel. Each map call is given a pair (*k1,v1*) and produces a list of (*k2,v2*) pairs. The output of the map calls is transferred to the reduce nodes (*shuffle* phase) in a way that guarantees that all the intermediate records with the same intermediate key (*k2*) are sent to the same reducer node. At each reduce node, the received intermediate records are sorted and grouped. Then, each formed group is processed in a single reduce call. Spark [2] is a more recent Big Data system that uses Resilient Distributed Datasets (RDDs) as its core data structure

© Springer Nature Switzerland AG 2019
G. Amato et al. (Eds.): SISAP 2019, LNCS 11807, pp. 212–220, 2019.
https://doi.org/10.1007/978-3-030-32047-8_19

and supports a wider range of operations (including variants of *map* and *reduce* as well as grouping, filtering and set operations). Spark operations are performed in a distributed fashion primarily using the main-memory resources of a computer cluster.

Grouping operations are among the most useful data analytics operations. The standard grouping operator (Group-by) [5] is extensively supported in relational databases and was extended to perform subset aggregation (where grouping is performed at various levels) via the Roll-Up and Cube database operators [6]. While Group-by is very fast, its use is limited to equality-based grouping (all the records with the same grouping attribute values form the same group). Multiple sophisticated clustering operators have been proposed in data mining [11], e.g., K-Means [7] and DBSCAN [8]. They seek to find more complex patterns in data, but often at a steep increase in execution time. Single-scan versions of the well-known clustering algorithms K-Means and Cobweb are proposed in [12] and [13]. CURE [14] is an alternative algorithm based on sampling. Extensions of common clustering algorithms have also been proposed for big data platforms. The work in [15] presents an adaptation of K-Means for the Hadoop framework. K-Means is also supported in the Spark framework [16]. Considering the advantages and limitations of grouping and clustering, previous work introduced the Similarity Grouping (SGB) operator for numeric data [9, 17] to integrate the advantages of both types of operators, i.e., fast execution times and similarity-based grouping. SGB allows the specification of the desired grouping using descriptive properties such as group size and compactness. Tang et al. extended SGB operators to support multi-dimensional data [10]. While SGB operators cannot identify all the complex forms of clusters identified by clustering algorithms, they identify groups that are useful in many scenarios.

Considering that SGB operators were proposed for single-node relational databases, this paper introduces the Distributed Similarity Grouping (DSG) operator, a highly parallel and scalable approach for identifying similarity groups in big datasets. DSG identifies a particularly useful type of similarity groups where all the elements of a group are within a given threshold from each other. The proposed algorithm can be used with any distance function and data type. We also present guidelines to implement DSG in Spark and Hadoop and extensively assess its performance and scalability properties. We show that DSG performs significantly better than K-Means for identifying similarity groups and maintains execution times that are close to those of standard grouping.

2   The Distributed Similarity Grouping Algorithm

This section presents the algorithmic details of DSG. The type of similarity groups identified by this algorithm are groups where given any pair of elements (r_1, r_2) of a group, their separation is no larger than a parameter threshold (ε), i.e., distance$(r_1, r_2) \leq \varepsilon$. Like many clustering algorithms, DSG is non-deterministic, i.e., it is possible that different executions of the algorithm on the same dataset could generate slightly different solutions. Unlike K-Means, DSG does not require advance knowledge of the number of clusters. DSG uses pivot-based data partitioning to distribute and parallelize the computational tasks. The goal is to divide a large dataset into partitions that can be processed independently and in parallel. The pivots are a subset of input data records and each pivot is associated with a partition. Each input record is assigned to the partition associated with its closes pivot. In addition, DSG replicates the records at the boundary between partitions to ensure that similarity groups located in these regions are properly detected. Figure 1 shows an example of how DSG partitions and identifies the similarity groups using two pivots (P_0 and P_1). The left image represents the initial dataset with seven clusters or similarity groups (G_1 to G_7). The right image represents the two generated partitions (*Part0* and *Part1*). Observe that regions A and C contain the records that are closer to P_0 than to P_1, while B and D the points that are closer to P_1 than to P_0. Regions A + C and B + D are referred to as *base* partitions. Observe that the records in A + C and B + D are assigned to partitions *Part0* and *Part1*, respectively. Also, the regions at the boundary between the base partitions (points within ε from the boundary) are replicated, i.e., region D is added to *Part0* and C to *Part1*. Regions C and D are referred to as *window* partitions. The only problem is that some of the groups (G_2 to G_6) are partially or fully contained in both partitions. Our approach needs a mechanism to output each similarity group only once and ensure that a group is outputted in the partition that contains the entire group. To this end, DSG applies the following guidelines: (1) during partitioning, each record r in a given partition P is augmented with information of its base partition (partition of its closes pivot) and assigned partition (P), and (2) given any generated similarity group g, the group will be outputted only in the partition matching the smallest base partition among all the records in g. In Fig. 1, G_1, G_2, G_3, and G_4 are outputted in *Part0* while G_5, G_6, and G_7 in *Part1*.

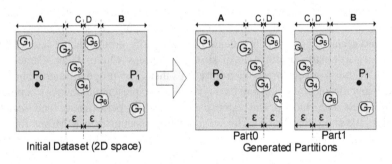

Fig. 1. Example of partitioning and similarity group generation using two pivots.

Algorithm 1 presents DSG's main algorithmic steps. After pivots are generated (line 1), the data is partitioned in parallel (lines 3–11). The partitioning phase is implemented using map operations in Spark and Hadoop. Each input record r is assigned to the partition of its closes pivot P_c (lines 4–5) and all the partitions p where r belongs to the window regions between the partitions of p and P_c (lines 6–10). In general, the records in the window region between two partitions should be a superset of the records whose distance to the hyperplane that separates the partitions is at most ε. Unfortunately, this hyperplane does not always explicitly exist in a metric space. Instead, the hyperplane is implicit and known as a generalized hyperplane. Since the distance of a record r to the generalized hyperplane between two partitions for pivots P_0 and P_1 cannot always be computed exactly, a lower bound is used [18] (line 7): $genHyperplaneDist(r, P_0, P_1) = (\text{distance}(r, P_0) - \text{distance}(r, P_1))/2$. This distance can be replaced by an exact distance when this can be calculated, e.g., with Euclidean distance, $genHyperplaneDist$ can be replaced by $euclideanHyperplaneDist$ $(r, P_0, P_1) = |(\text{distance}(r, P_0)^2 - \text{distance}(r, P_1)^2| /(2 \times \text{distance}(P_0, P_1)$. The partitioning phase records the information of base and assigned partitions of each record. The intermediate records generated during partitioning are grouped in the shuffle phase (line 12) such that all the records that belong to the same partition will form a single group. This is performed automatically in the shuffle phase of Hadoop and is implemented using the grouping operator in Spark. In the similarity group formation and output generation phase (lines 14–30), similarity groups are identified and outputted in each partition and in parallel. We first check if a partition is small enough to be efficiently processed in a single node (line 15–16). If this is not the case, the partition is stored for further processing using the same SGB algorithm. While this feature guarantees that the algorithm will be able to effectively partition datasets with high concentration of records in certain regions, in practice, we can increase the number of pivots and thus decrease the size of partitions to guarantee a single round. If a partition is small enough to be processed in a single node, the algorithm runs a single-node algorithm ($findSimGroups$) to identify the similarity groups of a single partition (line 18). The output of this algorithm is a set of clusters and each cluster is composed of its data records and information needed to ensure non-duplicated cluster generation ($flags$). The $flags$ component of a given cluster C maintains a sequence of flag arrays (one array per round where the cluster was processed). For instance, if four pivots are being used (P_0, P_1, P_2, P_3) and a single round is needed, the content of $C.flags$ has the form $\{[fo\_0, fo\_1, fo\_2, fo\_3]\}$. The content of this structure could be for example $\{[0, 1, 0, 1]\}$. A value of 1 at index i indicates that cluster C contains at least one record whose base partition is the one associated with pivot P_i. The $flags$ component is used to determine if a given cluster should be outputted while processing the current partition or not (lines 21–28). A given cluster of partition P_i will be outputted only if the minimum index on the corresponding flag array matches i (lines 25–27). The similarity group formation phase is performed using the reduce operations in Hadoop and Spark. The details of the $findSimGroups$ method are presented in Algorithm 2. For every record r, the algorithm tries to identify a suitable cluster among the ones that were already formed (lines 2–17). If Euclidean distance is used, a centroid-based filter is used to discard non-suitable clusters (line 5). For a potentially suitable cluster, the algorithm checks the similarity group condition between r and each element of the cluster (lines 9–13). If this check is successful, r can be added to this cluster (lines 14–19). If the process ends without finding a cluster, a new cluster is created and r is added to it. After processing the input records, the method generates the $flags$ components (line 28–30).

Algorithm 1 *DistSimGrouping*
Input: *inputData, eps, numPivots, memT*
Output: similarity groups in *inputData*
1 *pivots* = selectPivots(*numPivots, inputData*)

2 //**Partitioning** - *r*: ⟨*ID, value*,
3 *assignedPartitionSeq, basePartitionSeq*⟩
3 **for** each record *r* in a chunk of *inputData* **do**
4   *P_c* = getClosestPivot(*r, pivots*)
5   output ⟨*P_c, r*⟩ //intermediate output
6   **for** each pivot *p* in {*pivots-P_c*} **do**
7   **if** (dist(*r, p*) - dist(*r, P_c*))/2 ≤ *eps* **then**
8   output ⟨*p, r*⟩ //intermediate output
9   **end if**
10   **end for**
11 **end for**

12 //**Shuffle**: records with same key => partition

13 //**Group Formation**
14 **for** each partition *P_i* **do**
15   **if** size of *P_i* > *memT* **then**
16   store *P_i* for processing in subsequent round
17   **else**
18   *C_i* = findSimGroups(*P_i, eps*) //*C_i*:{*C_{i_k}*},
19   //*C_{i_k}*:⟨*records, flags*⟩, *flags*:{*F_m*}, *F_m*:{*f_{m_n}*}

20   //**Output Generation** (without duplication)
21   **for** each cluster *C_{i_k}* in partition *P_i* **do**
22   generate *minFlags* //minFlags[o]={index
          //of 1st element in *C_i.flags*[o] equal to 1}
23   *aPartitionSeq* = *r.assignedPartitionSeq*
24   //*r* is any record in *P_i*
25   **if** ∀o,minFlags[o]=aPartitionSeq[o] **then**
26   output *C_{i_k}* //final output
27   **end if**
28   **end for**
29   **end if**
30 **end for**

Alg. 1. Main DSG algorithm.

Algorithm 2 *findSimGroups*
Input: *S* (data records), *eps*
Output: *C* (list of clusters or similarity groups)
1 *C*={}
2 **for** each record *r* in *S* **do**
3   *clusterFound* = False
4   **for** each cluster *C_i* in *C* **do**
5   **if** distance(*r, C_i.getCentroid()*) > *eps* **then**
6   //*r* cannot belong to this cluster
7   **else** //*r* may or may not belong to *C_i*
8   *withinEps* = True // *r* is within *eps* of
        C_i's centroid, now verify all points in *C_i*
9   **for** each record *x* in *C_i* **do**
10   **if** distance(*r, x*) > *eps* **then**
          //*r* does not belong to *C_i*
11   *withinEps* = False
12   break
13   **end for**
14   **if** (*withinEps* =True) **then**//*r* belongs to *C_i*
15   *C_i*.add(*r*) //adds to *C_i.records*
16   *C_i.updateCentroid()*
17   *clusterFound*=True
18   break
19   **end if**
20   **end for**
21   **if** (*clusterFound* = False)
22   Create Cluster *Cnew*
23   *Cnew*.add(*r*)
24   *Cnew.updateCentroid()*
25   *C*.add(*Cnew*);
26   **end if**
27 **end for**
28 **for** each cluster *C_i* in *C* **do**
29   generateFlags(*C_i*) //updates *C_i.flags*
30 **end for**
31 **return** *C*

Alg. 2. findClusters method.

The implementation of DSG in Hadoop uses the job configuration (*jobConf*) object to distribute atomic parameters, e.g., *eps*, *memT*, and *numPivots*, to all the nodes. It also uses the random sampling and distributed cache facilities to generate the pivots and distribute them to the nodes. The partitioning and group formation phases are implemented through the *map* and *reduce* operations of Hadoop's MapReduce framework. Furthermore, customized MapReduce grouping and sorting operations were created to support the specific data structures of our solution during the shuffle stage. The Spark implementation uses the RDD API and is significantly shorter due to its robust support of data processing operations. In this case, the *sample* operation is used to select the pivots. Then, the *mapPartitionsToPair* operation is used to implement the partitioning phase and the *groupByKey* operation to group the records that belong to the same partition. Finally, the *flatMap* operation is used to perform the clustering generation phase.

3 Performance Evaluation

We implemented DSG, standard grouping (StandG), and K-Means clustering in Hadoop 2.9.1 and Spark 2.3.2. The experiments were executed on Google Cloud Platform. Unless otherwise stated, the cluster consisted of one master and ten worker nodes. Each node had 4 virtual CPUs, 15 GB of memory and 500 GB of disk space. The number of reducers per Hadoop job was set to 0.95 × (# of worker nodes) × (# of vCPUs per node - 1) and the number of splits per Spark job was 2 × (# of worker nodes) × (# of vCPUs). We implemented a parametrized synthetic dataset generator that enabled us to evaluate the algorithms under a variety of conditions. The datasets were composed of multidimensional vector-based similarity groups separated by 2ε. Given this dataset, DSG and K-Means were expected to have the same output while StandG only identified equality-based groups. Each data record consisted of an ID, an aggregation attribute, and a randomly generated multidimensional vector (100D–500D). The dataset for scale factor N (SFN) had 200,000 × N records. The SF1 datasets contained about 13,000 groups and each of them contained 50–100 records. Each record was duplicated between 1–3 times. We set DSG's *numPivots* = 40 × SF and *memT* = 50,000 based on preliminary tests.

Increasing Scale Factor. Figure 2 shows the execution time (lines) and the number of groups (bars) identified by DSG, StandG and K-Means as the scale factor increases. The execution times of DSG and StandG increase slowly as the scale factor increases. K-Means' execution time, on the other hand, is significantly larger than those of DSG and StandG. In Spark, DSG is about 13 times faster than K-Means while in Hadoop, about 8 times faster. While StandG generates a very large number of equality-based groups, DSG and K-Means identify the same similarity groups.

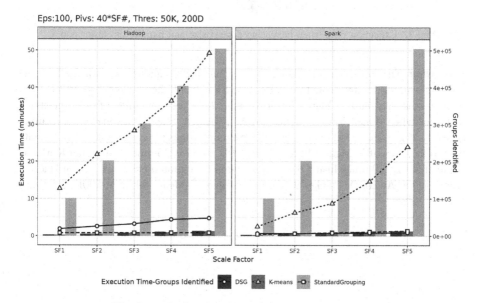

Fig. 2. Execution time when increasing dataset size.

Increasing Scale Factor and Number of Cluster Nodes. Figure 3 compares the execution time and the number of identified groups of DSG, StandG, and K-Means as the data size and number of nodes increase. In this experiment, we increase the scale factor and number of nodes available to the cluster from (SF1, 2 nodes) to (SF5, 10 nodes). DSG and StandG maintain near constant execution times while K-Means' performance increases significantly. In Hadoop, DSG's execution time for (SF5, 10 nodes) is approximately 1.25 times that of (SF1, 2 nodes). In Spark, it is practically constant.

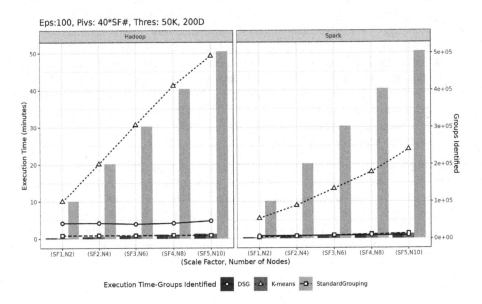

Fig. 3. Execution time when increasing dataset size and number of cluster nodes.

Increasing Number of Dimensions. We executed each algorithm with 200D-500D datasets while fixing the scale factor. Because SF is fixed, the number of groups in each dataset is nearly equal. Figure 4 shows that the execution time of all algorithms increases when dimensionality increases. As expected, StandG has the best execution times, K-Means the worst ones, and DSG execution times are closer to those of StandG than to the ones of K-Means. In Spark, at 200D, DSG is 26 times faster than K-Means while at 500D, 39 times faster. The difference in execution times witnessed in Hadoop is less acute. At 200D, DSG is 9 times faster than K-Means while at 500D, 3.5 times faster.

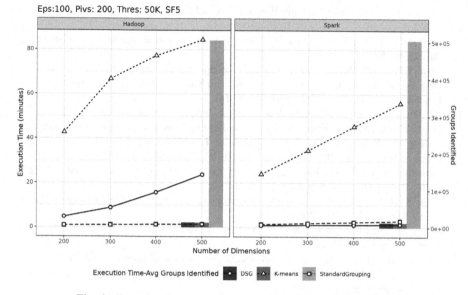

Fig. 4. Execution time when increasing the number of dimensions.

4 Conclusions

This paper introduces the Distributed Similarity Grouping (DSG) operator to efficiently identify similarity groups in very large datasets. The paper presents the general algorithmic details of DSG and the guidelines for its implementation in two popular big data systems. An extensive performance evaluation shows that DSG is successful at identifying similarity groups identified by the K-Means clustering algorithm while having small execution times that are in general very close to those of standard grouping. Future work in this area includes (1) the study of alternative ways of selecting the pivots, (2) the development of distributed similarity algorithms for other types of similarity groups, and (3) studying optimization techniques for non-vector data, e.g., text and sets.

References

1. Apache: Hadoop. https://hadoop.apache.org/
2. Apache: Spark. https://spark.apache.org/
3. Dean, J., Ghemawat, S.: MapReduce: simplified data processing on large clusters. In: OSDI (2004)
4. Chang, F., et al.: Bigtable: a distributed storage system for structured data. ACM Trans. Comput. Syst. **26**(2), 1–26 (2008)
5. Garcia-Molina, H., Ullman, J., Widom, J.: Database Systems: The Complete Book, 2nd edn. Pearson (2008)
6. Gray, J., Bosworth, A., Layman, A., Pirahesh, H.: Data cube: a relational aggregation operator generalizing group-by, cross-tab, and sub-totals. In: ICDE (1996)

7. Lloyd, S.P.: Least squares quantization in PCM. IEEE Trans. Inf. Theory **28**(2), 129–137 (1982)
8. Ester, M., Kriegel, H.P., Sander, J., Xu, X.: A density-based algorithm for discovering clusters. In: KDD (1996)
9. Silva, Y.N., Aref, W.G., Ali, M.: Similarity Group-by. In: ICDE (2009)
10. Tang, M., et al.: Similarity group-by operators for multi-dimensional relational data. IEEE Trans. Knowl. Data Eng. **28**(2), 510–523 (2016)
11. Berkhin, P.: Survey of clustering data mining techniques. Accrue Software (2002)
12. Li, M., Holmes, G., Pfahringer, B.: Clustering large datasets using Cobweb and K-means in tandem. In: Webb, G.I., Yu, X. (eds.) AI 2004. LNCS (LNAI), vol. 3339, pp. 368–379. Springer, Heidelberg (2004). https://doi.org/10.1007/978-3-540-30549-1_33
13. Farnstrom, F., Lewis, J., Elkan, C.: Scalability for clustering algorithms revisited. SIGKDD Explor. Newsl. **2**(1), 51–57 (2000)
14. Guha, S., Rastogi, R., Shim, K.: CURE: an efficient clustering algorithm for large databases. SIGMOD Rec. **27**(2), 73–84 (1999)
15. Anchalia, P.P., Koundinya, A.K., Srinath, N.K.: MapReduce design of K-means clustering algorithm. In: ICISA (2013)
16. Apache: Spark Clustering. https://spark.apache.org/docs/latest/ml-clustering.html
17. Silva, Y.N., Arshad, M., Aref, W.G.: Exploiting similarity-aware grouping in decision support systems. In: EDBT (2009)
18. Jacox, E.H., Samet, H.: Metric space similarity joins. ACM Trans. Database Syst. **33**(2), 7:1–7:38 (2008)

SIDEKICK: Linear Correlation Clustering with Supervised Background Knowledge

Maximilian Archimedes Xaver Hünemörder[(⊠)], Daniyal Kazempour,
Peer Kröger, and Thomas Seidl

Ludwig-Maximilians-Universität München, Munich, Germany
{huenemoerder,kazempour,kroeger,seidl}@dbs.ifi.lmu.de

Abstract. While explainable AI (XAI) is gaining in popularity, other more traditional machine learning algorithms can also benefit from increased explainability. A semi-supervised approach to correlation clustering opens up a promising design space that might provide such explainability to correlation clustering algorithms. In this work, semi-supervised linear correlation clustering is defined as the task of finding arbitrary oriented subspace clusters using only a small sample of supervised background knowledge provided by a domain experts. This work describes a first foray into this novel approach and provides an implementation of a basic algorithm to perform this task. We have found that even a small amount of supervised background knowledge can significantly improve the quality of correlation clustering in general. With confidence it can be stated, the results of this work have the potential to inspire several more semi-supervised approaches to correlation clustering in the future.

Keywords: Clustering · Subspace · Correlation · Semi-supervised · Background knowledge

1 Introduction

Explainable Artificial Intelligence (XAI) is rapidly gaining popularity among the data science community [6]. XAI is mainly motivated by the fact that neural networks, especially deep neural networks are often times treated as "black boxes", i.e. procedures and errors can be close to impossible to be comprehended by humans [8]. This creates problems that are interdisciplinary and manifold. For example, one major issue is the lack of trust in machine learning algorithms, both concerning the input and the results. XAI could help to build trust in AI, not only concerning the general population, but also researchers from domains other than computer science, i.e. users of such algorithms [8]. Even these domain experts want to be assured that the results they are seeing are direct and uncontaminated answers to their input and therefore require explainable results and

© Springer Nature Switzerland AG 2019
G. Amato et al. (Eds.): SISAP 2019, LNCS 11807, pp. 221–230, 2019.
https://doi.org/10.1007/978-3-030-32047-8_20

processes [13]. But outside of AI and deep learning, other data mining tasks and algorithms could also benefit from increased explainability [14].

Exploring the design space at the core of this work, a semi-supervised approach to correlation clustering, has the potential of producing explainable correlation clustering algorithms. In this context linear correlation clustering is defined as the task of finding arbitrary oriented subspace clusters [15]. This is, firstly, because quantitative models explaining the results of correlation clustering algorithms can easily be derived and made interpretable by domain experts [2]. Secondly, because semi-supervised algorithms such as constrained k-means clustering and similar algorithms allow for background knowledge driven machine learning, i.e. human ideas and opinions can influence the results of such algorithms. This allows domain experts to more accurately raise their intended queries towards the data [9]. Since to the best of our knowledge there has not been any semi-supervised algorithm that tackles the data mining task of correlation clustering, the goal of this work is to design, implement and evaluate an algorithm that detects linear correlated clusters given a small subset of a priori labeled data instances. The major contributions of this work are the introduction of SIDEKICK (SupervIseD Expert Knowledge Influenced Correlation Clustering), a first semi-supervised correlation clustering algorithm and a novel notion of correlation in the context of clustering (ϕ-correlated clusters).

2   Related Work

Since we are introducing the first linear correlation clustering algorithm that considers supervised background knowledge, we owe definitions as well as an overview on the related work in both fields, namely (a) linear correlation clustering and (b) clustering with supervised background knowledge. The task of linear correlation clustering is defined as finding clusters within a given data set that are located within interesting subspaces which are arbitrarily oriented [15]. It further means that the data objects within each of the clusters exhibit a linear correlation between a subset of their features. In this context a broad range of related work exists. ORCLUS [7] was the first of its kind tailored for detection of such clusters, followed by other algorithms such as 4C [10], HiCO [5], ERiC [4], COPAC [3] and CASH [1].

All named methods so far excel at certain aspects, but are not capable of dealing with supervised background knowledge. This is an aspect which our method is addressing. On the other hand there are semi-supervised methods for other types of clustering. Such clustering tasks, where results are influenced by semantic decisions by domain experts or data analysts are called *constrained clustering* tasks. The earliest algorithm tackling this task, Constrained K-means, performs a variation of the classic k-means algorithm under the restriction that instances have to be members of their corresponding clusters using a *must-link* constraint and two instances are not allowed to be in the same cluster, i.e. a *cannot-link* constraint [19]. There are multiple categories of constraints. At the time of this work we know of other contributions introducing instance-level, cluster-level and model-level constraints. Instance-level constraints are constraints specified between two instances, for example the must-link and cannot-link constraints from above. Cluster-level constraints are specified for multiple instances belonging to a cluster. For example ϵ- and δ-constraints [11]. Model-level constraints use a different approach. Here a user is shown the result of a certain clustering algorithm. Then the user can decide whether they like this specific result or not. If they do not like it, the algorithm is repeated but with automatic constraints that ensure that the result does not resemble the undesired results from earlier iterations [12].

2.1   Deriving Quantitative Models for Correlation Clusters

The main inspiration and basis of SIDEKICK is a 2006 paper with the title "Deriving Quantitative Models for Correlation Clusters" [2]. The primary idea behind this publication was to add an additional post-processing step to existing correlation clustering algorithms. In essence, a model is derived by using PCA on each cluster. The smallest number of Eigenvectors of this cluster C_i that explain a percentage higher than a predefined threshold α are called *strong Eigenvectors* $\check{V}_C$ of V_C. The remaining Eigenvectors are the *weak Eigenvectors* $\hat{V}_C$ of V_C. The the weak eigenvectors are used to derive a Hesse Normal Form hyperplane equation system, that can be solved by Gauß-Jordan Elimination for better readability. The paper also suggests a method of using the generated models to predict the cluster membership of additional instances. In order to predict the cluster membership of an arbitrary point p the distance between p and the correlation hyperplane of each Cluster C_i needs to be calculated. The distance is equal to the length of the vector between the orthogonal projection of p onto the correlation hyperplane and the point p itself. The projection can be calculated using the strong eigenvectors $s_1, s_2, ..., s_\lambda \epsilon \check{V}_{C_i}$ and the point, both normalized by the mean vector $\bar{c}_i$, in the following fashion:

$$d(p, C_i) = ||p - \sum_{j=1}^{\lambda} \langle p, s_j \rangle s_j ||$$

By calculating the distance of all cluster instances to the corresponding hyperplane the standard deviation σ_i can be derived. Assuming that the deviations of each cluster C_i fit to a Gaussian Distribution G_i, we can derive an equation to calculate the probability of a point p belonging to a distribution of a certain cluster C_i:

$$G_i(p, \sigma_i) = \frac{1}{\sigma_i \sqrt{2\pi}} e^{-\frac{(d(p, C_i))^2}{2\sigma^2}} \quad P(C_i|p, \sigma_i) = \frac{G_i(p, \sigma_i)}{\sum_{j=1}^{n} G_j(p, \sigma_j)}$$

3   Semi-supervised Correlation Clustering

As stated in the introduction, the goal of this work is to design an algorithm that provides a solution to a very specific problem. A domain expert has found correlations in a R-dimensional database D^R and specified certain instances which belong to these different correlations. The set of sets of the instances belonging to a correlation is called the background knowledge BK, while each single correlation cluster derived from the background knowledge is denoted as BK_i. The cardinality of BK_i is thought to be significantly smaller than the cardinality of D^R. While the dimensionality of BK is equal to R, the correlation dimensionality of each individual background knowledge cluster λ_{BK_i} has to be strictly smaller than R. Simply said the task is to utilize the background knowledge provided by a domain expert to decide which other unlabeled instances in D belong to each of the correlation clusters BK_i, while adhering to the restrictions set by the expert knowledge. To be more concise, the opinion of the domain expert is extrapolated and applied to $D\backslash\{BK\}$.

This task can be designated as semi-supervised, because it is an traditionally unsupervised task – correlation clustering – but includes preexisting supervised expert knowledge that influences the results. Additionally the resulting models can be simplified as described in Subsect. 2.1 and the domain expert can compare the models derived from the expert knowledge to the results. Thereby they receive feedback about how accurately the correlations can be extrapolated onto the unlabeled instances and how much the resulting model deviates from their provided background knowledge. Furthermore all subspace clustering algorithms are usually categorized as either top down or bottom up, because of the circular dependency inherent to the task [15]. In the case of semi-supervised correlation clustering the background knowledge already provides an a priori clustering. Thus the algorithm can work similarly to top-down algorithms using the background knowledge as a starting point. The task we are left with is finding the relevant subspace for each cluster and searching for other instances that are situated in these subspaces. In order to find these subspaces and predict additional points, this work contains a semi-supervised extension of the algorithm from Subsect. 2.1.

3.1 SIDEKICK

SIDEKICK follows four main steps. At the beginning we calculate the correlation hyperplanes for each background knowledge cluster using the algorithm from [2], which corresponds to derive_model(X). Additionally in accordance with Sect. 2.1 the standard deviation σ of distances between the cluster instances and the corresponding correlation hyperplane is calculated for each cluster. Each correlation cluster C_i is thereby clearly defined by its model, which is at this point defined as consisting of its eigenvectors – split into weak and strong (i.e.) – and the standard deviation σ_i and mean vector μ_i. The strong eigenvectors are needed to calculate the distance between an instance and the hyperplane. The standard deviation is needed to derive the normal distribution that is assumed to have produced the noise along the weak components. In summary:

1. Derive the underlying models for the each of the ground truth clusters
2. Predict the labels for all unlabeled instances

3. Assign either all or a subset of the predicted instances to their corresponding cluster
4. Deriving updated models for each resulting cluster and simplifying the hyperplane equation to highlight the underlying correlations

3.2   Unlimited SIDEKICK

The basic algorithm is called unlimited SIDEKICK since at step 3 all unlabeled instances are added to their corresponding cluster. This performs well with highly correlated clusters. Generally if a cluster is 100% correlated, i.e. it only possesses variance along the strong components SIDEKICK is expected to classify unlabeled instances correctly, even when only using a background knowledge consisting very few instances. However, missclassifications at the intersections of different clusters can occur, but only in cases where the correct membership of an instance is indeterminable anyways. Generally the Winner-Takes-All principle is applied, i.e. every instance is always classified as a member of the most probable cluster. On the other hand the biggest short-coming of the unlimited SIDEKICK algorithm is that whenever there are instances that were generated by a process not belonging to a cluster that was included in the background knowledge, almost all of these outliers are added to the cluster with the highest variance along the weak components, i.e. the lowest density along the weak components. This effect can be explained by the properties of the Gaussian distributions that are used to assign instances to a cluster. A slight difference of σ between two clusters has a huge impact on the normalized probability for a single instance and thereby it will be classified as highly likely to belong to the first cluster, even though it most probably does not belong to either correlation. In conclusion the resulting problem is how to find a suitable subset of the predicted points for each cluster to be assigned.

3.3 ϕ-correlated SIDEKICK

Our approach to solve this problem is to work with the definition of strong components and α from [2]. Then only using a new hyperparameter ϕ and the sum of the strong eigenvalues $\breve{e}$ of a background knowledge cluster, we can approximate the standard deviation l along the weak components of the resulting cluster:

$$l = \sqrt{\left(\frac{1}{\phi} - 1\right) \cdot \sum_{\forall e_i \in \breve{e}} e_i}$$

This cluster is then ϕ-correlated, in accordance to the original definition of α. Which is the minimum ratio of the total variance that can be explained by the strong components. Using the eigenvalues e of a cluster it holds:

$$\frac{\sum_{i=1}^{\lambda} e_i}{\sum_{i=1}^{d} e_i} \geq \alpha$$

Additionally, the number of strong components is equal to the correlation dimensionality λ of a cluster. That means, if the variance of the weak components becomes to high, they might become strong components. Therefore, the maximum variance $\hat{e}$ that the weak eigenvectors can explain while still resulting in a ϕ-correlated cluster is therefore equal to

$$\hat{e} = \frac{\check{e}}{\phi} - \check{e}$$

According to the Three Sigma Rule [17] about 68% of all instances belonging to a normal distribution are situated within a distance of one standard deviation from the mean and 99.7% at triple the standard deviation. That means if we only include the subset of instances for which the distances are lower than three times the square root of this variance, we can find about 99.7% of the instances that would be included in a normal distribution of a cluster that is ϕ-correlated. Which would mean that step 3 of SIDEKICK now involves only the subset of the predicted instances that are part of a certain correlation with a strength of ϕ. Therefore, as long as the direction of the hyperplane is close to the background knowledge, a domain expert can now specify the exact strength of correlation they are searching for.

3.4   Runtime Complexity

If we denote n as the total number of instances, c as the total number of clusters and bk as the amount of instances used as background knowledge the complexity of SIDEKICK for all clusters can be computed as $\mathcal{O}(c \cdot n)$, if bk is significantly smaller than n or $\mathcal{O}(c \cdot bk_2)$ if not. We came to this conclusion by simplifying the overall runtime complexity per cluster, that can be computed as the sum of the complexities of:

- Performing PCA on the background knowledge; Using Power Iteration [16] this would be $\mathcal{O}(bk^2)$
- Computing the standard deviation of the distances between the background knowledge objects and their corresponding clusters; This is $\mathcal{O}(bk)$
- Computing the probability of each object belonging to a cluster model; This is $\mathcal{O}(2(n - bk))$ for each unlabeled object, since we need to make two computations per object (c.f. Sect. 2.1)

In summary the overall complexity can be denoted as:

$$\sum_i^c (\overbrace{\mathcal{O}(bk_i^2)}^{\text{PCA}} + \overbrace{\mathcal{O}(bk_i)}^{\text{Compute standard deviation}} + \overbrace{\mathcal{O}(n - bk_i)}^{\text{Probability for single cluster}} + \overbrace{\mathcal{O}(n - bk_i))}^{\text{Normalized Probability over all clusters}}$$

The different variants only add a single step with a complexity of $\mathcal{O}(n)$ and therefore do not have a significant effect on the overall complexity.

4   Experiments and Discussion

In this section we evaluate SIDEKICK using two synthetic data sets. One data set contains 3000 three dimensional instances segmented into six linear correlated clusters, five with a correlation dimensionality of one and a single cluster with a correlation dimensionality of two. What makes this data set especially challenging is the fact that the linear correlated clusters exhibit different densities. The second data set is equal to the first one, except for a set of 1250 randomly generated outlier objects that were added to it. Using these two datasets we compared SIDEKICK against five established correlation clustering algorithms from the related work section (Sect. 2). Suitable Hyperparameter settings for each of the five algorithms have been determined through either a grid-based or a sequential scan (ERiC) for both data sets. The experiments were conducted using the ELKI [18] data mining framework. Furthermore, we have provided all the hyperparameter settings in Tables 1 and 2 and made the source code for SIDEKICK and the test data sets publicly available[1] to ensure reproducibility. Table 1 illustrates that the correlation clustering algorithm 4 C achieved the best results with an Adjusted Rand Index (ARI) of 78,93%. When we used the same hyperparameter settings on the data set that contains outliers, 4C remained the best performing correlation clustering algorithm. Its ARI decreased from 0.7893 to 0.7074 while other competing methods achieve a by far lower ARI score as seen in Table 2.

Table 1. ARI results and hyperparameter settings of competitive methods on the synthetic data set

Dataset	Algorithm	ARI	Hyperparameter settings
Without noise	CASH	0.5998	minpts: 370, maxlvl: 20, jitter: 2.5
Without noise	4C	**0.7893**	ε: 8.0, minpts: 15
Without noise	COPAC	0.5691	ε: 8.0, minpts: 15, kNN: 81
Without noise	ORCLUS	0.7351	k: 6, l:2
Without noise	ERiC	0.2260	k: 6

To evaluate SIDEKICK we started by using only unlimited SIDEKICK to cluster the data set without outliers, since all other variants were specifically designed to deal with outliers. We chose the background knowledge randomly from each cluster, mimicking a domain scientists expertise. To avoid the influence of the random choice, we repeated the experiment one hundred times using different random seeds. Our method achieved an average ARI of 95%. To cluster the data set that contains outliers, we used the ϕ algorithm, setting ϕ individually for each cluster. We started by using only 1% background knowledge per cluster, which equals to 3 instances per cluster. This yielded an average ARI of 69%

---

[1] https://github.com/huenemoerder/SIDEKICK.

Table 2. ARI results and hyperparameter settings of competitive methods on the synthetic data set with noise

Dataset	Algorithm	ARI	Hyperparameter settings
With noise	CASH	0.5348	minpts: 400, maxlvl: 30, jitter: 4.5
With noise	4C	**0.7074**	ε: 8.0, minpts: 15
With noise	COPAC	0.3786	ε: 8.0, minpts: 15, kNN: 89
With noise	ORCLUS	0.2465	k: 6, l:2
With noise	ERiC	0.146	k: 6

Table 3. ARI results and hyperparameter settings of SIDEKICK on the synthetic data set with and without noise

Setting	Algorithm	Average ARI	Max. ARI	Min. ARI	Variance ARI
Without noise, 0.99 BK	Unlimited	0.9569	1.0000	0.5634	0.0064
With noise, 0.99 BK	ϕ	0.6957	0.9644	0.2941	0.0248
With noise, 0.96 BK	ϕ	0.9381	0.9747	0.8568	0.0007

with a variance of 2% which means it was on average as good as 4C. When we increased the amount of background knowledge to 4%, which means that we used 20 instances as domain expert knowledge per cluster, the average ARI rose to 93%. This is superior to any of the state-of-the-art correlation clustering methods in our experiments. Even in the worst case, we got an ARI of 85%, which is still above any of the competing methods (Table 3).

Finally we want to highlight, that it was not our intention to show whether SIDEKICK is better than any of its competitors, since these competitors do not use any background knowledge. Rather the core message we want to convey is, that even a small amount of background knowledge is sufficient to boost the performance of solving a correlation clustering task significantly. Adding a further hyperparameter like ϕ can increase the robustness against noise and outliers.

5   Conclusion and Future Work

In conclusion the algorithm discussed in this work demonstrates the prospects of a semi-supervised approach to correlation clustering. As we have seen and discussed in the experiments using just a small amount of background knowledge can drastically improve the results of solving a correlation clustering task. Therefore any correlation clustering algorithm could theoretically benefit from such background knowledge.

Generally, SIDEKICK and its different variants should provide a useful toolkit for data exploration. In relation to the different variants of SIDEKICK itself we have learned that when working with a data set without outliers unlimited SIDEKICK is always the best choice. When working with a data set that contains many outliers choosing individual ϕ's for each cluster should be the best solution. The only complication that revealed itself is the dependence on correct background knowledge. But, firstly, this is intentional, because the goal of the algorithm was to trust the knowledge of the domain experts and only change it slightly at best. Secondly, this conclusion is somewhat diminished by the fact that during evaluation, the background knowledge for each cluster was sampled randomly from that cluster. Background knowledge provided by humans should usually be much closer to the truth and thereby provide better results even when using small amounts of instances as background knowledge.

Acknowledgement. This work has been funded by the German Federal Ministry of Education and Research (BMBF) under Grant No. 01IS18036A. The authors of this work take full responsibilities for its content.

References

1. Achtert, E., Böhm, C., David, J., Kröger, P., Zimek, A.: Global correlation clustering based on the hough transform. Stat. Anal. Data Min.: ASA Data Sci. J. **1**(3), 111–127 (2008)
2. Achtert, E., Böhm, C., Kriegel, H.P., Kröger, P., Zimek, A.: Deriving quantitative models for correlation clusters. In: Proceedings of the 12th ACM SIGKDD International Conference on Knowledge Discovery and Data Mining, pp. 4–13. ACM (2006)
3. Achtert, E., Böhm, C., Kriegel, H.P., Kröger, P., Zimek, A.: Robust, complete, and efficient correlation clustering. In: Proceedings of the 2007 SIAM International Conference on Data Mining, pp. 413–418. SIAM (2007)
4. Achtert, E., Böhm, C., Kriegel, H.P., Zimek, A., et al.: On exploring complex relationships of correlation clusters. In: Null, p. 7. IEEE (2007)
5. Achtert, E., Böhm, C., Kröger, P., Zimek, A.: Mining hierarchies of correlation clusters. In: 18th International Conference on Scientific and Statistical Database Management, pp. 119–128. IEEE (2006)
6. Adadi, A., Berrada, M.: Peeking inside the black-box: a survey on explainable artificial intelligence (XAI). IEEE Access **6**, 52138–52160 (2018)
7. Aggarwal, C.C., Yu, P.S.: Finding generalized projected clusters in high dimensional spaces, vol. 29. ACM (2000)
8. Goebel, R., et al.: Explainable AI: the new 42? In: Holzinger, A., Kieseberg, P., Tjoa, A.M., Weippl, E. (eds.) CD-MAKE 2018. LNCS, vol. 11015, pp. 295–303. Springer, Cham (2018). https://doi.org/10.1007/978-3-319-99740-7_21
9. Basu, S., Davidson, I., Wagstaff, K.: Constrained Clustering: Advances in Algorithms, Theory, and Applications. CRC Press, Boco Raton (2008)
10. Böhm, C., Kailing, K., Kröger, P., Zimek, A.: Computing clusters of correlation connected objects. In: Proceedings of the 2004 ACM SIGMOD International Conference on Management of Data, pp. 455–466. ACM (2004)
11. Davidson, I., Ravi, S.: Clustering with constraints: feasibility issues and the k-means algorithm. In: Proceedings of the 2005 SIAM International Conference on Data Mining, pp. 138–149. SIAM (2005)
12. Gondek, D., Vaithyanathan, S., Garg, A.: Clustering with model-level constraints. In: Proceedings of the 2005 SIAM International Conference on Data Mining, pp. 126–137. SIAM (2005)

13. Holzinger, A., Kieseberg, P., Weippl, E., Tjoa, A.M.: Current advances, trends and challenges of machine learning and knowledge extraction: from machine learning to explainable AI. In: Holzinger, A., Kieseberg, P., Tjoa, A.M., Weippl, E. (eds.) CD-MAKE 2018. LNCS, vol. 11015, pp. 1–8. Springer, Cham (2018). https://doi.org/10.1007/978-3-319-99740-7_1
14. Kazempour, D., Seidl, T.: Insights into a running clockwork: On interactive process-aware clustering. In: Proceedings of the 22nd International Conference on Extending Database Technology (EDBT) (2019, in press)
15. Kriegel, H.P., Kröger, P., Zimek, A.: Subspace clustering. Wiley Interdiscip. Rev.: Data Min. Knowl. Discov. 2(4), 351–364 (2012)
16. Mises, R., Pollaczek-Geiringer, H.: Praktische verfahren der gleichungsauflösung. ZAMM-J. Appl. Math. Mech./Zeitschrift für Angewandte Mathematik und Mechanik 9(2), 152–164 (1929)
17. Pukelsheim, F.: The three sigma rule. Am. Stat. 48(2), 88–91 (1994). http://www.jstor.org/stable/2684253
18. Schubert, E., Zimek, A.: ELKI: a large open-source library for data analysis - ELKI release 0.7.5 "heidelberg". CoRR abs/1902.03616 (2019). http://arxiv.org/abs/1902.03616
19. Wagstaff, K., Cardie, C., Rogers, S., Schrödl, S., et al.: Constrained k-means clustering with background knowledge. In: ICML, vol. 1, pp. 577–584 (2001)

Subspaces and Embeddings

Query Filtering with Low-Dimensional Local Embeddings

Edgar Chávez[1], Richard Connor[2(✉)], and Lucia Vadicamo[3]

[1] Centro de Investigación Científica y de Educación Superior de Ensenada
(CICESE), Ensenada, Mexico
elchavez@cicese.mx

[2] Division of Mathematics and Computer Science, University of Stirling,
Stirling, Scotland
richard.connor@stir.ac.uk

[3] Institute of Information Science and Technologies (ISTI), CNR,
Via Moruzzi 1, 56124 Pisa, Italy
lucia.vadicamo@isti.cnr.it

Abstract. The concept of *local pivoting* is to partition a metric space
so that each element in the space is associated with precisely one of a
fixed set of reference objects or pivots. The idea is that each object of
the data set is associated with the reference object that is best suited to
filter that particular object if it is not relevant to a query, maximising
the probability of excluding it from a search. The notion does not in itself
lead to a scalable search mechanism, but instead gives a good chance of
exclusion based on a tiny memory footprint and a fast calculation. It is
therefore most useful in contexts where main memory is at a premium,
or in conjunction with another, scalable, mechanism.

In this paper we apply similar reasoning to metric spaces which pos-
sess the four-point property, which notably include Euclidean, Cosine,
Triangular, Jensen-Shannon, and Quadratic Form. In this case, each element of the space can be associated with two reference objects, and a
four-point lower-bound property is used instead of the simple triangle
inequality. The probability of exclusion is strictly greater than with simple local pivoting; the space required per object and the calculation are
again tiny in relative terms.

We show that the resulting mechanism can be very effective. A consequence of using the four-point property is that, for m reference points,
there are $\binom{m}{2}$ pivot pairs to choose from, giving a very good chance of a
good selection being available from a small number of distance calculations. Finding the best pair has a quadratic cost with the number of references; however, we provide experimental evidence that good heuristics
exist. Finally, we show how the resulting mechanism can be integrated
with a more scalable technique to provide a very significant performance
improvement, for a very small overhead in build-time and memory cost.

Keywords: Metric search · Extreme pivoting · Supermetric space ·
Four-point property · Pivot based index

G. Amato et al. (Eds.): SISAP 2019, LNCS 11807, pp. 233–246, 2019.
https://doi.org/10.1007/978-3-030 32047-8_21

1   Introduction

In a metric space, the distances among a set of any three objects may be used to construct a triangle in Euclidean space, with corresponding vertices and edge lengths. If any two of these distances are known, the triangle inequality property can be used to determine upper and lower bounds for the third.

For any supermetric space [12], the distances among a set of any four objects can be used to construct a tetrahedron in Euclidean space, with corresponding vertices and edge lengths. An equivalent lower-bound calculation can be made for a final edge length, given any four objects, when any five of the six distances among them are known. In simple terms, the five distances can be used to fix two adjacent faces of a tetrahedron, and lower and upper bounds for the last edge can be easily determined by considering the rotation of the two triangles around their common edge.

We show a novel way of exploiting this situation, as follows. From a finite metric space S, a relatively small set of reference objects P is selected. For all $p_i, p_j \in P$, the distance $d(p_i, p_j)$ is calculated and stored. For each element s_i in S, a single pair of reference objects $\langle p_x, p_y \rangle$ is selected, and the the distances $d(s_i, p_x)$ and $d(s_i, p_y)$ are stored. Thus the space S is represented as a set of tuples $\langle x, y, d(s_i, p_x), d(s_i, p_y) \rangle$, indexed by i, therefore requiring only a few bytes per object.

When a query is executed, the distances $d(q, p_i)$ for each $p_i \in P$ are first calculated. At this point, considering the objects q, p_x, p_y and any $s_i \in S$, five of the six distances among them can be retrieved, leaving only $d(q, s_i)$ as an unknown. Therefore, for each element of s_i a lower-bound for $d(q, s_i)$ can be calculated, with a cheap geometric calculation, without any requirement to access the original value $s_i \in S$.

The approach we take is based on the observation that, for a selection of n reference points, there exist $\binom{n}{2}$ pairs from which the representation of each data point can be selected. This number, of course, becomes rapidly very large even with modest increases in n. If for each element of S we can find a particularly effective pair p_i, p_j, within this large space, then this tiny representation of S can be used as a powerful threshold query filter. This exclusion mechanism leads to a sequential scan, which is virtually unavoidable in light of a recent conditional hardness result in [18] for nearest neighbor search, even in the approximate setup, computing a $(1 + \epsilon)$-approximation to the nearest neighbor requires $\Omega(N - \delta)$ time, with N the size of the database.

The above hardness result has been suspected for a long time by the indexing community, and it has been named the *curse of dimensionality*. It is known, for example, that a metric inverted index [1] has high recall rates only if a substantial part of the candidate results is revised. We aim our approach at this final part of query filtering or re-ranking.

The contributions of this paper are as follows:

1. We show that the outline mechanism is viable. For the well-known SISAP benchmark data sets we show exclusion rates of over 99% can be achieved using our small memory footprint and cheap calculations.
2. We examine the problem of finding the best pair of reference points per datum; this can be done perfectly, but expensively, by an exhaustive search of the pair space. We show that much cheaper heuristics are also effective.
3. Finally, we show one example of how the mechanism can be combined with another, by describing its incorporation with the List of Clusters index. We use a pragmatic selection of pivot pairs to ensure that no new distances are measured at either construction or query time, and show a halving of overall query cost.

2   Related Work

Pivot based indexes have populated the metric indexing scene for long time. A pivot table, with the triangle inequality, is just the direct product of one dimensional projections obtained from a single pivot at a time. Each coordinate gives a lower bound to the actual distance from database points to the query. A safe choice is to take the maximum over all the available lower bounds. The most competitive algorithm published for searching is AESA [21], which proceeds as following. All the $O(n^2)$ distances between every object in the database of n elements is pre-computed. With this, every object in the database is a potential pivot. At query time a subset of the n pivots is selected, one at a time, using a heuristic which consist in selecting the $j+1$ pivot, the closest to the query, using as bound the j pivots known so far and the first pivot at random. The output of this heuristic is both a set of *good* pivots for the query, and the nearest object to it. Two things can be noticed from this basic approach, first, the number of pivots actually used is way smaller than n, and second, they are tailored for each query on the fly. Since the space usage is quadratic, the approach is impractical. Also notice that a sequential scan is implied to obtain the closest next pivot in the interaction. Linear space approaches of the same idea were used in [17], and a better heuristic for selecting the next pivot is proposed in [14]. The sequential scan can be avoided using a tree [2,5].

Selecting the best pivot for a given query is not possible offline. A weaker alternative is to select the best pivot for each database object, increasing the probability of exclusion at query time. Two options have been explored in the literature, in [6] each pivot in the pool only keep distances to objects in the extreme of the distribution, those objects near and far the pivot. This process is sub-optimal and may end with a few objects guarded by many pivots, and many objects guarded by a few or none pivots. A second alternative, ensuring some fairness in the coverage, was proposed in [19], this time each object can select the best pivot. This latter approach is called *extreme pivoting*. In those heuristics the gain is in filtering power, when the amount of available memory is fixed. Pivot tables are useful for post filtering in a hierarchical metric index, as in [20], or they can be used as a stand alone index using directly the table as in [7,9].

The exclusion based on the four point property was firstly proposed in [11], and generalized to $n + 1$ polytopes in [13]. The exclusion increased with the dimension of the polytope.

For post-filtering, when a primary index is applied to filter the data and only a small fraction of the databases should be checked against the original metric, a table is useful. A high rate of exclusion will prevent the use of the more expensive distance computation, and moreover, it will require to fetch a smaller number of objects from secondary memory. Hence a small table, with just a couple of coordinates, is an excellent trade-off because it can be kept in main memory. In the same spirit as the extreme pivots for unidimensional mapping, in this paper we are aiming at building a table of small memory footprint using the four point property.

2.1 The Four-Point Property and Supermetric Spaces

Much work on finite isometric embeddings was conducted in the 20th century, by e.g. Blumenthal [4], Wilson [22] and Menger [16]. Blumenthal uses the phrase *four-point property* to mean a space that is 4-embeddable in 3-dimensional Euclidean space: that is, that for any four objects in the original space it is possible to construct a distance-preserving tetrahedron.

More recently we have applied these results in theoretical mathematics to the practical domain of metric search [10–12]. For this context, the important result is that the four-point property applies to many commonly-used distance metrics, including Euclidean, Cosine[1], Jensen-Shannon, Triangular and Quadratic Form distances, all of which can be safely used in conjunction with the mechanisms described here.

2.2 The Four-Point Planar Lower Bound

For two points that have not been directly compared, q and s_i, it is shown in [12] how a lower bound of their distance can be established by comparing the distances between both points and two further reference points. For reference points p_1 and p_2, two triangles with a common base, $\triangle p_1 q p_2$ and $\triangle p_1 s_i p_2$, can be used to form two adjacent faces of a tetrahedron. Because of the four-point property, the unmeasured distance $d(q, s_i)$ must form the sixth edge of a tetrahedron. It is then clear, by consideration of the rotation of these triangles around the common baseline $\overline{p_1 p_2}$, that upper and lower bounds for the distance $d(q, s_i)$ can be determined as the two cases where the triangles lie in the same plane.

A lower bound of their distance can therefore be calculated by notionally plotting Cartesian points p_1' and p_2' arbitrarily on 2D axes, say at positions $(0, 0)$ and $(d(p_1, p_2), 0)$ respectively, and then plotting points q' and s_i', both above the X-axis, according to their respective distances from p_1 and p_2. Then the distance $\ell_2(q', s_i')$ is a lower bound of $d(q, s_i)$.

---

[1] for the correct formulation, see [10].

The value of this is that, independently of the size of individual data values and the cost of the distance metric, any value can be represented, for a fixed choice of reference points, as a small 2D coordinate, and compared with a cheap 2D ℓ_2 distance; the result of this comparison may mean that there is no requirement for the full comparison to be made. Of course, the value of the method depends heavily upon the probability of its success.

3   Distribution of Values in the 2D Plane

To visualise this property we use scatter diagrams constructed as follows. The two selected reference points are plotted on the X-axis according to the distance between them, and a data set is represented as points in the 2D space plotted above the X-axis, according to their respective distances from these reference points. The triangle inequality property gives the ability to create such a plot.

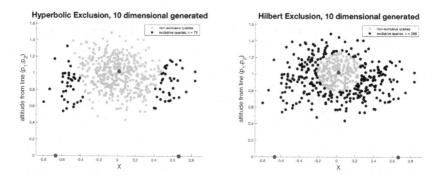

Fig. 1. 500 points from a generated Euclidean space plotted against randomly selected reference points. Left and right plots show the exclusion potential based on simple metric (left) and supermetric (right) properties. (Color figure online)

Figure 1 shows two versions of such a scatter plot created from a 10-dimensional Euclidean space, using the same data and reference points. Although the triangle inequality property guarantees the ability to create such a plot, the relationship among the plotted points is more subtle.

An example query point is selected from the centre of the diagram, coloured blue. For every other point plotted in the plane, we then consider whether it *might* be within a threshold distance t from this blue-coloured point, based only on the distances calculated to the two reference points. Here we have chosen $t = 0.24$, representing around one-millionth of the volume of the generated space.

The diagrams are then colour-coded so that those points which *may* be within that distance, i.e. those that cannot be excluded from a search, are highlighted, plotted in yellow. The four-point planar lower bound is illustrated on the right-hand side, clearly represented by a simple exclusion radius in the 2D plane. On the left-hand side, only the triangle inequality property is used, giving much wider hyperbolic bounds.

These diagrams represent a situation where only two reference points have been used, with respect to a single query. The left-hand side shows the effectiveness of local pivoting, where this very small amount of information allows 73 out of 500 data points to be excluded from the candidate solution space. It can be seen on the right-hand side that, if the four-point property can be used, then 298 ex 500 potential solutions can be excluded, using exactly the same information.

4   Independence of Reference Points

It appears that, for a given choice of reference points, the *distribution* of other points in the 2D plane with respect to these points is fairly predictable. However, where *individual* data points land within the scatter varies widely with the choice of reference points.

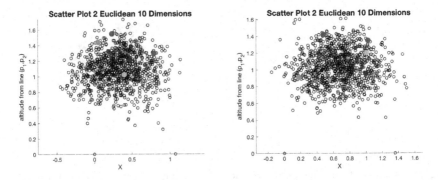

Fig. 2. 1,000 points plotted in the 2D plane based on two different, randomly selected, pairs of reference points. The data plotted is the same in each diagram, and the colour-coded points represent the same values. The (X, Y) scatter is similar in both cases, close to uncorrelated normal distributions on both axes, but it can be seen that where an individual point lands depends on the choice of reference points. (Color figure online)

Figures 2 and 3 show some diagrams to illustrate this. In each figure, a single set of data points is plotted in the XY plane according to their distances from two randomly-selected reference points; the left and right sides of each figure now represent the same data plotted against a *different* choice of reference points.

In the two figures, a random selection of five data points has been made and these are highlighted in colour in the charts; that is, the coloured spots in the left and right sides of the figure represent the same data point and its position with respect to the different reference points. It can be seen that there is a relationship among the positions where the coloured dots are plotted, but only a relatively weak one.

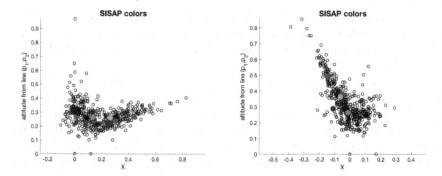

Fig. 3. 500 points from the *colors* data set plotted in the 2D plane, again with two randomly selected pairs of reference points. It can be seen how much the distribution changes in a non-uniform set with the choice of reference points. Again, where an individual point lands within the scatter depends on the choice of reference point. (Color figure online)

4.1   Choice of Reference Points

An underlying hypothesis in our work is that the distribution of queries within U will be similar to the distribution of S within U. Thus, looking at the scatter diagrams in Figs. 2 and 3, we could be viewing the distribution of either data or queries with respect to those same reference points. The probability of successful elimination, for a given q and s_i, therefore depends upon the choice of reference points, and the relative position of both s_i and q with respect to them.

If the hypothesis is correct, then the notion of a "good" pair of reference points for an individual $s_i \in S$ corresponds to the (inverse) *density* of the region where s_i lands, within a representative set. If query and datum lie further than the query threshold within the 2D plot, then the datum cannot be a solution to the query; this is most likely to occur when either query or datum lie within a sparsely populated region of the plane. If queries and data follow the same distribution patterns, then the best pair of reference points can be selected with reference to a representative set of data points from within S.

5   Selection and Query

5.1   Selection of Best Reference Pair

It is possible to use a statistical technique to select a good reference point pair per individual datum. A sample set of data is used, the *witness* set.

For a query over a finite metric space (S, d), first a set of n objects is taken from S and used to form a set P comprising numbered reference points p_i. For a given set of n reference objects, each of the $\binom{n}{2}$ pairs p_i, p_j is considered. For each, a 2D Euclidean space is built, exactly corresponding to those depicted in the earlier figures. Each space is built using the data from the witness set, according to the distances of each element to the pair of reference objects. These spaces may be efficiently searched using normal metric indexing techniques, and as the space is a genuine 2D space very efficient mechanisms such as the KD-Tree [3] can be used.

Each element of the data set is now considered as a query against each of these $\binom{n}{2}$ metric indexes, and the one with the least *local density* is selected to represent that element. There are various mechanisms for assessing local density, for example the smallest number of results for a threshold query, or the largest distance in the result set of a kNN query. We tested various ways over some different data sets and found relatively little difference in the cost or outcome, and settled on the strategy of picking the pair which gave the largest distance to the third-nearest 2D point.

While this mechanism is effective, it is of course extremely expensive, with a quadratic cost according to the number of reference points. In general, for high-dimensional queries, a relatively large number of reference points will be required. We discuss linear geometric approximations in Sect. 7.

5.2   Query

Having selected the most promising pair of reference points for each element $s_k \in S$, it is now represented as a tuple $\langle i, j, x, y \rangle$ where i and j are the identifiers of the reference points, and x and y are the 2D coordinates of the point where these reference points cause s_k to be projected onto the corresponding plane[2]. At query time, each distance $d(q, p_i)$ is first calculated; then for each tuple in the data, i and j are used to select the appropriate distances from which x_q and y_q can be calculated. Finally, the 2D Euclidean distance $\ell_2((x,y),(x_q,y_q))$ is calculated, which gives a lower bound to the distance $d(s_k, q)$ in the original space.

6   Initial Measurements

Table 1 shows the results of applying this strategy to the SISAP *colors* and *nasa* data sets [15]. The figures reported represent the proportion of the data set excluded when searched, using the reported technique, at each of the standard thresholds[3]. Note that the left-hand column reports the number of reference points used; while this represents of the number of distance calculations necessary, both per datum at build time and per query at query time, the number of available pairs is $\binom{n}{2}$ for n reference points, thus ranging from 45 to 11,175.

It is immediately apparent that the proposed mechanism is very effective. With only 10 reference points, already 97% of *colors* and 99% of *nasa* is successfully excluded at the smallest threshold. To put this in context, the top two rows of the table give the exclusion rates reported in [12] for the Distal SAT operating with both normal metric and supermetric exclusion mechanisms. However,

---

[2] as this is marginally more efficient than storing the distances to p_i and p_j.
[3] *colors*: 0.052, 0.083, 0.131; *nasa*: 0.12, 0.285, 0.53.

Table 1. Exclusion Rates for different numbers of reference points, taking the statistically best pair available. The top two rows give comparable figures for the Distal SAT index structure (see text).

	colors			nasa		
	t_0	t_1	t_2	t_0	t_1	t_2
DiSAT: 3pt	0.960	0.910	0.805	0.985	0.941	0.824
DiSAT: 4pt	0.980	0.943	0.840	0.991	0.964	0.851
no. of refs						
10	0.973	0.927	0.821	0.988	0.928	0.761
30	0.987	0.959	0.880	0.996	0.967	0.851
50	0.991	0.969	0.902	0.997	0.975	0.872
70	0.993	0.974	0.912	0.998	0.981	0.894
90	0.994	0.977	0.918	0.998	0.984	0.903
110	0.995	0.979	0.924	0.999	0.986	0.910
130	0.995	0.981	0.929	0.999	0.987	0.915
150	0.996	0.982	0.932	0.999	0.988	0.920

even although much better exclusion rates are achieved here, the mechanism is explicitly sequential.

Apparently, the value of the mechanism goes on increasing as the number of reference points is increased, with what appears to be a slow asymptotic approach towards perfect exclusion.

6.1 Build Cost

The dominant cost is in searching the 2D pair space at build time; the tables show results up to 150 reference points which of course also requires 150 distance calculations per datum. However these distance calculations are likely to be amortised within another search mechanism as shown in Sect. 8.

The cost of searching the pair space however increases quadratically with the number of reference points, making it infeasible for larger numbers. This cost is almost independent of the cost of distance calculations or size of data in the metric space: the cost of searching $\binom{n}{2}$ 2D spaces becoming quickly predominant as n increases. The cost is perfectly quadratic, in our experiments we have measured the cost $C(n) = 0.007n^2$ milliseconds for n pivots; even with only 150 reference points this is approaching 0.2 s per datum. In the context of searching a very large, high-dimensional, data set, then thousands of extra distance calculations are unlikely to be significant, but this would result in a huge potential space of reference point pairs that is intractable to search.

This leaves an interesting problem. The number of reference points does not typically constitute a performance problem in terms of distance calculations; the large cost is in the exhaustive search for the best pair of points. The reason the cost is high is because there are a huge number of potential pairs, which is the reason the mechanisms works so well. We have shown tremendous potential when the best pair of points is calculated from the very large number of pairs available. If we can find a way of finding these cheaply, ideally in a manner that scales linearly rather than quadratically with the number of reference points, the mechanism should become even more useful.

In the context of searching a very large, high-dimensional, data set, then thousands of extra distance calculations are unlikely to be significant, but this would result in a huge potential space of reference point pairs that is intractable to search; thus we seek linear-scaling solutions using geometric analysis instead. For once, it is not reasonable to assume an arbitrary amount of pre-processing time is acceptable in order to achieve a small improvement in query time.

7 Geometric Approach

A number of intuitively-derived methods for the selection of first and second reference points for were tested. In all cases, sets of 10, 50, 150 and 500 objects were chosen to act as reference points, and these were scanned linearly in two passes according to the following strategies. The intent is to find a strategy that gradually improves with respect to the number of reference points, but where the construction cost remains linear.

The strategies used for each of two linear-cost scans were as follows:

1. random, to act as a benchmark
2. for each data point, associate the closest reference point
3. for each data point, associate the farthest reference point
4. for each of the n reference points, associate it with the $\frac{1}{n}$ closest subset of the data (and do not consider these data points again)
5. for each of the n reference points, associate it with the $\frac{1}{n}$ farthest subset of the data (and do not consider these data points again)
6. having selected a first reference point, choose the second to minimise the altitude (Y-coordinate) of the plotted 2D apex point
7. having selected a first reference point, choose the second to minimise the horizontal displacement (X-coordinate) of the plotted 2D apex point

The first five strategies were tried for each of first and second reference point choice, whereas the last two were used only for the choice of the second point; thus a total of 35 different strategies were tested.

Methods (2) and (3) in any combination proved no better than random, and actually became slightly worse with a larger number of reference points; we believe this is because of non-uniformity within the sets and the presence of outliers in the reference points. This problem was fixed by use of methods (4) and (5), where the closest or farthest $\frac{1}{n}$ of the data is associated with each reference point.

Table 2 shows a few of the results. The first row shows a purely random choice for comparison. The second shows method (4) used for the first point, and

Table 2. Results shown only for the lowest threshold of the *colors* data set, other results are consistent. We give the build cost (msec per object) and exclusion rate for some of the strategies tested.

Pivot strategy			Number of pivots			
First	Second		10	50	150	500
Random	Random	build cost	0.0008	0.0008	0.0010	0.0016
		exclusion	0.929	0.925	0.926	0.924
Low dist	Low alt	build cost	0.014	0.022	0.045	0.124
		exclusion	0.930	0.958	0.967	0.966
Low dist	High dist	build cost	0.023	0.030	0.045	0.089
		exclusion	0.946	0.962	0.971	0.973

method (6) for the second. Finally the third row shows the use of method (4) for the first point and method (5) for the second, which gives the best compromise for these data sets and thresholds. The final effect of achieving 97% exclusion – as much as is achieved by a very sophisticated indexing structure over the full data set – through a linear cost construction of a 10-byte data representation is really a significant achievement. Note that in the cost comparisons, the "random" benchmark cost is effectively zero; at 500 pivots the cost of either mechanisms is restricted to around 0.1 ms per datum independent of the size of the data set, when the thorough search described in Sect. 5.1 would have cost 1.75 s.

8   Incorporation Within List of Clusters

Finally, we report results where our mechanism is incorporated with another, scalable, indexing mechanism. We have chosen a well-known indexing structure, and give a very simple technique which extends this using the four-point exclusion mechanism as a post-filter. That is, the mechanism is embedded within the original structure to act an internal filter, avoiding the calculation of original-space distances where the lower-bound calculation makes this unnecessary.

For this purpose we choose the List of Clusters [8], generally regarded as the most scalable mechanism known. We have measured this, with and without our optimisation, over the SISAP benchmark data sets *colors* and *nasa*, to perform threshold search using the three standard benchmark thresholds; we show a very significant improvement in performance.

As the list of clusters is built, at each node a pivot point is selected and a fixed number of objects, those being closest to this pivot point, are stored in an associated "bucket". Especially towards the start of this process, the cover radius of these objects from the pivot point is likely to be very small, therefore maximising the probability of the bucket being excluded from a search. When each cluster is constructed, the distances between every object in that cluster, and every pivot point from the root to that point in the list, will have been to be calculated as a part of the construction algorithm.

244   E. Chávez et al.

To this structure, we add only our small representations of the objects within each bucket, and cause no extra distance calculations at either build or query time. The local pivot point is used as the first reference point, and the furthest pivot from the so-far constructed spine of the tree as the second. This gives an approximation to the geometric technique (low dist, high dist) described in Sect. 7, and the only extra construction-time cost is the calculation of the 2D coordinate from these distances; in experiments, this was literally undetectable. The extra space cost is 10 bytes per object, for the *colors* data set representing an increase of around 1%.

At query time, the mechanism is used in the normal way based on the measured distance between the query and each pivot point down the spine of the list. In cases where the local "cluster" requires to be searched, then the four-point representations are first checked. The four-point representation of the query requires only the calculation of the 2D representative point, as all of the distances required have already been measured as the query algorithm progresses down the spine of the list. The lower-bound computation then comprises a 2-dimensional ℓ_2 distance. If the lower-bound distance is greater than the query threshold, there is no requirement to access the corresponding object and check its true distance against the query object. This saves not only an expensive distance calculation, but also the movement of the object within memory.

Table 3. Improvement shown on List of Clusters using four-point post-filtering. Values given are mean number of distance calculations per query.

Threshold	Standard			Optimised		
	t_0	t_1	t_2	t_0	t_1	t_2
SISAP colors	5645	11649	24401	2256	3987	10402
SISAP nasa	1381	3258	8790	1007	1402	3384

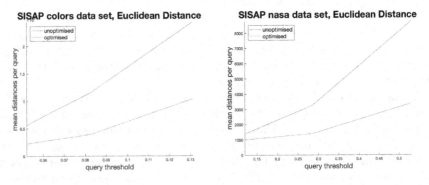

Fig. 4. SISAP benchmark space results with and without optimisation

8.1   Experimental Results

Table 3 shows the number of distance calculations made against the original data sets, along with the percentage improvement shown; the same values are plotted in Fig. 4. It can be seen in almost all cases that the query cost is better than halved, in return for only a small increase in memory size.

9   Conclusions and Future Work

We have shown how the four-point property can be used in conjunction with the concept of a pivot table in order to produce a minimally-sized table comprising only two reference objects identifiers, and two distances, per database object. These are used to construct a coordinate in a two-dimensional Euclidean space which gives a lower-bound on a query distance. The combination of the very large space of object pairs available from a relatively small set of reference objects, and the observation that each pair gives a significantly different projection of the space, combines to allow a very high rate of successful exclusion for a typical range search, with exclusion rates of 99.6 and 99.9% obtained for the SISAP benchmark *colors* and *nasa* data sets, with only 150 reference objects being used. For a data size of around 10 bytes per object and a cheap arithmetic check these results are impressive.

It is remarkable that a random selection of pairs of reference points produce exclusion rates quite close to the more expensive exhaustive search. Other linear cost pair selection heuristics are closer to the ground truth. There is room for trying to match the almost perfect exclusion rate with other heuristics.

Finally, since it is theoretically impossible to avoid a sequential scan for nearest neighbour search, even in the approximate sense, a cheap exclusion mechanism that is trivially parallelizable is competitive. We remark that this mechanism can be used in conjunction with probabilistic methods requiring postfiltering or re-ranking, like metric inverted files. We have given one successful example of this: for an almost immeasurably small increase in build cost and memory, the performance of the List of Clusters indexing structure has been shown to be radically improved. It is likely that many similar examples exist.

References

1. Amato, G., Gennaro, C., Savino, P.: Mi-file: using inverted files for scalable approximate similarity search. Multimedia Tools Appl. **71**(3), 1333–1362 (2014)
2. Baeza-Yates, R., Cunto, W., Manber, U., Wu, S.: Proximity matching using fixed-queries trees. In: Crochemore, M., Gusfield, D. (eds.) CPM 1994. LNCS, vol. 807, pp. 198–212. Springer, Heidelberg (1994). https://doi.org/10.1007/3-540-58094-8_18
3. Jon Louis Bentley: Multidimensional binary search trees used for associative searching. Commun. ACM **18**(9), 509–517 (1975)
4. Blumenthal, L.M.: A note on the four-point property. Bull. Am. Math. Soc. **39**(6), 423–426 (1933)

5. Burkhard, W.A., Keller, R.M.: Some approaches to best-match file searching. Commun. ACM **16**(4), 230–236 (1973)
6. Celik, C.: Priority vantage points structures for similarity queries in metric spaces. In: Shafazand, H., Tjoa, A.M. (eds.) EurAsia-ICT 2002. LNCS, vol. 2510, pp. 256–263. Springer, Heidelberg (2002). https://doi.org/10.1007/3-540-36087-5_30
7. Chávez, E., L Marroquín, J., Baeza-Yates, R.: Spaghettis: an array based algorithm for similarity queries in metric spaces. In: String Processing and Information Retrieval Symposium, 1999 and International Workshop on Groupware, pp. 38–46. IEEE (1999)
8. Chávez, E., Navarro, G.: A compact space decomposition for effective metric indexing. Pattern Recogn. Lett. **26**(9), 1363–1376 (2005)
9. Chavez, E., Ruiz, U., Tellez, E.: CDA: succinct spaghetti. In: Amato, G., Connor, R., Falchi, F., Gennaro, C. (eds.) SISAP 2015. LNCS, vol. 9371, pp. 54–64. Springer, Cham (2015). https://doi.org/10.1007/978-3-319-25087-8_5
10. Connor, R., Cardillo, F.A., Vadicamo, L., Rabitti, F.: Hilbert exclusion: improved metric search through finite isometric embeddings. ACM Trans. Inf. Syst. **35**(3), 17:1–17:27 (2016)
11. Connor, R., Vadicamo, L., Cardillo, F.A., Rabitti, F.: Supermetric search with the four-point property. In: Amsaleg, L., Houle, M.E., Schubert, E. (eds.) SISAP 2016. LNCS, vol. 9939, pp. 51–64. Springer, Cham (2016). https://doi.org/10.1007/978-3-319-46759-7_4
12. Connor, R., Vadicamo, L., Cardillo, F.A., Rabitti, F.: Supermetric search. Inf. Syst. **80**, 108–123 (2018)
13. Connor, R., Vadicamo, L., Rabitti, F.: High-dimensional simplexes for supermetric search. In: Beecks, C., Borutta, F., Kröger, P., Seidl, T. (eds.) SISAP 2017. LNCS, vol. 10609, pp. 96–109. Springer, Heidelberg (2017). https://doi.org/10.1007/978-3-319-68474-1_7
14. Figueroa, K., Chávez, E., Navarro, G., Paredes, R.: Speeding up spatial approximation search in metric spaces. J. Exp. Algorithmics (JEA) **14**, 6 (2009)
15. Figueroa, K., Navarro, G., Chávez, E.: Metric spaces library (2007). http://www.sisap.org
16. Menger, K.: Untersuchungen ber allgemeine metrik. Math. Ann. **100**, 75–163 (1928)
17. Micó, M.L., Oncina, J., Vidal, E.: A new version of the nearest-neighbour approximating and eliminating search algorithm (AESA) with linear preprocessing time and memory requirements. Pattern Recogn. Lett. **15**(1), 9–17 (1994)
18. Rubinstein, A.: Hardness of approximate nearest neighbor search. In: Proceedings of the 50th Annual ACM SIGACT Symposium on Theory of Computing, pp. 1260–1268. ACM (2018)
19. Ruiz, G., Santoyo, F., Chávez, E., Figueroa, K., Tellez, E.S.: Extreme pivots for faster metric indexes. In: Brisaboa, N., Pedreira, O., Zezula, P. (eds.) SISAP 2013. LNCS, vol. 8199, pp. 115–126. Springer, Heidelberg (2013). https://doi.org/10.1007/978-3-642-41062-8_12
20. Skopal, T., Pokorný, J., Snášel, V.: Nearest neighbours search using the PM-tree. In: Zhou, L., Ooi, B.C., Meng, X. (eds.) DASFAA 2005. LNCS, vol. 3453, pp. 803–815. Springer, Heidelberg (2005). https://doi.org/10.1007/11408079_73
21. Vidal, E.: New formulation and improvements of the nearest-neighbour approximating and eliminating search algorithm (AESA). Pattern Recogn. Lett. **15**(1), 1–7 (1994)
22. Wilson, W.A.: A relation between metric and Euclidean spaces. Am. J. Math. **54**(3), 505–517 (1932)

Characteristics of Local Intrinsic Dimensionality (LID) in Subspaces: Local Neighbourhood Analysis

Tahrima Hashem$^{(\boxtimes)}$, Lida Rashidi, James Bailey, and Lars Kulik

The University of Melbourne, Melbourne, Australia
tahrimah@student.unimelb.edu.au,
{rashidi.l,baileyj,lkulik}@unimelb.edu.au

Abstract. The local intrinsic dimensionality (LID) model enables assessment of the complexity of the local neighbourhood around a specific query object of interest. In this paper, we study variations in the LID of a query, with respect to different subspaces and local neighbourhoods. We illustrate the surprising phenomenon of how the LID of a query can substantially decrease as further features are included in a dataset. We identify the role of two key feature properties in influencing the LID for feature combinations: correlation and dominance. Our investigation provides new insights into the impact of different feature combinations on local regions of the data.

Keywords: Intrinsic dimension · Neighbourhood · Subspace

1 Introduction

Many core operations in data-mining and machine learning are dependent on the choice of similarity measure, as well as the choice of feature space. As the number of features in a dataset increases, the similarity between any pair of data points converges to the distribution mean and the similarity measure loses its discriminability power, i.e., the 'curse of dimensionality'. To overcome this challenge, a range of dimension reduction techniques [1–3] have been developed, to search for a lower dimensional representation that provides a good approximation of the data. A key concept in this context is a dataset's *intrinsic dimensionality* (ID), the minimum number of latent features required to represent the data. This is a natural measure to assess the complexity of a dataset.

In addition to considering the intrinsic dimensionality of an entire dataset, one can also consider intrinsic dimensionality with respect to a particular query object of interest. For this task, one can use local measures of ID [4,5], which focus on the k-nearest neighbor distances from a specific (query) location in the space. Recently developed *local intrinsic dimensionality* models, i.e., the expansion dimension (ED) [6], the generalised expansion dimension (GED) [7], and local continuous intrinsic dimension (LID) [8,9], quantify the ID in terms of the

© Springer Nature Switzerland AG 2019
G. Amato et al. (Eds.): SISAP 2019, LNCS 11807, pp. 247–264, 2019.
https://doi.org/10.1007/978-3-030-32047-8_22

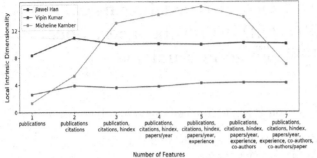

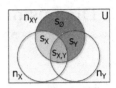

Fig. 2.   Neighbourhoods of a query along the individual distance variables, i.e., X and Y as well as their joint distance variable XY. (Color figure online)

Fig. 1. LID values (computed using the MLE estimator [4]) of three prominent researchers, i.e., *Jiawei Han*, *Micheline Kamber* and *Vipin Kumar* with respect to increasing number of features for the data-mining community of scholars from AMiner.

growth rate of objects with the expansion in distance from a specific query location. A wide range of applications, e.g., manifold learning, dimension reduction, similarity search [10], local density estimation [11] and anomaly detection [12], have benefited from the use of local ID measures.

In this paper, given a query object, our goal is to analyse how its LID estimates change with respect to different size feature sets. In particular, as more features are used, does the estimated LID of the query increase or decrease? Intuitively, one might expect that as one adds more features, the estimated LID of the query should either increase or remain stable. However, for some situations (in both real and synthetic data), we will demonstrate an unexpected and somewhat counterintuitive phenomenon, that the estimated LID of a query object can actually decrease as more features are used.

We provide a brief example to illustrate the idea. Figure 1 shows the estimated LID values of three researchers (queries): *Jiawei Han*, *Micheline Kamber* and *Vipin Kumar* from the data-mining community[1] of scholars in the AMiner[2] dataset. We observe that the LID trends are not always smooth as more features are considered. Importantly for researcher *M. Kamber*, there is significant drop in LID when going from 5 to 6 features, and going from 6 to 7 features.

Our purpose is to understand how such a drop in LID is possible and what factors might be responsible. Intuitively, the phenomenon is related to how outlying or inlying the query is within a given subspace, as well as relations between the features themselves, such as their degree of *correlation* and whether a property we call *feature dominance* is present. Developing such an understanding may lead to strategies for more effective feature engineering. Our contributions can be summarised as follows.

1. We identify and illustrate the counter intuitive phenomenon of how the estimated LID of a query object may decrease as more features are considered.

---

[1] https://aminer.org/lab-datasets/soinf/.

[2] https://aminer.org/data.

2. We identify the role of two key factors which can influence changes in LID and local neighbourhood for a query: feature dominance and correlation.
3. Given a query object, we study the estimated LID and neighbourhood variations within a feature space and its subspaces, using carefully controlled experiments.

2   Background and Preliminaries

We will first define local intrinsic dimensionality [9] and its estimator [4], and introduce the concept of neighbourhood.

Local Intrinsic Dimensionality: Classical expansion models [6,8] evaluate the growth rate of the number of data points as the distance to an object of interest increases. E.g., in Euclidean space, when the size of a d-dimensional ball increases by r, it's volume increases by r^d. It is possible to deduce the expansion dimension d from this growth rate of volume with respect to the size/distance as follows.

$$\frac{V_2}{V_1} = \left(\frac{r_2}{r_1}\right)^d \Rightarrow d = \frac{\ln(V_2/V_1)}{\ln(r_2/r_1)} \tag{1}$$

The notion of volume is analogous to the probability measure for continuous random variables. The expansion models can be adapted for distance distributions for a given query by replacing the ball set size with the probability of the lower tails of the distribution (Extreme Value Theory), providing a local view of the dimensional structure of the data, as their estimation is restricted to a neighbourhood around the object of interest. Houle et. al. [9] provides the formal definition of LID in light of this theory.

Definition 1. *Assume a reference object $q \in \mathbb{R}$. Let $X > 0$ be a random variable representing distances from q to other objects.[3] If $F(x)$ represents the cumulative distance distribution function of X such that $F(x)$ is continuously differentiable at distance $x \in X$, the local intrinsic dimensionality (LID) of the query q at distance x is defined as:*

$$LID_X(x) = \lim_{\epsilon \to 0} \frac{\ln(F((1+\epsilon)x))/F(x)}{\ln((1+\epsilon)x/x)} = \lim_{\epsilon \to 0} \frac{\ln(F((1+\epsilon)x))/F(x)}{\ln(1+\epsilon)} \tag{2}$$

whenever the limit exists.

Applying L'Hopital's rule to the limits of Eq. 2, LID can be expressed as follows [9].

Theorem 1 ([9]). *If $F(x)$ represents the cumulative distribution function for a distance variable X and $F(x)$ is continuously differentiable such that $F(x) > 0$ for $x > 0$, then*

$$LID_X(x) = \frac{x \cdot F'(x)}{F(x)} \tag{3}$$

---

[3] Suppose $q = 0 \in \mathbb{R}$ and $x_1 = 2 \in X$ are 1 dimensional data values. Then, x_1 directly represents a distance value from q to itself along the X axis.

Thus, when $x \in X$ tends to zero, the LID of q can be defined in terms of the limit:

$$LID_X = \lim_{x \to 0} LID_X(x) \tag{4}$$

LID gives a rough indication of the dimensionality of the submanifold containing q that would best fit the distribution of data in the vicinity of q. Comprehensive theory regarding the LID model can be found in [8,9,13,14].

LID Estimation: The k nearest neighbour distances can be considered as extreme events associated with the lower tail of the distance distribution according to the Extreme Value Theory. The tails of the continuous probability distributions converge to the Generalized Pareto Distribution (GPD), under some reasonable assumption [15]. Amsaleg et. al. [4,5] developed several estimators of LID to heuristically approximate the actual underlying distance distribution by a transformed GPD. The Maximum Likelihood Estimator (MLE) has showed a useful trade-off between efficiency and complexity. For a query object q from a data distribution, the MLE estimator of LID(q) is,

$$\widehat{LID}(q) = -\left(\frac{1}{k} \sum_{i=1}^{k} \log \frac{r_i(q)}{r_k(q)}\right)^{-1} \tag{5}$$

where $r_i(q)$ denotes the distance between q and its i-th nearest neighbour in the sample.

Neighbourhood: Given two features F_X, F_Y[4] and a query object q, we define random variables, X, Y that represent the distance distributions from q to other objects using either F_X or F_Y. The joint distribution XY represents the distance distribution from q in the joint space $\{F_X, F_Y\}$. Let LID_X, LID_Y and LID_{XY} be the estimates of the LID for q using X, Y and XY, respectively. The nearest neighbours, n_X, n_Y and n_{XY} that are used to estimate the individual and joint LIDs, are shown as circles in Fig. 2. n_{XY} is a mixture of data objects from n_X, n_Y and U (the whole region).

We use s_X to represent the nearest neighbours within X, that are common with the neighbours in the joint space XY and not with the neighbours in Y (shown in yellow color in Fig. 2). Similarly for s_Y. Thus, $s_X = (n_X \cap n_{XY}) \setminus n_Y$ and $s_Y = (n_Y \cap n_{XY}) \setminus n_X$. Also, $s_{X,Y}$ (the green region) represents the neighbours that are common in both the individual and joint dimensions, $s_{X,Y} = n_X \cap n_Y \cap n_{XY}$. The nearest neighbours in the joint space XY that are not common with any of the neighbours in the individual dimensions are represented as s_ϕ, $s_\phi = n_{XY} \setminus (s_X \cup s_Y \cup s_{X,Y})$ (the pink region).

For the rest of the paper, we refer the **estimate of the LID value** using Eq. 5 as the **LID** of a query.

---

[4] In fact, our model allows F_X (or F_Y) to be a set of features, rather than a single feature, but for simplicity we will present in the context of being a single feature.

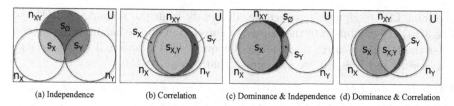

(a) Independence   (b) Correlation   (c) Dominance & Independence   (d) Dominance & Correlation

Fig. 3. Four different scenarios of the neighbourhoods for a given query along two distance variables, i.e., X and Y as well as their joint distance variable XY.

3 Research Questions

The local intrinsic dimensionality (LID) of the query in the joint space XY varies with respect to changes in the local joint neighbourhood (n_{XY}). We next characterise the relationship between the nearest neighbours in the joint space XY and the nearest neighbours in the individual variables, i.e., X and Y, w.r.t. the following two properties:

- **Correlation**: When the two distance variables, X and Y, are positively correlated, one expects that a significant portion of the nearest neighbours in the joint space XY overlap with the nearest neighbours in both X and Y. One also expects that this phenomenon is absent when X and Y are not correlated. i.e. $\mid s_{X,Y}^{cor.} \mid \gg \mid s_{X,Y}^{uncor.} \mid$.
- **Dominance**: A dominant distance variable is one which has a strong influence in determining the nearest neighbours of the query in the joint space XY. If X dominates Y, then a major portion of the nearest neighbours in the joint space XY overlap with the nearest neighbours in X as compared to Y. i.e., $\mid s_X \mid \gg \mid s_Y \mid$.

We will assess in what circumstances LID_{XY} can be less than the individual estimated LID values, LID_X and LID_Y. We particularly focus on the role of a dominant distance variable and/or the presence of a strong correlation between X and Y. We consider the following four research questions (RQ1–RQ4):

RQ1: *Given a query, when two distance variables are independent (uncorrelated), how can LID_{XY} and n_{XY} be characterised with respect to LIDs and neighbourhoods of the individual dimensions (LID_X, LID_Y, n_X, n_Y)?*

For RQ1, we will analyse a query's characteristics, i.e., inlyingness/outlyingness, in terms of its estimated LID in 2D spaces, when the individual distance variables have no dependency between them. Figure 3(a) illustrates this scenario, where we observe $\mid s_{X,Y} \mid \cong 0$ and $\mid s_X \mid \cong \mid s_Y \mid$.

RQ2: *Given a query object, when two distance variables are dependent (correlated), how can LID_{XY} and n_{XY} be characterised with respect to LIDs and neighbourhoods of the individual dimensions (LID_X, LID_Y, n_X, n_Y)?*

Correlation between two distance variables can lead to significant changes in the joint neighbourhood in comparison to the uncorrelated case and we expect the joint LID (estimated) to behave differently from the scenario in RQ1 (see Fig. 3(b)). To demonstrate the impact of correlation on the neighbourhood, consider the top 100 nearest neighbours of a query in XY space, if the correlation between X and Y is 1.0, we can expect that $|s_{X,Y}| \cong 100$.

RQ3: *How are LID_{XY} and n_{XY} influenced when one of the distance variables dominates the other? (X dominates Y or vice versa)*

A dominating distance variable can strongly influence the formation of neighbourhood in the joint space. In Fig. 3(c) we note, a significant part of n_{XY} overlaps with n_X and a small part of it overlaps with n_Y. In this case, we have assumed that the distance variables are independent, i.e., $|s_{X,Y}| \cong 0$. In this case, we expect the query to have neighbourhood characteristics for the joint space that are similar to those for the individual variable X, due to the dominance property of X.

RQ4: *In the presence of both correlation and dominance, how can LID_{XY} and n_{XY} be characterised in terms of (LID_X, LID_Y, n_X, n_Y)?*

Figure 3(d) illustrates this scenario where we observe a positive correlation between X and Y as $|s_{X,Y}| \gg 0$. We find $|s_X| > |s_Y|$, meaning that X still dominates Y.

4   Experimental Study Using Synthetic Data

We observe the behaviour of LID in multiple univariate (Sect. 4.1) and bivariate (Sects. 4.2–4.4) synthetic datasets that are generated to model the scenarios in the research questions *RQ1-RQ4*. We will later investigate a real dataset in Sect. 5. For our experiments, we model the **distance distribution** instead of the actual data distribution, i.e., the generated data values represent the distances from a query that is located at the origin. Note that the query is not generated by the data generation process. Since we always ensure that the generated values of the synthetic datasets are greater than or equal to 0, the data values along each dimension directly represent the distances from the query to themselves. The Euclidean Norm ($\|\cdot\|_2$) is used to measure distance. We use $k = 100$ neighbours in the MLE estimator of LID (see Eq. 5). Unless otherwise stated, *z-score* normalisation has been applied on both synthetic and real data (i.e. on the raw feature values for F_X or F_Y) before estimating the LIDs of a given query.

4.1   LID in Univariate Synthetic Datasets

To model the distance distributions, we have selected the Weibull distribution as it is lower bounded (given x≥0). Equation 6 shows the Weibull probability density and cumulative distribution functions. We generate three uniscaled ($\lambda = 1$) Weibull distributions for different values of shape parameter (κ) in Fig. 4. For shape values, $1 < \kappa < 2.6$, the Weibull pdf is positively skewed (right tail), for $2.6 < \kappa < 3.5$ its coefficient of skewness approaches zero (no tail) and for $\kappa > 3.5$ it is negatively skewed (left tail) [16].

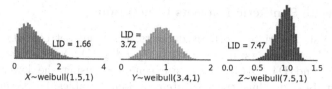

Fig. 4. Histograms of three $Weibull(\kappa,\lambda)$ distributed distance variables, i.e., X, Y and Z.

$$f_w(x; \kappa, \lambda) = \frac{\kappa}{\lambda}\left(\frac{x}{\lambda}\right)^{\kappa-1} \exp^{-(\frac{x}{\lambda})^{\kappa}}$$

$$F_w(x; \kappa, \lambda) = 1 - \exp^{-(\frac{x}{\lambda})^{\kappa}} \quad F_w^{-1}(x; \kappa, \lambda) = \lambda[-\ln(1-x)]^{\frac{1}{\kappa}} \tag{6}$$

$$LID_X = \lim_{x \to 0} \frac{x \cdot f_w(x; \kappa, \lambda)}{F_w(x; \kappa, \lambda)} = \lim_{x \to 0} \frac{\frac{d}{dx}(x \cdot f_w(x; \kappa, \lambda))}{\frac{d}{dx}(F_w(x; \kappa, \lambda))} = \kappa \tag{7}$$

The theoretical LID of a Weibull distributed distance variable is derived in Eq. 7 and is equal to the shape value. Also, experimentally the query (at origin) obtains LID values 1.66, 3.72 and 7.47, corresponding to κ values, i.e., 1.5, 3.4, 7.5, respectively. Thus, the larger the shape value of the Weibull distribution, the higher the LID and the more outlying the query is relative to other objects.

Table 1. Description of distribution and correlation parameters of synthetic bivariate datasets.

Scenarios	Title	Name of datasets	Distribution type	Description of the parameters	Rank correlation
Scenario 1	D1.	ND-Independent	Weibull	$\kappa_X = 4, \lambda_X = 1,$ $\kappa_Y = 6, \lambda_Y = 1$	$\alpha_s = 0$
	D2.	ND-Correlated	same as D1	same as D1	$\alpha_s = 0.89$
Scenario 2	D3.	D-Independent	same as D1	$\kappa_X = 4, \lambda_X = 8,$ $\kappa_Y = 6, \lambda_Y = 1$	$\alpha_s = 0$
	D4.	D-Correlated	same as D1	same as D3	$\alpha_s = 0.89$
	D5.	ED-Independent	Pearson	$\mu_X = 7, \sigma_X^2 = 0.5,$ $\beta_X = -1.75,$ $\gamma_X = 9 \; \mu_Y = 8,$ $\sigma_Y^2 = 1, \beta_Y = 0,$ $\gamma_Y = 3$	$\alpha_s = 0$
	D6.	ED-Correlated	same as D5	same as D5	$\alpha_s = 0.89$

4.2   Bivariate Synthetic Datasets Generation

We generate six bivariate synthetic datasets. Each dataset consists of $10,000$ data points. Four datasets, i.e., D1:*ND-Independent*, D2:*ND-Correlated*, D3:*D-Independent*, and D4: *D-Correlated*, are generated using the lower bounded *Weibull* distribution. Thus, the generated random variables, i.e., X and Y, can be treated as continuous distance variables for a query at origin $(0,0)$. To achieve control over the *mean* (μ), *variance* (σ^2), *skewness* (β) (i.e., measure of symmetry), and *kurtosis* (γ) (i.e., measure of whether the data is heavy-tailed or not in relation to the normal distribution) of the distance distributions, we generate two further datasets, D5:*ED-Independent* and D6:*ED-Correlated*, using the *Pearson* distribution family, which is effective in modelling skewed observations [17]. In this case, we ensure that all data values are greater or equal to 0, so that the generated data values correspond to distances from the query.

We consider two scenarios, *Scenario 1* and *Scenario 2*, where we model *dominance* and *non-dominance* between two distance variables. For each scenario we generate two different types of datasets, i.e., uncorrelated and correlated, using a Gaussian Copula (described below). Datasets D1 and D2, model the scenarios stated in *RQ1* and *RQ2*, respectively, whereas D3 and D4, model the scenarios in *RQ3* and *RQ4*, respectively. Datasets D5 and D6 illustrate the LID behaviour for the same phenomena as D3 and D4, using extremely skewed and heavy-tailed distance distributions.

When generating the bivariate distance distributions, our goal is to illustrate the circumstances where the query has different LID values in individual dimensions. We investigate the properties of the 2D local neighbourhood around the query, where LID_{XY} may show an expected increase or unexpected decrease with respect to the individual LIDs LID_X and LID_Y. To ensure that we have different LID values in X and Y dimensions, we use smaller values for the shape parameter in X than Y, making $LID_X < LID_Y$, leveraging our observations in Sect. 4.1.

We use a copula [18,19] to generate both the correlated and uncorrelated datasets. Copulas (C) provide a way to model correlated multivariate data. According to Sklar's Theorem [18], any multivariate cumulative distribution function can be expressed in terms of the marginal cumulative distribution functions of the random variables, together with a copula describing their dependence structure (α) (see Eq. 8).

$$F(x,y) = C(F^1(x), F^2(y); \alpha) \tag{8}$$

The *Gaussian Copula* (C_g) generates correlated uniformly distributed values from a multivariate normal distribution with a given linear correlation (α_p). Thus, a correlated multivariate distribution with the same or different marginal distributions can be obtained by applying the desired inverse cumulative distribution functions $(ICDF)$ to the corresponding uniform variables. We follow this technique to generate the four bivariate datasets, i.e., D1, D2, D3 and D4, using the following steps.

- **Step 1:** We use a Gaussian copula C_g with selected linear correlation parameter to sample bivariate uniformly distributed values $U = [U_1, U_2]$ for $U \in [0, 1]$.
- **Step 2:** We apply the *ICDF* of the Weibull distribution (F_w^{-1}) to U_1 and U_2, with the given parameters for each dimension i.e., κ_X, κ_Y, λ_X, and λ_Y, and obtain the desired marginal (Weibull) distributions for X and Y; the process is known as inverse transform sampling [20].

Though we need to provide the linear correlation (α_p) as an input to C_g, this linear correlation is not preserved during the inverse sampling because F_w^{-1} is a non-linear function (see Eq. 6). However, $F_w^{-1}(u)$ is monotonically increasing for $u \in U$ and $\kappa, \lambda > 0$, and under any monotonic transformation, rank correlation, e.g., *Spearman*'s correlation coefficient (α_s), is preserved [18]. There remains a one-to-one mapping between α_p and α_s for normally distributed data [21] (see Eq. 9). Hence, the value of α_s between the Weibull distributed variables is almost identical to the initial value of α_p specified in C_g, since C_g is constructed from normally distributed data.

$$\alpha_s = (6/\pi) * sin^{-1}(\alpha_p/2) \qquad (9)$$

For generating the uncorrelated datasets, i.e., D1 and D3, we use $\alpha_p = 0$ $(\alpha_s = 0)$ for C_g. We use the same scale $(\lambda = 1)$ and different shapes (κ), i.e., 4 and 6, for X and Y in the D1 dataset. In D3, we use a larger value of scale for X $(\lambda = 8)$ than Y $(\lambda = 1)$, so that X can be treated as a dominating distance variable. In fact, we intend to observe how X with its heavily-tailed neighbours, dominates Y in selecting the neighbours in the 2D space (XY). On the other hand, we use $\alpha_p = 0.9$ $(\alpha_s = 0.89)$ to C_g for the generation of correlated datasets in both non-dominance and dominance cases, i.e., D2 and D4. We use the same Weibull parameters as D1 and D2, for D3 and D4, respectively.

We model extreme scenarios of dominance in both the absence and presence of correlation using D5 and D6 datasets, respectively. The data values are sampled from a *Pearson* distribution family [17]. In D5, X follows a negatively skewed $(\beta_X = -1.75)$ heavy-tailed $(\gamma_X = 9)$ distribution whereas Y models symmetric $(\beta_Y = 0)$ light-tailed $(\gamma_Y = 0)$ distribution (see Table 1). Since both of them are independently sampled, they are uncorrelated. Due to the very skewed distribution along X, we are able to see the drop of joint LID even after applying the *z-score normalisation* in these datasets. Since it is not straightforward to obtain the *ICDF* for the X dimension with the given Pearson parameters, we generate the correlated Pearson numbers in the following step[5].

- **Step 1:** We generate independent (uncorrelated) Pearson values, P_1 and P_2 using the same parameters as D5 and sort them in ascending order.
- **Step 2:** We generate the correlated uniform values, i.e., U_1 and U_2, with $\alpha_p = 0.9$ (equivalent to $\alpha_s = 0.89$) from the Gaussian copula C_g.

---

[5] https://au.mathworks.com/help/stats/generate-correlated-data-using-rank-correlation.html.

- **Step 3:** After sorting the uniform values in ascending order, we obtain two indices, in_1 and in_2, describing the rearranged order of U_1 and U_2, respectively.
- **Step 4:** We position the sorted values of P_1 and P_2 in the same order as the indices, in_1 and in_2, to obtain the final Pearson variables, P_1^c and P_2^c for dimensions X and Y, respectively, in D6 dataset.

4.3   Scenario 1 (Non-dominance)

D1 and D2 are generated in a setting where there is no dominant feature (Table 1). Figure 5(a) and (b) provide the scatter plots and nearest neighbour distance graphs for D1 and D2, respectively. It is clearly notable from the scatter plots that the data values are correlated in D2 (elliptical shape) whereas in D1 they are not (circular shape).

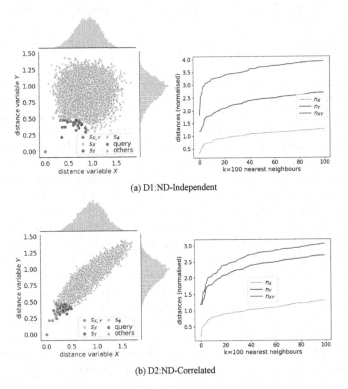

(a) D1:ND-Independent

(b) D2:ND-Correlated

Fig. 5. Scatter plots and **normalised** distance graphs of datasets D1 and D2 modelling the *uncorrelated* and *correlated* variables, respectively, in the *non-dominance* setting. (Color figure online)

In Fig. 5(a), we observe that the distances of the local neighbours with respect to X variable are relatively smaller than that of Y. This happens because Y has a larger shape than X. Hence the distribution of Y is skewed more to the left in comparison to X. As a result, LID_Y is larger than LID_X. We further note that there is no common neighbour between the joint space and the individual dimensions, i.e., $s_{X,Y}=0$ (no green dots). The distances of the neighbours along the joint variable XY are far in comparison to the individual variables. Hence, the query obtains a very high LID value, $LID_{XY} = 10.18$, that is approximately the summation of the individual LIDs, $LID_X = 4.87$ and $LID_Y = 6.35$, which matches with results mentioned in [8].

Figure 5(b) corresponds to the correlated variables X and Y of dataset D2. We observe that 48% of the neighbours, n_{XY} are overlapped with both n_X and n_Y, i.e., $s_{X,Y} = 48$ (green dots). We found $LID_X = 3.78$, $LID_Y = 6.35$ and $LID_{XY} = 6.46$. Note that the joint LID in the correlated case is 6.46 which is smaller than the joint LID($=10.18$) of the uncorrelated case with the same parameter settings. Thus, if the continuous distance variables are positively correlated, the query finds its 2D neighbours to be more common with the neighbours of the individual dimensions and thus obtains a smaller LID in comparison with the independent case. **This observation answers $RQ1$ and $RQ2$ for the independent and correlated distance variables in the non-dominance setting.**

4.4   Scenario 2 (Dominance)

For the dominance scenario, we consider D3:*D-Independent* and D4:*D-Correlated* datasets. In order to observe the dominance property of a distance variable, we do not standardise these datasets. We note from the scatter plots that the data shows greater variance in X compared to Y and the distances of the nearest neighbours for Y remain almost constant, whereas there is a steady increase in the distances for X as the number of nearest neighbours grows (see Fig. 6). For the query at the origin, LID_Y is 6.35 for both D3 and D4 datasets while LID_X is 4.87 and 3.80 in D3 and D4, respectively.

For D3 dataset, we observe a drop in LID value with respect to the joint distance variable XY compared to Y, i.e., $LID_{XY} = 5.70$ while $LID_Y = 6.35$. We note that a major portion of the neighbours in XY overlap with the neighbours from X, i.e., $s_X = 95$ (the orange dots). There is no overlapping between the neighbours of XY and the individual dimensions X and Y, i.e., $s_{X,Y} = 0$. As a result, the distances along XY are following the similar trend of along X (Fig. 6(a)). Thus in scenarios where one of the features X is dominant and has a lower LID value in comparison to the non-dominant feature Y, the LID value in the joint space LID_{XY} becomes smaller than that of LID_Y. However, if the dominant variable does not have such property (low LID), we do not observe this reduction of LID value in the joint space (**answering question $RQ3$**).

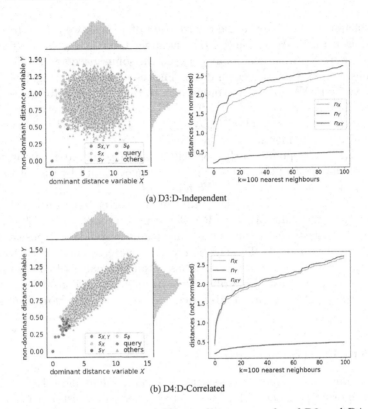

(a) D3:D-Independent

(b) D4:D-Correlated

Fig. 6. Scatter plots and nearest neighbours distance graphs of D3 and D4 datasets modelling the *uncorrelated* and *correlated* variables, respectively, in the *dominance* setting. (Color figure online)

We demonstrate the correlated scenario of D4 in Fig. 6(b). We find 99% $(s_X + s_{X,Y} = 51\% + 48\%)$ of n_{XY} are overlapped with n_X and 48% of them are common with n_Y (green dots). The uniformity of the distances of n_X has a significant influence on the distances of the neighbours in the joint space (XY), causing significant reduction in the LID of XY, i.e., $LID_{XY}(=3.93) \ll LID_Y(=6.35)$. n.b. the joint LID in D4 (3.93) is smaller than the joint LID of the uncorrelated case in D3 (5.70) (**answering question RQ4**).

Extreme Distributions: D5:*ED-Independent* and D6:*ED-Correlated* illustrate the extreme case of dominance for uncorrelated and correlated Pearson random variables, respectively (see Fig. 7). Here, X has a negative long tailed asymmetric distribution whereas Y follows a short tailed symmetric distribution. The query is an outlier in both dimensions, but it obtains smaller LID in X ($LID_X = 7.63$) than Y ($LID_Y = 16.59$) for both datasets. This phenomenon occurs since the query is surrounded by a group of outliers in X, whereas all the nearest neighbours are quite far away from the query in Y.

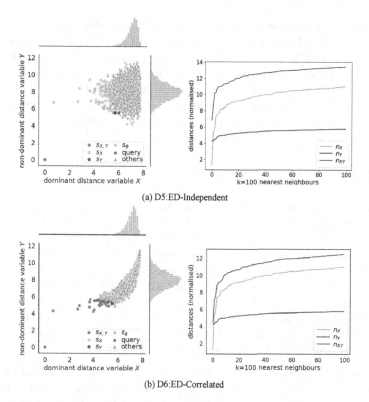

(a) D5:ED-Independent

(b) D6:ED-Correlated

Fig. 7. Scatter plots and nearest neighbours distance graphs of D5 and D6 datasets modelling the *uncorrelated* and *correlated* variables, respectively, in the *dominance* setting. (Color figure online)

Table 2. Changes in local neighbourhood and LID estimation w.r.t. the distance variables X, Y and XY for all six synthetic datasets in absence(/presence) of dominance and correlation.

Scenarios	Title	Name of the datasets	Neighbourhood in (XY)				LID estimates		
			s_X	s_Y	$s_{X,Y}$	s_ϕ	LID_X	LID_Y	LID_{XY}
Scenario 1	D1.	ND-Independent	9	35	0	56	4.87	6.35	10.18
	D2.	ND-Correlated	13	38	48	01	3.78	6.35	6.46
Scenario 2	D3.	D-Independent	95	01	0	04	4.87	6.35	5.70
	D4.	D-Correlated	51	01	48	0	3.80	6.35	3.93
	D5.	ED-Independent	75	04	00	21	7.63	16.59	13.74
	D6.	ED-Correlated	44	05	51	0	7.63	16.59	9.80

In D5, the uncorrelated dataset, the neighbours along XY mostly intersect with the neighbours along X as we find $s_X = 75$. Since they are uncorrelated there is no overlap among the neighbours in the joint space XY and the individual dimensions X and Y. Here, X has a bigger influence on LID_{XY} since the nearest neighbours of the query in X are having the dominant distances. We find $LID_{XY} = 13.74$ which is smaller than $LID_Y(=16.59)$. We obtain a similar LID behaviour in D6. However, due to the correlation between X and Y, 51% of n_{XY} overlaps with both n_X and n_Y in D6. A significant portion of n_{XY} is coming only from n_X, i.e., $s_X = 44\%$, which causes a drop in $LID_{XY} = 9.80$ as compared to $LID_Y(=16.59)$. Note that for the correlated case the reduction of LID value in the joint space is much greater than the uncorrelated case.

Our results for synthetic datasets are summarised in Table 2.

5   Experiments with Real Data

The AMiner dataset is a large academic social network comprising 1.7M authors, 2.1M papers and 4.3M coauthor relationships. We consider 7 numerical features: publications (*pub*), citations (*ct*), h-index (*hi*), papers/year (*ppy*), co-authors, co-authors/paper (*avgco*), and research experience. We analysed the LID behaviour in this dataset by considering different authors as the query and estimating the LID value for various combinations of features. We consider two prominent researchers, i.e., *Micheline Kamber* and *Jeffrey Xu Hu* from the data-mining community consisting of 641 researchers, as queries to model different phenomena described in Sects. 4.3–4.4. We use k = 100 in the MLE estimator, but obtained similar results (not reported) for $k = 30, 60$.

Case Study 1- Dominance: Given *M. Kamber* as the query, and the three features, *ct*, *hi* and *avgco*, we illustrate how dominance influences the LID value. In this scenario, the dominant variable X corresponds to the distances from the query to other authors on the feature *avgco*. While the non-dominant variable Y corresponds to the distances between query to others with respect to the two features *citations* and *hindex*. We find that $LID_X = \text{LID}(avgco) = 3.39$ and $LID_Y = \text{LID}(ct, hi) = 10.85$.

Figure 8 provides the 3D scatter plot and the normalised distance graph of 100 nearest neighbours that are used to estimate the LID along the distance

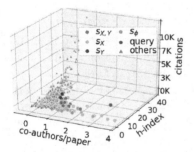
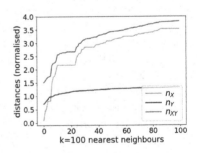

Fig. 8. The left figure is a scatter plot of the data-mining community of researchers from AMiner for the query *M. Kamber* and the three features *ct*, *hi* and *avgco*. Nearest neighbours distance graph is shown on the right. *M. Kamber* obtains LID(*ct*, *hi*) = 10.85 < LID(*ct*, *hi*, *avgco*) = 5.2. (Color figure online)

variables, X, Y and XY. It is evident that the query is an outlier in both X and Y. We find that 81% of the neighbours (the orange dots) in XY are coming only from X, i.e., $s_X = 81$ (see Table 3), which is the reason for obtaining a smaller LID value of 5.2 for the 3D feature-set (*ct, hi, avgco*), i.e., $LID_{XY} = 5.2$, after adding *avgco* to the 2D feature-set (*ct, hi*).

Case Study 2- Correlation: We observe the LID behaviour on AMiner dataset in terms of the correlation of the features. In our experiments, we also explored the effect of decorrelation on the LID value. Consider the query: *Jeffrey* and the two features papers/year and publications. Here, X represents the distances

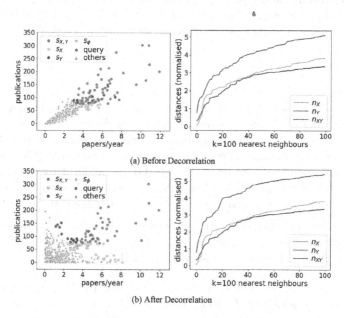

(a) Before Decorrelation

(b) After Decorrelation

Fig. 9. Scatter plots of the data-mining community of researchers, and distance graphs of the query *Jeffrey Xu Hu* before and after decorrelation of the features papers/year and publications. *Jeffrey* obtains LID values of 3.4 and 4.2 before and after decorrelation, respectively. (Color figure online)

Table 3. Joint neighbourhood and LID values for the distance variables X, Y and XY in the dominance, correlation and decorrelation scenarios. The second column describes the query values for the features.

Query	Features	Scenarios	Correlation coefficients	n_{XY} s_X	s_Y	$s_{X,Y}$	s_ϕ	LID_X	LID_Y	LID_{XY}
M. Kamber	X:*avgco* = 3.6, Y:*ct* = 1546, *hi* = 4	Dominance	-	81	3	8	8	3.39	10.85	5.2
Jeffrey Xu Hu	X : *ppy* = 10.9, Y : *pub* = 228	Correlation	$\alpha_s = 0.96$	16	7	77	0	2.5	3.1	3.4
		Decorrelation	$\alpha_s = 0.30$	31	18	44	7	2.5	3.1	4.2

from the query to others on *ppy* while Y on *pub*. *Jeffrey* obtains LID values of 2.5 and 3.1 with respect to *ppy* and *pub*, respectively (see Table 3).

The features *ppy* and *pub* are highly correlated, i.e., $\alpha_s(X,Y) = 0.96$, where *Jeffrey* obtains $LID_{XY}^{corr} = 3.4$. After removing the correlation [22], i.e., by random permutation of objects in X and Y, yielding $\alpha_s(X,Y) = 0.3$, *Jeffrey* obtains $LID_{XY}^{decor} = 4.2$ which is larger than LID_{XY}^{corr}. Figure 9(a) and (b) display the scatter and distance plots before and after the decorrelation, respectively. We note in Fig. 9(b), that the no. of common neighbours between XY and the individuals, i.e., X and Y, ($s_{X,Y}$) decreases (green dots) while the neighbours in s_X and s_Y increases (orange and blue dots) as compared to Fig. 9(a). The distance plot in Fig. 9(a) shows a more uniform distance distribution, compared to Fig. 9(b) which shows an abrupt increase at multiple locations of the plot.

6   Discussion

As a default, one might expect that the local intrinsic dimensionality of a query should increase as more features are used. However, our studies using both real and synthetic data indicate that under certain conditions such as dominance of a feature or presence of correlation between features, the estimated LID of a query can instead decrease. During the expansion of an existing feature space, significant changes might occur in the neighborhood local to the query. Our studies found, when a query's local neighborhoods are dissimilar with respect to different features, this phenomenon could occur. Some general observations are:

- **Independence**: When the features are independent, the LID in merged space is approximately the summation of the LIDs of the individual features. It matches with the theoretical observation of LID in joint space as stated in [8, 13].
- **Dominance**: When a dominant feature with low LID (LID_X), is combined with a feature with high LID (LID_Y), the LID in the joint space will be lie between LID_X and LID_Y ($LID_X < LID_{XY} \leq LID_Y$).
- **Correlation**: In the presence of a positive correlation, when a feature with low LID (LID_X) is combined with another feature with high LID (LID_Y), the joint LID is much smaller than the summation of the LIDs of the individual dimensions ($LID_{XY} \ll (LID_X + LID_Y)$). The stronger the correlation, the larger the reduction in LID_{XY}.
- **Dominance and Correlation**: In the presence of positive correlation, when a dominant feature with low LID (LID_X) is combined with another feature with high LID (LID_Y), the joint LID is between LID_X and LID_Y ($LID_X < LID_{XY} \ll LID_Y$). The stronger the dominance and correlation, the larger the reduction in LID_{XY}.

7   Conclusions

We have analysed the behaviour of local intrinsic dimensionality (LID) for changes in the feature-space as well as the neighbourhood of a query. We considered two key factors, correlation and dominance, that can cause the LID to decrease when more features are considered. Thus, increasing the number of features may not always result in an increase in the (local) complexity of the data around a query object. Our observations may provide insights into the feature selection and enumeration process, as well as object inlyingness/outlyingness across subspaces. For the future, it will be interesting to develop further theory to understand these findings.

References

1. Bouveyron, C., Celeux, G., Girard, S.: Intrinsic dimension estimation by maximum likelihood in probabilistic PCA. Pattern Recogn. Lett. **32**, 1706–1713 (2011)
2. Tenenbaum, J.B., Silva, V., Langford, J.C.: A global geometric framework for nonlinear dimensionality reduction. Science **290**, 2319–2323 (2000)
3. Roweis, S.T., Saul, L.K.: Nonlinear dimensionality reduction by locally linear embedding. In: ICCV, vol. 290, pp. 2323–2326 (2000)
4. Amsaleg, L., et al.: Extreme-value-theoretic estimation of local intrinsic dimensionality. DMKD **32**(6), 1768–1805 (2018)
5. Amsaleg, L., et al.: Estimating local intrinsic dimensionality. In: SIGKDD, pp. 29–38 (2015)
6. Karger, D.R., Ruhl, M.: Finding nearest neighbors in growth-restricted metrics. In: Proceedings of the Thirty-Fourth Annual ACM STOC, pp. 741–750 (2002)
7. Houle, M.E., Kashima, H., Nett, M.: Generalized expansion dimension. In: ICDMW, pp. 587–594 (2012)
8. Houle, M.E.: Dimensionality, discriminability, density and distance distributions. In: ICDMW, pp. 468–473 (2013)
9. Houle, M.E.: Local intrinsic dimensionality I: an extreme-value-theoretic foundation for similarity applications. In: Beecks, C., Borutta, F., Kröger, P., Seidl, T. (eds.) SISAP 2017. LNCS, vol. 10609, pp. 64–79. Springer, Heidelberg (2017). https://doi.org/10.1007/978-3-319-68474-1_5
10. Houle, M.E., Ma, X., Nett, M., Oria, V.: Dimensional testing for multi-step similarity search. In: ICDM, pp. 299–308 (2012)
11. Von Brünken, J., Houle, M., Zimek, A.: Intrinsic dimensional outlier detection in high-dimensional data. NII Technical Reports, pp. 1–12 (2015)
12. Houle, M.E., Schubert, E., Zimek, A.: On the correlation between local intrinsic dimensionality and outlierness. In: Marchand-Maillet, S., Silva, Y.N., Chávez, E. (eds.) SISAP 2018. LNCS, vol. 11223, pp. 177–191. Springer, Cham (2018). https://doi.org/10.1007/978-3-030-02224-2_14
13. Houle, M.E.: Inlierness, outlierness, hubness and discriminability: an extreme-value-theoretic foundation. NII Technical Reports, pp. 1–32 (2015)
14. Houle, M.E.: Local intrinsic dimensionality II: multivariate analysis and distributional support. In: Beecks, C., Borutta, F., Kröger, P., Seidl, T. (eds.) SISAP 2017. LNCS, vol. 10609, pp. 80–95. Springer, Heidelberg (2017). https://doi.org/10.1007/978-3-319-68474-1_6
15. Coles, S.G.: An Introduction to Statistical Modeling of Extreme Values, vol. 208. Springer, London (2001). https://doi.org/10.1007/978-1-4471-3675-0
16. Rousu, D.N.: Weibull skewness and kurtosis as a function of the shape parameter. Technometrics **15**(4), 927–930 (1973)
17. Pearson, K.: Contributions to the mathematical theory of evolution. II. skew variation in homogeneous material. Philos. Trans. R. Soc. Lond. Ser. A **186**, 343–414 (1895)

18. Nelsen, R.B.: An Introduction to Copulas. Springer, New York (2006). https://doi.org/10.1007/0-387-28678-0
19. Takeuchi, T.: Constructing a bivariate distribution function with given marginals and correlation: application to the galaxy luminosity function. Mon. Not. R. Astron. Soc. **406**, 1830–1840 (2010)
20. Kendall, M.G., Stuart, A., Ord, J.K. (eds.): Kendall's Advanced Theory of Statistics. Oxford University Press Inc., Oxford (1987)
21. Kendall, M.G.: Rank and product-moment correlation. Biometrika **36**(1/2), 177–193 (1949)
22. Gionis, A., Mannila, H., Mielikäinen, T., Tsaparas, P.: Assessing data mining results via swap randomization. ACM TKDD **1**(3), 14 (2007)

Metric Embedding into the Hamming Space with the n-Simplex Projection

Lucia Vadicamo[1](✉), Vladimir Mic[2], Fabrizio Falchi[1], and Pavel Zezula[2]

[1] Institute of Information Science and Technologies (ISTI), CNR, Pisa, Italy
{lucia.vadicamo,fabrizio.falchi}@isti.cnr.it
[2] Masaryk University, Brno, Czech Republic
{xmic,zezula}@fi.muni.cz

Abstract. Transformations of data objects into the Hamming space are often exploited to speed-up the similarity search in metric spaces. Techniques applicable in generic metric spaces require expensive learning, e.g., selection of pivoting objects. However, when searching in common Euclidean space, the best performance is usually achieved by transformations specifically designed for this space. We propose a novel transformation technique that provides a good trade-off between the applicability and the quality of the space approximation. It uses the *n-Simplex projection* to transform metric objects into a low-dimensional Euclidean space, and then transform this space to the Hamming space. We compare our approach theoretically and experimentally with several techniques of the metric embedding into the Hamming space. We focus on the applicability, learning cost, and the quality of search space approximation.

Keywords: Sketch · Metric search · Metric embedding · n-point property

1 Introduction

The *metric search* problem aims at finding the most similar data objects to a given query object under the assumption that there exists a metric function assessing the dissimilarity of any two objects. The broad applicability of the metric space similarity model makes the metric search a challenging task, since the distance function is the only operation that can be exploited to compare two objects. One way to speed-up the metric searching is to transform the space to use a cheaper similarity function or to reduce data object sizes [4,9,14,19]. Recently, Connor et al. proposed the *n-Simplex projection* that transforms the metric space into a finite-dimensional Euclidean space [8,9]. Here, specialised similarity search techniques can be applied. Moreover, the Euclidean distance is more efficient to evaluate than many distance functions.

Another class of metric space transformations is formed by *sketching techniques* that transform data objects into short bit-strings called *sketches* [4, 17,19]. The similarity of sketches is expressed by the Hamming distance, and

© Springer Nature Switzerland AG 2019
G. Amato et al. (Eds.): SISAP 2019, LNCS 11807, pp. 265–272, 2019.
https://doi.org/10.1007/978-3-030-32047-8_23

sketches are exploited to prune the search space during query executions [18,19]. While some sketching techniques are applicable in generic metric spaces, others are designed for specific spaces [4]. The metric-based sketching techniques are broadly applicable, but their performance is often worse than that of the vector-based sketching approaches when dealing with the vector spaces [4,17].

We propose a novel sketching technique *NSP_50* that combines advantages of both approaches: wide applicability and good space approximation. It is applicable to the large class of metric spaces meeting the *n-point property* [3,7], and it consists of the projection of the search space into a low-dimensional Euclidean space (*n-Simplex projection*) and the binarization of the vectors. The NSP_50 technique is particularly advantageous for expensive metric functions, since the learning of the projection requires a low number of distance computations. The main contribution of the NSP_50 is a better trade-off between its applicability, quality of the space approximation, and the pre-processing cost.

2   Background and Related Work

We focus on the similarity search in domains modelled by the *metric space* (D, d), with the domain of objects D and the metric (*distance*) function d : $D \times D \to \mathbb{R}^+$ [21] that expresses the dissimilarity of objects $o \in D$. We consider the data set $S \subseteq D$, and the so-called *kNN queries* that search for the k closest objects from S to a *query* object $q \in D$. Similarity queries are often evaluated in an approximate manner since the slightly imprecise results are sufficient in many real-life applications and they can be delivered significantly faster than the precise ones. Many metric space transformations have been proposed to speed-up the approximate similarity searching, including those producing the Hamming space [4,5,11,18,19], Euclidean space [9,16] and Permutation space [1,6,20]. We further restrict our attention to the metric embedding into the Hamming space.

2.1   Bit String Sketches for Speeding-Up Similarity Search

Sketching techniques $sk(\cdot)$ transform the metric space (D, d) to the Hamming space $(\{0,1\}^\lambda, h)$ to approximate it with smaller objects and more efficient distance function. We denote the produced bit strings as *sketches* of length λ. Many sketching techniques were proposed – see for instance the survey [4]. Their main features are: (1) Quality, i.e., the ability to approximate the original metric space; (2) Applicability to various search spaces; (3) Robustness with respect to data (intrinsic) dimensionality; (4) Cost of the object-to-sketch transformation; (5) Cost of the transformation learning. In the following, we summarise concepts of three techniques that we later compare with the newly proposed NSP_50 technique. They all produce sketches with *balanced bits*, i.e. each bit i is set to 1 in one half of the sketches $sk(o), o \in S$. This is denoted by the suffix _50 in their notations.

GHP_50 technique [18] uses λ pairs of reference objects (*pivots*), that define λ instances of the *Generalized Hyperplane Partitioning* (GHP) [21] of the dataset S. Therefore, each GHP instance splits the dataset into two parts according to the closer pivot, and these parts define values of one bit of all sketches $sk(o), o \in S$. The pivots are selected to produce balanced and low correlated bits [18]: (1) an initial set of pivots $P_{sup} \in D$ is selected in random, (2) the balance of the GHP is evaluated for all pivot pairs using a sample set T of S, (3) set P_{bal} is formed by pivot pairs that divide T into parts balanced to at least 45 % to 55 %, and corresponding sketches sk_{bal} are created, (4) the correlation matrix M with absolute values of the Pearson correlation coefficient is evaluated for all pairs of bits of sketches sk_{bal}, and (5) a heuristic is applied to select rows and columns of M which form its sub-matrix with low values and size $\lambda \times \lambda$. (6) Finally, the λ pivot *pairs* that produce the corresponding low correlated bits define sketches $sk(o), o \in S$.

BP_50 uses the *Ball Partitioning* (BP) instead of the GHP [18]. BP uses one pivot and a radius to split data into two parts, that again define the values in one bit of sketches $sk(o), o \in S$. Pivots are selected again via a random set of pivots P_{sup}, for which we evaluate radii dividing the sample set T into halves. The same heuristic as in case of the technique GHP_50 is than employed to select λ pivots that produces low correlated bits.

PCA_50 is a simple sketching technique surprisingly well approximating the Euclidean spaces [4,12,13,15,17]. It uses the *Principal Component Analysis* (*PCA*) to shrink the original vectors, which are then rotated using a random matrix and binarized by the thresholding. The i-th bit of sketch $sk(o)$ thus expresses whether the i-th value in the shortened vector is bigger then the median computed on a sample set T. If sketches longer than the original vectors are desired, we propose to apply the PCA and to rotate transformed vectors using independent random matrices. Then we concatenate corresponding binarized vectors.

Sketching techniques applicable to generic metric spaces, e.g., GHP_50 and BP_50, are usually of a worse quality than vector-based sketching techniques when dealing with the vectors spaces [4,17]. Moreover, they require an expensive learning of the transformation. We propose the sketching technique NSP_50 to provide a better trade-off between the quality of the space approximation, applicability of the sketching, and the pre-processing cost.

2.2 The n-Simplex Projection

The *n-Simplex projection* [9] associated with a set of n pivots $\mathcal{P}_n$ is a space transformation $\phi_{\mathcal{P}_n} : (D, d) \rightarrow (\mathbb{R}^n, \ell_2)$ that maps the original metric space to a n-dimensional Euclidean space. It can be applied to any metric space with the *n-point property*, which states that any n points $o_1, .. o_n$ of the space can be *isometrically* embedded in the $(n-1)$-dimensional Euclidean space. Many often used metric spaces such as Euclidean spaces of any dimension, spaces with the Triangular or Jensen-Shannon distances, and, more generally, any Hilbert-embeddable spaces meet the n-point property [7]. The n-Simplex projection is properly described in [9]. Here, we sketch just the main concepts.

First, the n-point property guarantees that there exists an isometric embedding of the n pivots into $(\mathbb{R}^{n-1}, \ell_2)$ space, i.e., it is possible to construct the vertices $v_{p_i} \in \mathbb{R}^{n-1}$ such that $\ell_2(v_{p_i}, v_{p_j}) = d(p_i, p_j)$ for all $i, j \in \{1, \ldots, n\}$. These vertices form the so-called *base simplex*. Second, for any other object $o \in D$, the $(n + 1)$-point property guarantees that there exists a vertex $v_o \in \mathbb{R}^n$ such that $\ell_2(v_o, v_{p_i}) = d(o, p_i)$ for all $i = 1, \ldots, n$. The n-Simplex projection assigns such v_o to o, and Connor et al. [9] provide an iterative algorithm to compute the coordinates of the vertices v_{p_i} of the simplex base as well as the coordinates of the vector v_o associated to $o \in D$. The base simplex is computed once and reused to project all data objects $o \in S$. Moreover, the Euclidean distance between any two projected vectors $v_{o_1}, v_{o_2} \in \mathbb{R}^n$ is a lower-bound of their actual distance, and this bound becomes tighter with increasing number of pivots n [9].

3   The n-Simplex Sketching: Proposal and Comparison

We propose the sketching technique *NSP_50* that transforms metric spaces with the n-point property to the Hamming space. It uses the n-Simplex projection with λ pivots to project objects into λ-dimensional Euclidean space; the obtained vectors are then randomly rotated and binarized using the median values in each coordinate. These medians are evaluated on the data sample set. The random rotation is applied to distribute information equally over the vectors, as the n-Simplex projection returns vectors with decreasing values along the dimensions.

For each data set S, there exists a finite number of pivots $\tilde{n}$ such that $\phi_{\mathcal{P}_{\tilde{n}}}$ is an isometric space embedding[1]. The identification of the minimum $\tilde{n}$ with this property is still an open problem. The convergence is achieved when all the projected data points have a zero value in their last component, so the NSP_50 technique as described above cannot produce meaningful sketches of length $\lambda > \tilde{n}$. We overcome this issue by a concatenation of smaller sketches obtained using different rotation matrices.

The proposed NSP_50 technique is inspired by the PCA_50 approach, but provides significantly broader applicability, as it can transform all the metric spaces with the n-point property. This includes spaces with very expensive distance functions, as mentioned in Sect. 2.2. Sketching techniques also require transformation learning of a significantly different complexity. We compare the novel NSP_50 technique with the GHP_50, BP_50 and PCA_50 approaches and we provide the table summarising the main features of these sketching techniques, including the costs of the learning and object to sketch transformations in terms of floating point operations and distance computations. This table is provided online[2], due to the paper length limitation.

The GHP_50 and BP_50 techniques require an expensive pivot learning. Specifically, the GHP_50 requires (1) to examine the balance of the GHPs defined by various pivot pairs to create long sketches with the balanced bits, (2) an analysis of the pairwise bit correlations made for these sketches, and (3) a selection

---

[1] The proof is made trivially by a selection of all objects from the data set S as pivots.
[2] http://www.nmis.isti.cnr.it/falchi/SISAP19SM.pdf.

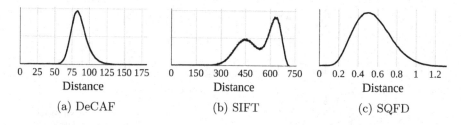

0 25 50 75 100 125 150 175	0 150 300 450 600 750	0 0.2 0.4 0.6 0.8 1 1.2
Distance	Distance	Distance
(a) DeCAF	(b) SIFT	(c) SQFD

Fig. 1. Distance densities for DeCAF, SIFT and SQFD data sets

of low correlated bits. The learning of the BP_50 is cheaper, since the proper radii are selected for a set of pivots directly. The rest of the learning is the same as in case of the GHP_50. The cost of the PCA_50 learning is given by the PCA learning cost and evaluation of the medians over the transformed vectors. We compute the PCA matrix using the Singular Value Decomposition (SVD) over the centred data. The learning of the NSP_50 is the cheapest one; it consists of the n-Simplex projection that has the quadratic cost with respect to the number of pivots n, and the binarization, which consists of the medians evaluations over coordinates of vectors in the sample set T.

4 Experiments

We evaluate the search quality of the NSP_50 technique on three data sets and we compare it with the sketching techniques PCA_50, GHP_50 and BP_50. We use three real-life data sets of visual features extracted from images:

SQFD: 1 million *adaptive-binning feature histograms* [2] extracted from the *Profiset collection*[3]. Each signature consists of, on average, 60 cluster centroids in a 7-dimensional space. A weight is associate to each cluster, and the signatures are compared by the Signature Quadratic Form Distance [2]. Note that this metric is a cheaper alternative to Earth Movers Distance, nevertheless, the cost of the Signature Quadratic Form Distance evaluation is quadratic with respect to the number of cluster centroids.

DeCAF: 1 million deep features extracted from the *Profiset collection* using the Deep Convolutional Neural Network described in [10]. Each feature is a 4,096-dimensional vector of values from the last hidden layer ($fc7$) of the neural network. The deep features use the ReLU activation function and are not ℓ_2-normalised. These features are compared with the Euclidean distance.

SIFT: 1 million SIFT descriptors from the *ANN data set*[4]. Each descriptor is a 128-dimensional vector. The Euclidean distance is used for the comparison.

Figure 1 shows particular distance densities. We express the quality of the sketching techniques by the *recall* of the k-NN queries evaluated using a simple

---

[3] http://disa.fi.muni.cz/profiset/.
[4] http://corpus-texmex.irisa.fr/.

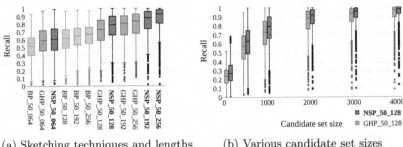

(a) Sketching techniques and lengths   (b) Various candidate set sizes

Fig. 2. SQFD data set: Quality of 3 sketching techniques varying sketch lengths (2a), comparison of 128bit sketches using various candidate set sizes (2b).

sketch-based filtering. More specifically, sketches are applied to select the candidate set $CandSet(q)$ for each query object $q \in D$ that consists of a fixed number of the most similar sketches to the query sketch $sk(q)$; then, the candidate set is refined by the distance $d(q, o)$, $o \in CandSet(q)$ to return the k most similar objects o to q with the sketches in the candidate set $CandSet(q)$. This approximate answer is compared with the precise one that consists of the k closest objects $o \in S$ to q. The candidate sets consist of 2,000 sketches in the case of DeCAF and SIFT data sets, and 1,000 sketches in the case of the SQFD data set.

We evaluate experiments using 1,000 randomly selected query objects $q \in D$, and we depict results by *Tukey box plots* to show distributions of the recall values for particular query objects: the lower- and upper-bounds of the box show the quartiles, and the lines inside the boxes depict the medians of the recall values. The ends of the whiskers represent the minimum and the maximum non-outliers, and dots show the outlying recall values. In all cases, we examine 100 nearest neighbours queries to investigate properly the variance of the recall values over particular query objects. We use sketches of lengths $\lambda \in \{64, 128, 196, 256\}$.

Results. Figure 2a shows results for the SQFD data set. The colours of the box plots distinguish particular sketching techniques, the suffix of the column names denotes the length of sketches. The proposed NSP_50 technique significantly outperforms both, GHP_50 and BP_50 techniques, fixing the sketch length. The PCA_50 approach is not applicable for this data set, as we search different than the Euclidean space. The BP_50 technique performs worst and provides the median recall just 0.67 in case of 256bit sketches. The NSP_50 and GHP_50 approaches achieve a solid median recall of 0.88 and 0.81, respectively, even in case of 192bit sketches. We show also a coherence of the results when varying the candidate set size. Figure 2b reports the recalls for the candidate set sizes $c \in \{100, 500, 1000, 2000, 3000, 4000\}$ and sketches of length 128 bits made by the sketching techniques NSP_50 and GHP_50. This figure shows that a given recall value can be achieved by the NSP_50 technique using a smaller candidate set than in case of the GHP_50.

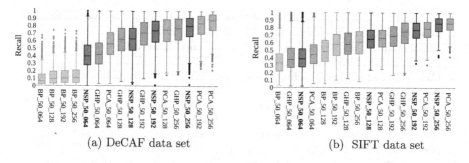

(a) DeCAF data set   (b) SIFT data set

Fig. 3. Quality of sketching techniques varying sketch lengths

The recall values for the DeCAF and SIFT data sets are depicted in Fig. 3. The BP_50 technique is less robust concerning the dimensionality of the data, so it achieves poor recalls in case of DeCAF descriptors, but it is still reasonable for the SIFT data set. The quality of the newly proposed NSP_50 technique is slightly better then that of the GHP_50 technique in case of the DeCAF data set. Both are, however, outperformed by the PCA_50 technique, which is specialised for the Euclidean space. This interpretation is valid for all the sketch lengths λ we have tested. The differences between the NSP_50 and PCA_50 techniques practically dismiss in case of the SIFT data set. Both these techniques achieve significantly better recall than the BP_50 and the GHP_50 techniques.

5 Conclusions

We contribute to the area of the metric space embeddings into the Hamming space. We propose the NSP_50 technique that leverages the n-Simplex projection to transform metric objects into bit-string sketches. We compare the NSP_50 technique with three other state-of-the-art sketching techniques designed either for the general metric space or the Euclidean vector space. The experiments are conducted on three real life data sets of visual features using four different sketch lengths. We show that our technique provides advantages of both metric-based and specialised vector-based techniques, as it provides a good trade-off between the quality of the space approximation, applicability, and transformation learning cost.

Acknowledgements. The work was partially supported by VISECH ARCO-CNR, CUP B56J17001330004, and AI4EU project, funded by the EC (H2020 - Contract n. 825619). This research was supported by ERDF "CyberSecurity, CyberCrime and Critical Information Infrastructures Center of Excellence" (No. CZ.02.1.01/0.0/0.0/ 16_019/0000822).

References

1. Amato, G., Gennaro, C., Savino, P.: MI-File: using inverted files for scalable approximate similarity search. Multimed. Tools Appl. **71**(3), 1333–1362 (2014)

2. Beecks, C., Uysal, M.S., Seidl, T.: Signature quadratic form distance. In: Proceedings of the ACM-CIVR 2010, pp. 438–445. ACM (2010)
3. Blumenthal, L.M.: Theory and Applications of Distance Geometry. Clarendon Press, Oxford (1953)
4. Cao, Y., et al.: Binary hashing for approximate nearest neighbor search on big data: a survey. IEEE Access **6**, 2039–2054 (2018)
5. Charikar, M.S.: Similarity estimation techniques from rounding algorithms. In: Proceedings of ACM-STOC 2002. ACM (2002)
6. Chávez, E., Figueroa, K., Navarro, G.: Effective proximity retrieval by ordering permutations. IEEE Trans. Pattern Anal. Mach. Intell. **30**(9), 1647–1658 (2008)
7. Connor, R., Cardillo, F.A., Vadicamo, L., Rabitti, F.: Hilbert exclusion: improved metric search through finite isometric embeddings. ACM Trans. Inf. Syst. **35**(3), 17:1–17:27 (2016)
8. Connor, R., Vadicamo, L., Cardillo, F.A., Rabitti, F.: Supermetric search. Inf. Syst. **80**, 108–123 (2018)
9. Connor, R., Vadicamo, L., Rabitti, F.: High-dimensional simplexes for supermetric search. In: Beecks, C., Borutta, F., Kröger, P., Seidl, T. (eds.) SISAP 2017. LNCS, vol. 10609, pp. 96–109. Springer, Heidelberg (2017). https://doi.org/10.1007/978-3-319-68474-1_7
10. Donahue, J., et al.: DeCAF: a deep convolutional activation feature for generic visual recognition. In: Proceedings of ICML 2014, vol. 32, pp. 647–655 (2014)
11. Douze, M., Jégou, H., Perronnin, F.: Polysemous codes. In: Leibe, B., Matas, J., Sebe, N., Welling, M. (eds.) ECCV 2016. LNCS, vol. 9906, pp. 785–801. Springer, Cham (2016). https://doi.org/10.1007/978-3-319-46475-6_48
12. Gong, Y., Lazebnik, S., Gordo, A., Perronnin, F.: Iterative quantization: a procrustean approach to learning binary codes for large-scale image retrieval. IEEE Trans. Pattern Anal. Mach. Intell. **35**(12), 2916–2929 (2013)
13. Gordo, A., Perronnin, F., Gong, Y., Lazebnik, S.: Asymmetric distances for binary embeddings. IEEE Trans. Pattern Anal. Mach. Intell. **36**(1), 33–47 (2014)
14. Indyk, P., Motwani, R.: Approximate nearest neighbors: towards removing the curse of dimensionality. In: Proceedings of ACM STOC, pp. 604–613 (1998)
15. Jégou, H., Douze, M., Schmid, C., Pérez, P.: Aggregating local descriptors into a compact image representation. In: Proceedings of CVPR 2010, pp. 3304–3311. IEEE (2010)
16. Kruskal, J.B.: Multidimensional scaling by optimizing goodness of fit to a nonmetric hypothesis. Psychometrika **29**(1), 1–27 (1964)
17. Mic, V., Novak, D., Vadicamo, L., Zezula, P.: Selecting sketches for similarity search. In: Proceedings of ADBIS, pp. 127–141 (2018)
18. Mic, V., Novak, D., Zezula, P.: Designing sketches for similarity filtering. In: Proceedings of IEEE ICDM Workshops, pp. 655–662 (2016)
19. Mic, V., Novak, D., Zezula, P.: Binary sketches for secondary filtering. ACM Trans. Inf. Syst. **37**(1), 1:1–1:28 (2018)
20. Novak, D., Zezula, P.: PPP-codes for large-scale similarity searching. In: Hameurlain, A., Küng, J., Wagner, R., Decker, H., Lhotska, L., Link, S. (eds.) Transactions on Large-Scale Data- and Knowledge-Centered Systems XXIV. LNCS, vol. 9510, pp. 61–87. Springer, Heidelberg (2016). https://doi.org/10.1007/978-3-662-49214-7_2
21. Zezula, P., Amato, G., Dohnal, V., Batko, M.: Similarity Search: The Metric Space Approach, vol. 32. Springer, New York (2006). https://doi.org/10.1007/0-387-29151-2

On coMADs and Principal Component Analysis

Daniyal Kazempour$^{(\boxtimes)}$, M. A. X. Hünemörder, and Thomas Seidl

Ludwig-Maximilians-Universität München, Munich, Germany
{kazempour,huenemoerder,seidl}@dbs.ifi.lmu.de

Abstract. Principal Component Analysis (PCA) is a popular method for linear dimensionality reduction. It is often used to discover hidden correlations or to facilitate the interpretation and visualization of data. However, it is liable to suffer from outliers. Strong outliers can skew the principal components and as a consequence lead to a higher reconstruction loss. While there exist several sophisticated approaches to make the PCA more robust, we present an approach which is intriguingly simple: we replace the covariance matrix by a so-called coMAD matrix. The first experiments show that PCA based on the coMAD matrix is more robust towards outliers.

Keywords: Covariance · coMAD · Principal Component Analysis

1 Introduction

When dealing with vast amounts of data and a large number of features performing principal component analysis (PCA) [5] is a common approach. PCA yields the principal components, i.e. the directions of highest variance in the data. Furthermore, PCA can be used to reveal hidden correlations and is sometimes used to detect arbitrary oriented linear correlated clusters. For example it is used in correlation clustering algorithms like 4C [2] or ORCLUS [1]. However PCA based on the covariance matrix is highly sensity towards outliers, particularly strong ones, can have an impact on the resulting principal components. This is due to the fact that outliers can influence the mean for each of the features of a data set. That is why in statistics the median is used as a robust measure against outliers. For the measure of dispersion of a feature the so called median absolute deviation from the median, short MAD, is the method of choice. In this work we propose to use a coMAD matrix instead of a covariance matrix on which the eigenvalues and eigenvectors are computed. We will elaborate on the MAD and the coMAD matrix in detail. In our first tests it can be seen that with heavy noise the principal components of a PCA are heavily deflected, while those resulting from a PCA based on the coMAD remain stable.

© Springer Nature Switzerland AG 2019
G. Amato et al. (Eds.): SISAP 2019, LNCS 11807, pp. 273–280, 2019.
https://doi.org/10.1007/978-3-030-32047-8_24

2   Related Work

There are many approaches to make the PCA more robust towards noise. With the term robust, we understand the following: Given data where a significant amount of the objects exhibits a linear correlation and some objects are outliers. The method is considered as robust, if the increasing number as well as increasing distance of outliers, does not significantly affect the direction of computed principal components in comparison to the case where there would be no outliers in the data set. In the work of [6] the authors develop a theory of Robust Principal Component Analysis (RPCA) and describe a robust M-estimation algorithm for capturing linear multivariate representations of high dimensional data, exemplary on images. M-estimators are a class of extremum estimators which can be regarded as a generalization of maximum-likelihood estimation. The authors further state that while methods such as RANSAC and Least Median Squares are more robust compared to M-estimation, it is not clear how to apply the techniques efficiently on high-dimensional data. In another work [4] the authors propose the ROBPCA method which combines the concept of a so-called 'projection pursuit' with a robust scatter matrix estimation on which the eigenvectors and eigenvalues are computed. Their method relies on several criteria and definitions. They use e.g. for the computation of outlierness the so called Stahel-Donoho affine-invariant outlyingness. They further compute a reweighted mean and covariance matrix based on the Rousseeuw and Van Driessen consistency factor. In another work [7] a generative RPCA model is proposed which relies on the Bayesian framework in which data noise is modelled as a mixture of Gaussians (MoG).

While all of the mentioned methods rely on various more complex and sophisticated methods, we challenge the task of robust PCA by asking: What if we exchange the covariance matrix against a coMAD matrix? Since the simple median and MAD is robust against outliers, the coMAD and MAD-PCA should be, too. In the following section we first define the coMAD and contrast it against the covariance. Then we support our claim in the experimental evaluation section, in which we also elaborate briefly on a metric we use to measure the robustness of our method in which we compare the reconstruction errors from a coMAD-based PCA against the covariance-based solution. We later on critically question the appropriateness of the MAE evaluation. Our approach is based on the MAD and the comedian as defined in the work of [3]. However we have taken the liberty of renaming the comedian to coMAD since the covariance is traditionally not named the 'comean'.

3   MAD and coMAD

Given a data matrix D where each of its rows represents a data record and its columns represent the features $(A_1, ..., A_d)$. The first step of performing PC is computing the covariance matrix Σ, which is defined as:

$$\Sigma_D = \begin{pmatrix} var(A_1) & \cdots & cov(A_1, A_d) \\ \vdots & \ddots & \vdots \\ cov(A_d, A_1) & \cdots & var(A_d) \end{pmatrix}$$

With covariance being

$$cov(A_i, A_j) := E(((A_i - E(A_i))(A_j - E(A_j)))$$

For the case that $A_i = A_j$ it is defined as the variance $var(A_i) = E((A_i - E(A_i))^2) = E((A_j - E(A_i))^2) = var(A_j)$. The covariance is thus a generalization of the variance. At this point it can be seen that from each of the features the expected value E (mean) is subtracted. The mean however is sensitive towards outliers which skew the mean value significantly. A more robust measure is the median. The analogon to the variance is the median absolute deviation from the median (MAD) which is defined as:

$$mad(A_i) = med(|A_i - med(A_i)|)$$

We can now generalize MAD like the covariance is a generalization of the variance. Then the coMAD can be defined as:

$$com(A_i, A_j) := med((A_i - med(A_i))(A_j - med(A_j)))$$

Like $cov(A_i, A_j)$, $com(A_i, A_j)$ is also a measure of covariance. Therefore, building on the definition of the coMAD, we can define a coMAD matrix Λ:

$$\Lambda_D = \begin{pmatrix} com(A_1, A_1) & \cdots & com(A_1, A_d) \\ \vdots & \ddots & \vdots \\ com(A_d, A_1) & \cdots & com(A_d, A_d) \end{pmatrix}$$

Now we can perform PCA using Λ_D instead of Σ_D, i.e. the eigenvalues and their corresponding eigenvectors of the coMad-matrix are computed.

While we shall see later in the experimental section the effects of noise on a covariance and a coMAD-based PCA, we further ask in this work-in-progress if there is a way to quantify the quality of the different methods. For this purpose we shall elaborate first on what happens after the eigenpairs (eigenvector, eigenvalue) of a PCA are computed. The eigenvectors are put into an eigenvector matrix U where each column corresponds to an eigenvector. What we do now is to use the eigenvectors with the k-largest eigenvalues, where k is widely noted in the literature as the number of principal components which cover more than 85% of the variance as a rule of thumb. The percentage of variance in a given dataset is explained by the following ration for each of the eigenvalue λ_i:

$$\varphi(k) = \frac{\sum_{i=1}^{k} \lambda_i}{\sum_{i=1}^{d} \lambda_i}$$

We discard as an example the second principal component of a performed PCA, which yields $U_{k=1}$. We project now the data D down to a lower dimensional representation Y through:

$$Y = D \cdot U_{k=i}$$

If we now want to reconstruct the data back to its original two-dimensional representation we achieve this with:

$$Z = Y \cdot U_{k=i}$$

This procedure of projecting the data (D) down to a lower dimension (Y) and reconstructing (Z) it gives us the opportunity to compute the Mean Absolute Error (MAE) which is defined as:

$$MAE(D, Z) = \frac{\sum_{i=1}^{n} |d_i - z_i|}{n},$$

where $d_i \in D$, $z_i \in Z$ and n denoted the number of objects for which holds $n = |D| = |Z|$.

4   Experiments and Discussion

To provide an intuition, we apply this method on the following toy examples as seen in Fig. 1[1]. In Fig. 1(a) we have a data set with a subset of objects which clearly exhibits a linear correlation. To this data set we added three outliers. On the left side it can be observed that the principal components of a PCA using the covariance matrix are deflected towards the direction of the three outliers. In contrast the principal components from the coMAD-based PCA remain with barely noticeable deflection in the direction of the objects belonging to the linear correlation. If we add now a fourth outlier which is even more apart, one can observe in Fig. 1(b) that the deflection of the principal components of a covariance-based PCA increases, exhibiting an by far larger eigenvalue. The coMAD variant remains again barely affected. In Fig. 1(c) and (d) we increase the distance of the fourth outlier, being more distant to the other outliers as well as to the linear correlated objects. In Fig. 1(d) the deflection is massive in the case of the covariance-based PCA, while principal components of the coMAD variant remain robust. These simple synthetic experiments reveal that a coMAD-based PCA excels regarding robustness against noise.

In a next step we conduct experiments on different data sets comparing the resulting MAE for the covariance and for the coMAD variant. For each of the data sets we repeat the MAE computation for choosing a range from $k = 1, ..., k = d$ principal components. The reason for this approach is twofold: first, one can observe how the MAE decreases per data set by increasing the number of principal components. Second: one can observe certain number of

---

[1] We used python with several libraries. The code can be found at: https://github.com/huenemoerder/MAD-PCA.

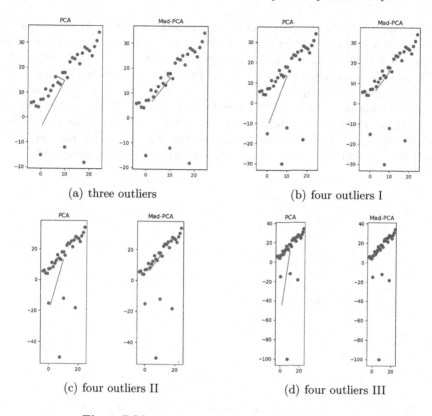

(a) three outliers   (b) four outliers I

(c) four outliers II   (d) four outliers III

Fig. 1. PCA using covariance vs. coMAD matrices

eigenvectors which result in an increase or decrease of the MAE. For all the experiments we used the data sets offered by the sklearn[2] library.

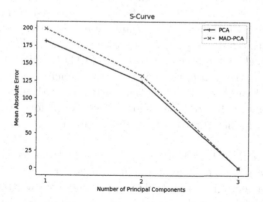

Fig. 2. MAE with increasing number of principal components on the synthetic s-curve data set. orange: coMAD variant; blue: covariance variant (Color figure online)

---

[2] https://scikit-learn.org/stable/modules/classes.html#module-sklearn.datasets.

We begin as a first experiment on a synthetic data set. We generated an s-curve (50 samples, 3 features, 20 gaussian noise, random state = 42) without outliers. By that we have a null-case where we would not expect the coMAD to surpass the covariance approach at all. This data set has been chosen for its inherently non-linear shape. In Fig. 2 we see on the horizontal axis the number of principal components and on the vertical axis the MAE. The coMAD approach is mostly the same like the covariance method, since we do not have outliers but just an increase of noise. However taking the first two principal components, the MAE is by around 25 units lower for the covariance method compared to the coMAD.

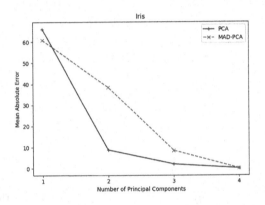

Fig. 3. MAE with increasing number of principal components on the iris data set. orange: coMAD variant; blue: covariance variant (Color figure online)

The second experiment is conducted on the iris data set (450 samples, 4 features). In Fig. 3 it can be seen significant differences in the MAE between the covariance and coMAD PCA variants. Taking the first principal component yields an lower MAE for the coMAD variant compared to the covariance version. For the second principal component we get already a visible difference where the covariance excels in comparison to the coMAD-based PCA. However, the improvement of the MAE from the covariance approach reduces drastically with the third principal component. It remains for future work to further investigate the reasons for why in the coMAD yields a lower MAE for the second principal component and a higher for the third PC compared to the covariance-based approach.

Since iris is small with regards to number of features as well as number of samples, we move in our next experiment to a larger scale. We test now both approaches on the pendigit data set which represents handwritten digits containing over 1797 samples in total, and 64 features. The results can be observed in Fig. 4. Here we observe that the MAE of the coMAD variant is marginally higher compared to the covariance version. This observation raises several questions: 1. Is the coMAD PCA inferior to the covariance PCA? The answer is: it depends.

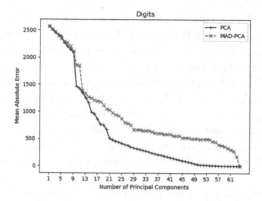

Fig. 4. MAE with increasing number of principal components on the digits data set. orange: coMAD variant; blue: covariance variant (Color figure online)

In the synthetic experiments one could clearly observe that the coMAD based method is superior. It is resilient towards outliers. But why do we get worse MAE results compared to the covariance variant? If we think of what happens in the case of the reconstruction error we recognize the following: It turns out that the MAE is lower for the covariance based approach since we obtain principal components which are heavily deflected. While this deflection is bad, since it means that it does not represent the direction of linear correlated objects in the data set, it is good for the reconstruction, since it means that the deflected principal component minimizes the distance from the outliers to the principal components, and as such, minimizes the error. The observed behaviors regarding the reconstruction are therefore expected. However, we may question at this point whether the MAE itself is a good measure for our purpose. The MAE is suitable if we want to evaluate of how well the detected principal components minimize the distance to each data object in the data set. However, it does not reflect how well the principal components maintain a small distance of the objects belonging to a linear correlation.

5   Conclusion and Future Work

In this work-in-progress we have presented the coMAD in context of PCA. Based on the idea that a median is more robust compared to mean we defined a coMAD matrix on which a PCA is performed. The idea is intriguingly simple compared to competing methods. Experimental results show that the coMAD approach yields principal components which seem unaffected by any noise, while the covariance-based PCA experiences heavy deflection of its resulting principal components.

The potential of the coMAD for PCA may reveal in future work, when for correlation clustering methods such as 4 C the coMAD is used instead of the covariance. Especially in the 4 C scenario where for each data object within an ε-radius a PCA is conducted, we have a small sample size of objects, making it especially prone to just a small number of noise. It also remains to future work to further study those cases where the coMAD approach delivers higher as well as a lower MAE compared to the covariance variant. Detecting and characterizing those cases opens the scene for the development of novel approaches dealing with such special cases. Further it remains of special interest to develop criteria to evaluate the quality of a coMAD-based PCA against a covariance variant, since the MAE does not satisfy the task of determining the resilience of principal components against noise, but rather quantifies the quality of the principal components with regards to the minimization of the distance to all objects in the data set. We hope to kindle with this work future research on the coMAD.

Acknowledgement. This work has been funded by the German Federal Ministry of Education and Research (BMBF) under Grant No. 01IS18036A. The authors of this work take full responsibilities for its content.

References

1. Aggarwal, C.C., Yu, P.S.: Finding generalized projected clusters in high dimensional spaces, vol. 29. ACM (2000)
2. Böhm, C., Kailing, K., Kröger, P., Zimek, A.: Computing clusters of correlation connected objects. In: Proceedings of the 2004 ACM SIGMOD International Conference on Management of Data, pp. 455–466. ACM (2004)
3. Falk, M.: On mad and comedians. Ann. Inst. Stat. Math. **49**(4), 615–644 (1997)
4. Hubert, M., Rousseeuw, P.J., Vanden Branden, K.: Robpca: a new approach to robust principal component analysis. Technometrics **47**(1), 64–79 (2005)
5. Pearson, K.: Liii. On lines and planes of closest fit to systems of points in space. London, Edinburgh, Dublin Philos. Mag. J. Sci **2**(11), 559–572 (1901)
6. De la Torre, F., Black, M.J.: Robust principal component analysis for computer vision. In: Proceedings Eighth IEEE International Conference on Computer Vision, ICCV 2001, vol. 1, pp. 362–369. IEEE (2001)
7. Zhao, Q., Meng, D., Xu, Z., Zuo, W., Zhang, L.: Robust principal component analysis with complex noise. In: International Conference on Machine Learning, pp. 55–63 (2014)

Subspace Determination Through Local Intrinsic Dimensional Decomposition

Ruben Becker[1], Imane Hafnaoui[2], Michael E. Houle[3(✉)], Pan Li[4], and Arthur Zimek[5]

[1] Gran Sasso Science Institute, L'Aquila, Italy
ruben.becker@gssi.it
[2] Ecole Polytechnique de Montréal, Montréal, Canada
imane.hafnaoui@polymtl.ca
[3] National Institute of Informatics, Tokyo, Japan
meh@nii.ac.jp
[4] Saarland Informatics Campus, Saarbrücken, Germany
panli1989@gmail.com
[5] Department of Mathematics and Computer Science,
University of Southern Denmark, Odense, Denmark
zimek@imada.sdu.dk

Abstract. Axis-aligned subspace clustering generally entails searching through enormous numbers of subspaces (feature combinations) and evaluation of cluster quality within each subspace. In this paper, we tackle the problem of identifying subsets of features with the most significant contribution to the formation of the local neighborhood surrounding a given data point. For each point, the recently-proposed Local Intrinsic Dimension (LID) model is used in identifying the axis directions along which features have the greatest local discriminability, or equivalently, the fewest number of components of LID that capture the local complexity of the data. In this paper, we develop an estimator of LID along axis projections, and provide preliminary evidence that this LID decomposition can indicate axis-aligned data subspaces that support the formation of clusters.

Keywords: Intrinsic dimensionality · Estimation · Subspace

1 Introduction

In data mining, machine learning, and other areas of AI, we are often faced with datasets that contain many more attributes than needed, or that can even be helpful for tasks such as clustering or classification. Problems associated with such high dimensional data are for example the concentration effect of distances [6,9] or irrelevant features [13,26]. For clustering [17,23] and outlier detection [26], researchers have made use of various techniques to identify relevant subspaces, as defined by subsets of features that are informative for a particular

© Springer Nature Switzerland AG 2019
G. Amato et al. (Eds.): SISAP 2019, LNCS 11807, pp. 281–289, 2019.
https://doi.org/10.1007/978-3-030-32047-8_25

task. Examples of how relevant subspaces can be determined for individual clusters or outliers include local density estimation in a systematic search through candidate subspaces, or the adaptation of distance measures based on the distribution within local neighborhoods. For sufficiently tight local neighborhoods, the underlying local data manifold can be regarded as approaching a linear form [21], an assumption that further justifies the determination of locally relevant features for subspace determination.

In this paper, we present a novel technique for the identification of subsets of features with the most significant contribution to the formation of the local neighborhood surrounding a given data point, using the recently introduced Local Intrinsic Dimensionality (LID) [10,11] model. LID is a distributional form of intrinsic dimensional modeling in which the volume of a ball of radius r is taken to be the probability measure associated with its interior, denoted by $F(r)$. The function F can be regarded as the cumulative distribution function (cdf) of an underlying distribution of distances. Theoretical properties of LID in multivariate analysis have been studied recently [12]. LID has also seen practical applications in such areas as similarity search [7], dependency analysis [20], and deep learning [18,19].

To make use of the LID model to identify locally-discriminative features, we develop an estimator of LID decomposed along axis projections that compensates for the bias introduced during projection. We also provide preliminary experimental evidence that LID decomposition can indicate axis-aligned data subspaces that support the formation of clusters, by implementing a simple two-stage technique whereby points are first assigned to relevant subspaces, and then clustered. As the relevant features can be different for each cluster, feature relevance is assessed cluster-wise or even point-wise (as the clusters are not known in advance). It is not our intent here to propose a complete subspace clustering strategy; rather the goal in this preliminary investigation is to provide some guidance as to how subspace identification could be done as an independent, initial step as part of a larger clustering strategy.

In Sect. 2, we give some preliminaries on intrinsic dimensionality, before discussing LID decomposition and its estimation. In Sect. 3, to illustrate how LID decomposition could be used within subspace clustering, we propose as an example a simple method using LID to determine eligible subspaces within which DBSCAN is used for clustering. In this preliminary version, only a brief summary of the experimentation is given; more details can be found in [4]. We conclude the paper in Sect. 4.

Preliminaries on ID. Let $X \in \mathbb{R}^m$ be an m-variate random variable, let $F : \mathbb{R}^m \to \mathbb{R}$ be its joint probability distribution, and let $\|\cdot\|$ denote an arbitrary norm. The ID of F at a non-zero point x is defined as follows.

Definition 1 ([12])**.** *Let $x \in \mathbb{R}^m_{\neq 0}$ such that $F(x) \neq 0$. Assume that the partial derivatives $\frac{\partial f}{\partial x_i}(x)$ at x exist for all $i \in [m] = \{1, \ldots, m\}$, the ID of F at x is defined as $\mathrm{ID}_F(x) := x^T \nabla F(x)/F(x)$.*

It is well-known that, under suitable mild continuity assumptions, the ID of F at x is equivalent to both the indiscriminability and the intrinsic dimensionality of F at x, see [12, Theorem 1]. Local intrinsic dimensionalities have also been shown to satisfy the following useful decomposition rule.

Theorem 1 ([12]). *Let $x \in \mathbb{R}^m_{\neq 0}$ and let $I \subseteq \mathbb{R}$ with $0 \in I$ be an open interval such that F is non-zero and its partial derivatives exist and are continuous at $(1 + \varepsilon)x$ for all $\varepsilon \in I$. Assume that $x_i \neq 0$ for each $i \in [m]$. Then $\mathrm{ID}_F(x) = \sum_{i=1}^m \mathrm{ID}_{F_{i,x}}(x_i)$, where $F_{i,x}(t) := F(x_1, \ldots, x_{i-1}, t, x_{i+1}, \ldots, x_m)$ for $i \in [m]$.*

2   Decomposed LID Estimation

Definition and Properties. We now define $N_\delta := \{x \in \mathbb{R}^m : 0 < \|x\|_\infty < \delta\}$, and assume that F is non-zero and that its partial derivatives exist and are continuous at every $x \in N_\delta$. Then, for every $x \in N_\delta$, there is an interval I with $0 \in I$ such that F is non-zero and its partial derivatives exist and are continuous at $(1+\varepsilon)x$ for every $\varepsilon \in I$. Following [12], we define $\mathrm{ID}_F^* := \lim_{x \to 0, \|x\|_\infty \leq \delta} \mathrm{ID}_F(x)$ as the *local intrinsic dimensionality of F*.

Definition 2. *Let I_δ be the 'hollow' open interval $(-\delta, \delta) \setminus \{0\}$. For $x \in N_\delta$, we define the functions $F_{i,x} : I_\delta \to \mathbb{R}$ and $g_i : I_\delta \times I_\delta^{m-1} \to \mathbb{R}$ as $F_{i,x}(t) := F(x_1, \ldots, x_{i-1}, t, x_{i+1}, \ldots, x_m)$ and $g_i(t, x_{-i}) := t \cdot F_{i,x}'(t)/F_{i,x}(t)$, where $x_{-i} = (x_1, \ldots, x_{i-1}, x_{i+1}, \ldots, x_m) \in I_\delta^{m-1}$ for some $x \in N_\delta$.*

Using the Moore-Osgood theorem to interchange the order of limits, we obtain a decomposition rule for LID. For the precise statement of the Moore-Osgood theorem, see for example [16].

Theorem 2. *Assume that for every $i \in [m]$, it holds that (1) $\lim_{t \to 0} g_i(t, y)$ exists for every $y \in I_\delta^{m-1}$, (2) $\lim_{y \to 0} g_i(t, y)$ exists for every $t \in I_\delta$, and (3) at least one of the two limits exists uniformly. Then the limits $\mathrm{ID}_{F,i}^* := \lim_{x \to 0} x_i \cdot F_{i,x}'(x_i)/F_{i,x}(x_i)$ exist for all $i \in [m]$, and thus*

$$\mathrm{ID}_F^* = \sum_{i=1}^m \mathrm{ID}_{F,i}^* = \sum_{i=1}^m \lim_{x \to 0} \frac{x_i \cdot F_{i,x}'(x_i)}{F_{i,x}(x_i)} = \sum_{i=1}^m \lim_{y \to 0} \lim_{t \to 0} g_i(t, y). \tag{1}$$

We refer to $\mathrm{ID}_{F,i}^*$ as the *local intrinsic dimensionality of F in direction i*.

Estimating $\mathrm{ID}_{F,i}^*$. Now let $\phi : \mathbb{R} \to \mathbb{R}$ be a univariate function and assume that $\mathrm{ID}_\phi^* := \mathrm{ID}_\phi(0) = \lim_{t \to 0} t \cdot \phi'(t)/\phi(t)$ exists. We note that Theorems 2 and 3 in [11] yields that, as w approaches 0, it holds that $\phi(t) \approx \phi(w) \cdot (t/w)^{\mathrm{ID}_\phi^*}$. Moreover, differentiating this quantity yields $(\phi(w)/w) \cdot \mathrm{ID}_\phi^* \cdot (t/w)^{\mathrm{ID}_\phi^* - 1}$ as an approximation of $\phi'(t)$. We now apply this observation to the estimation of $\mathrm{ID}_{F,i}^*$ for some $i \in [m]$. Let us fix some $x \in \mathbb{R}^m_{\neq 0}$ and let us denote $\mathrm{ID}_i^* := \mathrm{ID}_{F_{i,x}}^*$ for $i \in [m]$. Given $p^{(1)}, \ldots, p^{(k)} \in \mathbb{R}^m$ following the joint distribution F, we are now in a position to state the log-likelihood function for the parameter ID_i^* under the observations $p^{(1)}, \ldots, p^{(k)}$. Assume that we associate a weight $\omega(p_i^{(j)})$ to the projection $p_i^{(j)}$ of each observation $p^{(j)}$—for the standard unweighted case of the log-likelihood function, all weights are set to 1. We may regard these weights as assigning a-priori likelihoods to the observations, by which an individual observation $p_i^{(j)}$ is accounted as having occurred $\omega(p_i^{(j)})$-many times. The weighted log-likelihood function can then be derived as

$$\mathcal{L}(\mathrm{ID}_i^* : p^{(1)}, \ldots, p^{(k)}) = \sum_{j=1}^{k} \omega(p_i^{(j)}) \cdot \log \left(\frac{F_{i,x}(w)}{w} \cdot \mathrm{ID}_i^* \cdot \left(\frac{|p_i^{(j)}|}{w} \right)^{\mathrm{ID}_i^* - 1} \right).$$

We are now interested in the parameter ID_i^* that maximizes $\mathcal{L}(\mathrm{ID}_i^* : p^{(1)}, \ldots, p^{(k)})$. For this purpose, we form its derivative w.r.t. ID_i^* and set it to zero. A straightforward derivation shows that the likelihood is maximized at

$$\widehat{\mathrm{ID}_i^*} = \left(-\frac{1}{\sum_{j=1}^{k} \omega(p_i^{(j)})} \sum_{j=1}^{k} \omega(p_i^{(j)}) \log \left(\frac{|p_i^{(j)}|}{w} \right) \right)^{-1}, \tag{2}$$

which has the form of a weighted variant of the Hill estimator with threshold w.

Note that we have now developed an estimator for ID_i^*. Assuming however, that for a reference point $x_0 \in \mathbb{R}^m$, the considered neighborhood from which the points $p^{(1)}, \ldots, p^{(k)}$ are chosen is sufficiently small, it is reasonable to use the same estimator for $\mathrm{ID}_{F,i}^*$ as well, as the outer limit in (1) can be neglected.

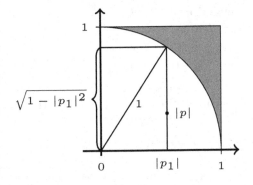

Neighborhood Weighting. In the previous subsection, we have developed an estimator for $\mathrm{ID}_{F,i}^*$; however, we have not yet stated how to determine a neighborhood for x_0. This turns out to be a delicate question, for which the use of observation weighting will become essential.

Note that the estimator for $\mathrm{ID}_{F,i}^*$ that we developed above assumes that neighborhood points $p^{(j)}$ with projections $|p_j|$ stem from the interval $[0, w]$. If we pick a 'box neighbor-

Fig. 1. When considering a circular neighborhood, points p with projections $\|p_1\|$ close to one are much less likely than points with small projections, since the blue region is not accounted for. Such a neighborhood can however still be employed, by associating a weight $\omega(p_1)$ with p that is proportional to 1 over the length of the line segment that contains all points with this projection $\|p_1\|$. (Color figure online)

hood' of x_0 consisting of the k closest points to x_0 with respect to the L_∞ norm (defined as $\|v\|_\infty := \max\{|v_i| : i \in [m]\}$ for $v \in \mathbb{R}^m$), the points p with projections $|p_i|$ close to zero are equally likely to be neighbors as points with projections close to one. This is however, not the case if we pick the neighborhood as the k closest points with respect to the Euclidean norm. In this case, points p with projections $|p_i|$ close to zero will be much more likely to be neighbors than points with $|p_i|$ close to one. As the Euclidean norm is however much more common in practical applications, due to its rotational invariance, we would still like to be able to handle this situation as well. In order to compensate for the bias that results from the fact that points with large projections are less likely than points with small projection, we will use the weighting scheme introduced in the previous subsection. When estimating $\text{ID}^*_{F,i}$, an observation p with projection $|p_i|$ must be weighted according to the ratio of the volume of the $m-1$-dimensional sphere with radius $(1-|p_i|^2)^{1/2}$ on the one hand, and the volume of its bounding hypercube on the other. This leads to the definition of weights $\omega(p_i) := 1/(1-|p_i|^2)^{(m-1)/2}$ for the case of the Euclidean norm. See Fig. 1 for an illustration of the 2D-case.

Verification of $\text{ID}^*_F = \sum_{i=1}^m \text{ID}^*_{F,i}$. In this paragraph we report on an experiment that aims at verifying the equation $\text{ID}^*_F = \sum_{i=1}^m \text{ID}^*_{F,i}$ from Theorem 2 for the case of a uniform distribution in a space equipped with the Euclidean distance metric. For the purpose of estimating ID^*_F, we use the MLE (Hill) estimator proposed in [3] (hill_distances). We compare its output value $\widehat{\text{ID}^*_F}$ on a hyperspherical neighborhood of radius 1 with the sum $\sum_{i=1}^m \widehat{\text{ID}^*_i}$, where we consider two different ways of obtaining the estimates $\widehat{\text{ID}^*_i}$. In the first case (sum_hill_projections), we pick a (unit-)hypercubical neighborhood, while in the second case (sum_w_hill_projections), we use the weighted estimator for the hyperspherical neighborhood compensating for bias using weights.

In our experiment, we create neighborhoods of 100 points for dimensions $m = 2, 4, 8, \ldots, 1024$. Note that in this example of a uniform distribution in m dimensions, the true LID value is m. The experiments show that the two decomposition-based estimators, when summed over all components, do match the total intrinsic dimensionality m, as does the MLE estimator, see Fig. 2.

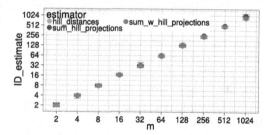

Fig. 2. Results for the three estimators. Errorbars denote 95% confidence intervals. Every measurement is an average of 5 runs.

3   Subspace Clustering Based on LID Decomposition

We now consider some of the issues surrounding the use of LID-decomposition ranking to support subspace clustering. It is not our intent here to propose a single full subspace clustering strategy; rather, the goal is to provide some guidance as to how subspace identification could be done as an independent, preliminary step as part of a larger clustering strategy.

We rely on the LID decomposition to determine relevant attributes for the cluster to which the neighborhood of q belongs. The subspace dimensionality of a point q is determined by searching for attributes with low ID estimates. A common way of doing this is by locating a gap in the sequence of LID estimates that best separates relevant attributes from irrelevant ones, much in the same way as a projective basis is found in PCA decompositions through gaps in the sequence of eigenvalues or variances. We track the relative difference in ID from attributes with low ID to high ID and fix the cut-off that determines the subspace dimensionality at the attribute that exhibits the highest relative difference.

Cluster Membership. To better define the local subspace preference vectors, we propose an additional refinement step. We use a sample of data points $\tilde{\mathcal{X}}$ to build a profile from their subspace preference vectors $\mathcal{P} = \{\mathcal{S}(x) \mid x \in \tilde{\mathcal{X}}\}$. The local subspace preference is refined by determining the membership of points $\mathcal{M}$ to the collected subspace profiles. Given the ordered attributes vector $\mathcal{O}(q)$, $\mathcal{M}(q)$ is selected as the subspace whose attributes are present in the first elements of $\mathcal{O}(q)$. Inside a subspace, points with preference towards that subspace are clustered using a traditional algorithm such as DBSCAN [8].

Experimental Evaluation

Besides the recall, we rely on three other metrics that are widely used in the literature to measure the performance of clustering techniques, namely the Adjusted Rand-Index (ARI) [15], the Normalized Mutual Information (NMI) [24], and the Adjusted Mutual Information (AMI) [25].

Synthetic Data. We synthetically generated three datasets (T1, T2, T3) with 30, 50, and 100 attributes, respectively, each consisting of 5 standard Gaussian clusters with each attribute value from a given cluster generated according to $\mathcal{N}(c,r)$, with c and r having been selected uniformly at random from $[-1,1]$ and $(0,0.2]$, respectively. For T1 and T2, each cluster was generated in its own *distinct* subspace (with no attributes in common between clusters). For the purpose of studying the resilience of the approach to noise, the data was augmented with attributes whose

Table 1. Synth. datasets: description.

	d	$\|\mathcal{S}\|$	Noisy $\mathcal{A}_i$
T1	30	$\{5,5,5,5,5\}$	5
T2	50	$\{3,5,7,7,11\}$	17
T3	100	$\{3,5,7,7,11\}$	67

Table 2. Synth. datasets: results.

		NMI	AMI	ARI	Recall
T1	DiSH	0.535	0.362	0.264	0.582
	CLIQUE	0.431	0.275	0.303	0.635
	LID-DBSCAN	**0.801**	**0.734**	**0.803**	**0.726**
T2	DiSH	0.568	0.396	0.532	0.7
	CLIQUE	0.644	0.473	0.568	**0.78**
	LID-DBSCAN	**0.779**	**0.695**	**0.716**	0.765
T3	DiSH	0.570	0.397	0.412	0.702
	CLIQUE	0.644	0.473	0.568	**0.78**
	LID-DBSCAN	**0.749**	**0.671**	**0.699**	0.76

values were drawn uniformly at random from $[-1, 1]$. T3 was generated from T2 by adding 50 additional attributes with uniform noise. The details are summarized in Table 1. Table 2 summarizes the clustering performance for these datasets comparing our approach against DiSH [1] and CLIQUE [2]. We chose DiSH as it also relies on a point-wise determination of relevant attributes (essentially comparing the spread of distances of nearest neighbors in all attributes) and could be seen as closely related to our approach. In addition, we test against the classical method CLIQUE, as it is arguably the best-known subspace clustering method. In most cases, our approach shows a superior performance.

Manifold Data. For the purpose of further validating the efficiency of the approach to detect significant subspaces on more complex datasets, we relied on the manifold generator proposed in [22] and generated manifolds of differing distributions in different dimensions. We compared the performance of the LID decomposition approach with the one of DiSH with respect to two different metrics (RNIA and ARR) that are generally used to judge clustering algorithms. With respect to both metrics, LID decomposition outperforms DiSH for each of the datasets considered, particularly for D4 (the set with highest average manifold dimension). We refer the reader to the full version for details.

4 Conclusion

Using decomposed LID as a new primitive for estimating the local relevance of a feature, future work could explore more refined subspace clustering approaches. Clustering approaches can be tailored to this new primitive but presumably many existing subspace clustering methods could be adapted to using the new primitive instead of conventional building blocks such as density-estimates, analysis of variance, or distance distributions. Beyond subspace clustering, many more applications can be envisioned, for example in subspace outlier detection [26] or in subspace similarity search [5,14].

Variance-based measures of feature relevance, such as those underlying PCA and its variants, have an advantage over LID in that sample variances decompose perfectly across the coordinates within a Euclidean space. However, although the theoretical values within an LID decomposition are guaranteed to be additive, their estimates are not. Although the experimental results shown in Fig. 2 indicate for the case of uniform distributions that MLE estimates for decomposed LID do sum to the overall LID estimate within reasonable tolerances, it is not clear how well additivity is conserved for real data. Since the additivity of estimators for LID decomposition may depend significantly on their accuracy, future research in this area could benefit from the further development of LID estimators of good convergence properties.

Acknowledgments. M. E. Houle was supported by JSPS Kakenhi Kiban (B) Research Grant 18H03296.

References

1. Achtert, E., Böhm, C., Kriegel, H.-P., Kröger, P., Müller-Gorman, I., Zimek, A.: Detection and visualization of subspace cluster hierarchies. In: Kotagiri, R., Krishna, P.R., Mohania, M., Nantajeewarawat, E. (eds.) DASFAA 2007. LNCS, vol. 4443, pp. 152–163. Springer, Heidelberg (2007). https://doi.org/10.1007/978-3-540-71703-4_15

2. Agrawal, R., Gehrke, J., Gunopulos, D., Raghavan, P.: Automatic subspace clustering of high dimensional data for data mining applications. In: Proceedings of SIGMOD, pp. 94–105 (1998)

3. Amsaleg, L., et al.: Estimating local intrinsic dimensionality. In: Proceedings of KDD, pp. 29–38 (2015)

4. Becker, R., Hafnaoui, I., Houle, M.E., Li, P., Zimek, A.: Subspace determination through local intrinsic dimensional decomposition: theory and experimentation. arXiv e-prints arXiv:1907.06771 (2019)

5. Bernecker, T., et al.: Subspace similarity search: efficient k-NN queries in arbitrary subspaces. In: Gertz, M., Ludäscher, B. (eds.) SSDBM 2010. LNCS, vol. 6187, pp. 555–564. Springer, Heidelberg (2010). https://doi.org/10.1007/978-3-642-13818-8_38

6. Beyer, K., Goldstein, J., Ramakrishnan, R., Shaft, U.: When is "Nearest Neighbor" meaningful? In: Beeri, C., Buneman, P. (eds.) ICDT 1999. LNCS, vol. 1540, pp. 217–235. Springer, Heidelberg (1999). https://doi.org/10.1007/3-540-49257-7_15

7. Casanova, G., et al.: Dimensional testing for reverse k-nearest neighbor search. PVLDB 10(7), 769–780 (2017)

8. Ester, M., Kriegel, H.P., Sander, J., Xu, X.: A density-based algorithm for discovering clusters in large spatial databases with noise. In: Proceedings of KDD, pp. 226–231 (1996)

9. François, D., Wertz, V., Verleysen, M.: The concentration of fractional distances. IEEE TKDE 19(7), 873–886 (2007)

10. Houle, M.E.: Dimensionality, discriminability, density and distance distributions. In: Proceedings of the ICDM Workshops, pp. 468–473 (2013)

11. Houle, M.E.: Local intrinsic dimensionality I an extreme-value-theoretic foundation for similarity applications. In: Beecks, C., Borutta, F., Kröger, P., Seidl, T. (eds.) SISAP 2017. LNCS, vol. 10609, pp. 64–79. Springer, Cham (2017). https://doi.org/10.1007/978-3-319-68474-1_5

12. Houle, M.E.: Local intrinsic dimensionality II: multivariate analysis and distributional support. In: Beecks, C., Borutta, F., Kröger, P., Seidl, T. (eds.) SISAP 2017. LNCS, vol. 10609, pp. 80–95. Springer, Cham (2017). https://doi.org/10.1007/978-3-319-68474-1_6

13. Houle, M.E., Kriegel, H.-P., Kröger, P., Schubert, E., Zimek, A.: Can shared-neighbor distances defeat the curse of dimensionality? In: Gertz, M., Ludäscher, B. (eds.) SSDBM 2010. LNCS, vol. 6187, pp. 482–500. Springer, Heidelberg (2010). https://doi.org/10.1007/978-3-642-13818-8_34

14. Houle, M.E., Ma, X., Oria, V., Sun, J.: Efficient algorithms for similarity search in axis-aligned subspaces. In: Traina, A.J.M., Traina, C., Cordeiro, R.L.F. (eds.) SISAP 2014. LNCS, vol. 8821, pp. 1–12. Springer, Cham (2014). https://doi.org/10.1007/978-3-319-11988-5_1

15. Hubert, L., Arabie, P.: Comparing partitions. J. Classif. 2(1), 193–218 (1985)

16. Kadelburg, Z., Marjanović, M.: Interchanging two limits. Teach. Math. 8(1), 15–29 (2005)

17. Kriegel, H.P., Kröger, P., Zimek, A.: Clustering high dimensional data: a survey on subspace clustering, pattern-based clustering, and correlation clustering. ACM TKDD **3**(1), 1–58 (2009)
18. Ma, X., et al.: Characterizing adversarial subspaces using local intrinsic dimensionality. In: Proceedings of ICLR, pp. 1–15 (2018)
19. Ma, X., et al.: Dimensionality-driven learning with noisy labels. In: Proceedings of ICML, pp. 3361–3370 (2018)
20. Romano, S., Chelly, O., Nguyen, V., Bailey, J., Houle, M.E.: Measuring dependency via intrinsic dimensionality. In: ICPR 2016, pp. 1207–1212, December 2016
21. Roweis, S.T., Saul, L.K.: Nonlinear dimensionality reduction by locally linear embedding. Science **290**, 2323–2326 (2000)
22. Rozza, A., Lombardi, G., Ceruti, C., Casiraghi, E., Campadelli, P.: Novel high intrinsic dimensionality estimators. Mach. Learn. **89**(1–2), 37–65 (2012)
23. Sim, K., Gopalkrishnan, V., Zimek, A., Cong, G.: A survey on enhanced subspace clustering. Data Min. Knowl. Disc. **26**(2), 332–397 (2013)
24. Strehl, A., Ghosh, J.: Cluster ensembles - a knowledge reuse framework for combining multiple partitions. J. Mach. Learn. Res. **3**, 583–617 (2002)
25. Vinh, N.X., Epps, J., Bailey, J.: Information theoretic measures for clustering comparison: variants, properties, normalization and correction for chance. J. Mach. Learn. Res. **11**, 2837–2854 (2010)
26. Zimek, A., Schubert, E., Kriegel, H.P.: A survey on unsupervised outlier detection in high-dimensional numerical data. Stat. Anal. Data Min. **5**(5), 363–387 (2012)

Applications

Leveraging Feature Similarity for Earlier Detection of Unwanted Feature Interactions in Evolving Software Product Lines

Seyedehzahra Khoshmanesh[(✉)] and Robyn R. Lutz[iD]

Iowa State University, Ames, IA 50011, USA
{zkh,rlutz}@iastate.edu
http://web.cs.iastate.edu/rlutz/

Abstract. Software product lines enable reuse of shared software across a family of products. As new products are built in the product line, new features are added. The features are units of functionality that provide services to users. Unwanted feature interactions, wherein one feature interferes with another feature's operation, is a significant problem, especially as large software product lines evolve. Detecting feature interactions is a time-consuming and difficult task for developers. Moreover, feature interactions are often only discovered during testing, at which point costly re-work is needed. This paper proposes a similarity-based method to identify unwanted feature interactions much earlier in the development process. It uses knowledge of prior feature interactions stored with the software product line's feature model to help find unwanted interactions between a new feature and existing features. The paper describes the framework and algorithms used to detect the feature interactions using three path similarity measures and evaluates the approach on a real-world, evolving software product line. Results show that the approach performs well, with 83% accuracy and 60% to 100% coverage of feature interactions in experiments, and scales to a large number of features.

Keywords: Feature interaction · Similarity measures · Software product lines

1 Introduction

A software product line (SPL) is a family of software products that share a set of basic features as a core and differ in other alternative or optional features [18]. Software product lines are widely used in industry to reduce the cost and time-to-market of new products. A *feature* is defined in a software product line as a unit of functionality that provides service to users [7,10,19] (i.e., different from a feature in machine learning or statistics). In a software product line, features are combined in various configurations to form a growing set of new products [9,18].

© Springer Nature Switzerland AG 2019
G. Amato et al. (Eds.): SISAP 2019, LNCS 11807, pp. 293–307, 2019.
https://doi.org/10.1007/978-3-030-32047-8_26

Feature interaction in a software product line refers to a situation in which individual features separately behave as expected, but when merged in a new product, one or more no longer operates as desired and may even be dangerous [3]. Batory et al. describe a classical example of unintended feature interaction, attributed to Kang. A building with a fire-control feature has sensors that, when they detect fire, activate water sprinklers. A building with a flood-control feature has water sensors that, when they detect standing water, turn off the water main. Either feature operates correctly alone; however, if the flood-control feature is added to a building having the fire-control feature, the features interfere with each other and create a hazardous situation [5].

The feature interaction problem is a challenging one that hinders the development of dependable product lines [11,19]. As the number of features grows, the problem typically grows.

Potential feature interactions can increase exponentially with the number of features, making the task even more difficult. In safety-critical systems, unplanned feature interactions can be hazardous. Detecting unwanted feature interactions is difficult, and developers currently depend primarily on testing or code analysis to locate them, both of which occur late in the development process [13].

In small product lines, class similarity can detect some feature interactions [15]; however, real-world product lines typically have very many features [8] and weak traceability from features to code [5].

Our work is based on two observations: first, that a new feature being added is often *similar* to an existing feature, and second, that a product line repository typically contains a *feature model*, a directed acyclic graph that specifies the features, and a representation or list of the known constraints on feature interactions. The constraints encode lessons learned from experience with prior products.

We are especially interested in mutually exclusive features (*alternative* constraints, where only one child feature among several can be selected, and cross-tree *excludes* constraints, where both features cannot be part of the same product). We focus here on these feature interactions because unwanted ones have proven especially troublesome unless discovered early on. A recent Dagstuhl seminar [1] described the need to improve such detection of feature interactions. Moreover, features involved in excludes constraints are often implicated in unsatisfactory user experience and anomalous performance [6]. Widely reported examples include recent iPhone and safety-critical Prius car problems, both of which involved *excludes* constraints on features [12].

We have observed that if there is a mutual exclusion constraint between Feature i and Feature j, and a new product introduces a new Feature k having high similarity with Feature j, then it is likely that Feature k also has a mutually exclusive relationship with Feature i. Detecting mutually exclusive features in code is difficult. Nadi et al. [16] reported that, while they could recover 28% of the existing constraints in a feature model, their approach detected only 3 of its 32 mutex constraints. Our method instead computes the similarity between those features known to interact in unwanted ways in the software product line and the planned new features. We apply this knowledge to discover related feature interactions in the new product much earlier in the development process.

We thus investigate three research questions:

RQ1: *How effectively can we measure the similarity between a new feature and existing features?*

RQ2: *To what extent can path similarity between two features detect unintended feature interaction at an early stage of development?*

RQ3: *To what extent can we effectively predict new, unwanted feature interactions in a new product in a software product line?*

We study these research questions by implementing our approach and evaluating it in a large software product line. Results from evaluation showed an accuracy of 83% and coverage of 60–100% in detecting feature interactions. This indicates that the use of similarity measures between features in a software product line can detect potential feature interactions in the design phase of a newly added product.

The rest of the paper is structured as follows. Section 2 provides an overview of the approach for detecting feature interactions in a software product line. Section 3 specifies the three feature-based similarity measures that were used. Section 4 describes results from the evaluation. Section 5 reviews related work, and Sect. 6 gives concluding remarks.

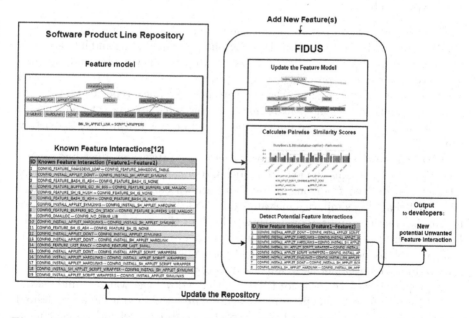

Fig. 1. FIDUS: Feature Interaction Detection Using Similarity in Software Product Lines

2   Overview

In this section we motivate our work and introduce our proposed framework.

2.1   Problem Statement

The problem that we address is detecting unwanted feature interactions. When features are combined to build a new product but do not work together as expected, an unplanned feature interaction occurs. Currently, testing or formal verification are used by developers to detect feature interactions. However, testing all combinations of features is impractical due to its combinatorial complexity in real-world product lines [14]. Moreover, formal verification requires developers to create a formal model, which is rarely done in practice, and then only for small software product lines. Thus, although methods which apply testing and formal methods to detect feature interaction in software product lines are helpful, these methods detect them only late in development, are often infeasible due to combinatorial explosion, or pre-suppose the existence of formal models.

These limitations motivated us to investigate an approach to detect unwanted feature interactions in an *earlier stage of development by using feature similarity measures*. We call our resulting method **FIDUS**, short for "Feature Interaction Detection Using Similarity."

2.2   Approach

Since detecting feature interactions at the code level is costly and occurs late in development, we focus instead on the early-phase feature model and introduce an efficient framework using similarity measures to detect new feature interactions in an evolving software product line. Our proposed framework, FIDUS, is based on three main components: (1) the software product line's repository of known feature interactions, (2) the feature model for the software product line, and (3) a set of similarity measures to understand how close new features are to existing features.

FIDUS is shown in Fig. 1. As shown there, the software product line repository on the left contains the feature model and documentation of known feature interactions, derived from bug reports and known constraints. Figure 1 shows that FIDUS applies similarity measures both to features participating in known feature interactions and to new features added for the next product in the SPL. FIDUS then uses the results of the similarity measurements to identify similar features and predict new feature interactions to be avoided. This information is reported to the developer as well as used to update the repository.

A proposed usage scenario for FIDUS is that a developer wants to understand whether a new product having a new feature $Fnew$ will interact in an undesired way with any existing features. FIDUS uses the feature model with its capture of known constraints (e.g., $Fi \oplus Fj$) to help answer this question by applying feature similarity measures to $Fnew$ with features in known, unwanted interactions. FIDUS then reports to the developer the extent to which the new feature $Fnew$

may potentially participate in any known feature interactions. This technique has the potential to reduce feature interactions inadvertently introduced in a new product, thereby reducing risk as well as saving time and effort in debugging.

3   Similarity Measures to Detect Unwanted Feature Interactions in Software Product Lines

This section first introduces the software product line used in our evaluation so that it can be used for illustrative purposes in the following discussion. The section then describes, specifies, and discusses the three similarity measures that we investigated for detecting feature interactions in a new product or version in an evolving software product line. Table 1 shows the similarity measures used in our study to measure the distance between two features in a tree-based feature model of a large software product line.

3.1   SPL Case Study

We selected three versions to study as products in the very large (600 features) BusyBox software product line [20]. BusyBox is a binary executable and highly configurable system consisting of small versions of many common GNU shell tools such as file utilities and shell utilities. BusyBox can be configured with configurator tools such as menuconfig. As shown in Table 1, we studied features introduced in the configuration part of BusyBox versions 1.7, 1.8, and 1.17, and in the shells part of BusyBox version 1.17. We used the LocMetrics tool[1] for counting lines of code. On average 14% to 17% of the source lines of code are kconfig code, which is a configuration language used to configure BusyBox. The number of features is the total number of features added to or deleted from the prior version. For example, in version 1.8, four features were added and two features were deleted, resulting in 42 features in version 1.8 compared to 40 features in version 1.7.

Table 1. Different parts of different versions of BusyBox studied in this paper

SPL name	Num. features	Num. interactions	#LOC
BusyBox v1.7 (Configuration)	40	16	14.8K
BusyBox v1.8 (Configuration)	42	22	15.5K
BusyBox v1.17 (Configuration,Shells)	72	28	19K

---

[1] http://www.locmetrics.com/index.html.

3.2   Similarity Measures

We chose three path-based metrics which had worked well on a very large as-is hierarchy, WordNet, introduced in [17]. However, here we used them to measure the pairwise similarity between features in a large software product line's feature model. The three similarity metrics are path, lch, and wup [17]. A description of each of these path-based similarity measures together with examples from BusyBox version 1.8 follows. Figure 2 shows the feature model and the position of two features, $SYMLINKS$ and $SCRIPT\_WRAPPER$. We calculate the similarity scores between them below.

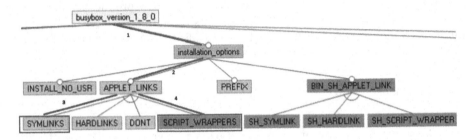

Fig. 2. A piece of the feature model for BusyBox version 1.8 (Color figure online)

Path length (path): In path measure, the similarity score is inversely proportional to the number of nodes along the shortest path between two features in the feature model, which is here treated as an undirected acylic graph. The similarity score is 0 to 1, inclusive. If the two features are identical, the path similarity score is 1. Two sibling features have a path score of 0.5 since the shortest path between them is of length two in terms of counting nodes. Here, Shortest path $(SYMLINKS, SCRIPT\_WRAPPER) = |3, 4| = 2$.

Leacock & Chodorow (lch): The lch similarity score proposed by Leacock and Chodorow (lch) is $-log(length/(2 * D))$, where length is the length of the shortest path between the two features using node-counting and D is the maximum depth of the feature model. Here, $D = maximum\_depth\_feature\_diagram = 4$, and $lch = -log(length/(2 * D)) = -log(2/(2 * 4)) = 0.6$.

Wu & Palmer (wup): The Wu & Palmer measure (wup) calculates the similarity score of two features by considering the depths of the two features in the feature model, along with the depth of the LCS, the least common subsumer (LCS) of the two features, which is the most distinct feature they share as an ancestor. The similarity score formula is $2 * depth(LCS)/(depth(F1) + depth(F2))$. The score can never be zero because the depth of the LCS is never zero since the depth of the root of a feature model is one. The score is 1 if the two features are the same. Here, $depth(LCS) = depth(APPLET\_LINKS) = 3$, and $wup = 2 * depth(LCS)/(depth(F1) + depth(F2)) = 2 * 3/(4 + 4)) = 0.75$.

3.3   Algorithms

Algorithms 1, 2 and 3 show how the path, lch and wup similarity measures are calculated. Algorithm 4 does pairwise calculation of similarity scores between features in a software product line. Lines 2 to 8 execute nested for loops and call Algorithms 1, 2 and 3 to calculate the similarity. Algorithm 5 shows how FIDUS recommends new possible feature interactions when a new feature is added to the existing software product line.

Algorithm 1. Calculate the path Similarity

Input: $Fi, Fj \in F$ (Two Nodes in Feature Model)
Output: $PathSim$(Path Similarity)

1: **function** PATH(Fi, Fj)
2:     $Length \leftarrow |Shortest\_Path(Fi, Fj)|$
3:     $PathSim \leftarrow \frac{1}{Length}$
4:     **return** $PathSim$
5: **end function**

Algorithm 2. Calculate the lch Similarity

Input: $Fi, Fj \in F$ (Two Nodes in Feature Model), D (maximum depth of FM)
Output: $LchSim$(lch Similarity)

1: **function** LCH(Fi, Fj)
2:     $Length \leftarrow |Shortest\_Path(Fi, Fj)|$
3:     $LchSim \leftarrow -log(\frac{Length}{2*D})$
4:     **return** $LchSim$
5: **end function**

Algorithm 3. Calculate the wup Similarity

Input: $Fi, Fj \in F$ (Two Nodes in Feature Model)
Output: $WupSim$(wup Similarity)

1: **function** WUP(Fi, Fj)
2:     $depth(LCS) \leftarrow |LCS(Fi, Fj)|$
3:     $WupSim \leftarrow 2 * depth(LCS)/(depth(F1) + depth(F2))$
4:     **return** $WupSim$
5: **end function**

Algorithm 4. Pairwise calculation of similarity scores between features in a software product line

Input: $F_1 \ldots F_N$
Output: $PathSim[\ ][\ ], LchSim[\ ][\ ], WupSim[\ ][\ ]$ (Pairwise Similarity Matrix)

1: **function** SIMMATRIX($F_1 \ldots F_N$)
2:    **for** $i \leftarrow 1$ to N **do**
3:       **for** $j \leftarrow i + 1$ to N **do**
4:          $PathSim[i][j] \leftarrow PATH(Fi, Fj)$
5:          $LchSim[i][j] \leftarrow LCH(Fi, Fj)$
6:          $WupSim[i][j] \leftarrow WUP(Fi, Fj)$
7:       **end for**
8:    **end for**
9:    **return** $PathSim[\ ][\ ], LchSim[\ ][\ ], WupSim[\ ][\ ]$
10: **end function**

Lines 2 to 19 of Algorithm 5 identify which features in the feature model have the highest similarity values to a new feature. Next, it checks all features in known feature interactions pairs to find whether the most similar features to a new feature are involved in any known unwanted feature interactions. If the algorithm matches the new feature to any known feature interactions, FIDUS recommends it/them as potential unwanted feature interactions. We implemented the algorithms in Python and Java in order to automate the calculation of pairwise similarity scores[2].

Time Complexity. Algorithms 1 and 2 depend only on finding the shortest path between two features in a feature model. Here, the length of the shortest path is defined as the number of nodes in the shortest path between the two features. Finding the shortest path in an undirected acyclic graph has linear time complexity of $O(|E + N|)$ where $|E|$ is the number of edges between features and $|N|$ is the number of features in the feature model. Algorithm 2 needs the maximum depth of the DAG which is linear in N. Thus, the time complexity is $\max(O(|E + N|), O(|N|))$ which is $O(|E + N|)$. The time complexity of Algorithm 3 is also linear in N since it only depends on line 2, which finds the depth of the common ancestor between two features and the depth of each feature. The time complexity of Algorithm 4, which calculates the pairwise similarity between features, is $O(N^3)$. Lines 2 to 8 of Algorithm 4 implement a nested for loop which runs $N(N-1)/2$ times. In lines 4, 5 and 6 of Algorithm4, we call Algorithms 1, 2 and 3, all of which are linear in the number of features, to calculate the similarity scores. Thus, the final time complexity of Algorithm 4 is $O(N^2(E + N))$. Algorithm 5 which reports out the potential new feature interactions that have been discovered has the same time complexity as Algorithm4 since it calculates the pair similarity of new features with existing features in known feature interactions.

---

[2] https://github.com/zahrakhoshmanesh/FIDUS.

Algorithm 5. Recommend new Feature Interaction using path similarity

Input: $F[\ ], FI[\ ], F_{new}$ (FI: list of Known pairwise Feature Interactions in the SPL)
Output: $NewFI[\ ]$ (list of new suggested feature interactions)

```
 1: function DETECT(F[N], FI[K], F_new)
 2:     FSimilarPath ← F1, FSimilarLch ← F1, FSimilarWup ← F1
 3:     PATHMax ← PATH(F1, F_new)
 4:     LCHMax ← LCH(F1, F_new)
 5:     WUPMax ← LCH(F1, F_new)
 6:     for i ← 2 to N do
 7:         PATHSim[i] ← PATH(Fi, F_new)
 8:         LCHSim[i] ← LCH(Fi, F_new)
 9:         WUPSim[i] ← WUP(Fi, F_new)
10:         if (PATH[i] > PATHMax) then
11:             FSimilarPath ← Fi
12:         end if
13:         if (LCHSim[i] > LCHMax) then
14:             FSimilarLch ← Fi
15:         end if
16:         if (WUPSim[i] > WUPMax) then
17:             FSimilarWup ← Fi
18:         end if
19:     end for
20:     for i ← 1 to k do
21:         Fi + FsimilarPath ← FindFI(FsimilarPath)
22:         NewFI[] ← Fi + FsimilarPath
23:         Fi + FsimilarLch ← FindFI(FsimilarLch)
24:         NewFI[] ← Fi + FsimilarLch
25:         Fi + FsimilarWup ← FindFI(FsimilarWup)
26:         NewFI[] ← Fi + FsimilarWup
27:     end for
28:     return NewFI[]
29: end function
```

3.4 Illustrative Example

Figure 2 shows a portion of the feature model for BusyBox versions 1.7 and 1.8. In Fig. 2, the new features added in version 1.8, are highlighted in green in the feature model of BusyBox version 1.7. FIDUS will calculate the similarity between them and those existing features which participate in known unwanted feature interactions of BusyBox version 1.7. As depicted in Fig. 2, there are six features in the installations_option section of BusyBox 1.7, including optional features INSTALL_NO_USR, APPLET_LINK, and PREFIX, and alternative features for APPLET_LINK feature including SYMLINKS, HARD LINK and DONT with the constraint that only one of these three features can be selected. Selecting the two features SYMLINKS and HARD LINK would cause an unwanted feature interaction since these two features have an XOR constraint. An XOR constraint is a mutually exclusive selection of two features such that selecting one feature or configuration must exclude the second feature from the product. XOR constraints are handled as unwanted feature interactions in FIDUS since the two features cannot both be selected for a product.

Feature SCRIPT_WRAPPER is added to BusyBox version 1.8. In other words, SCRIPT_WRAPPER is a new feature beyond BusyBox version 1.7.

We use a list of known unwanted feature interactions of version 1.7 which are shown in the software product line repository of Fig. 1. From that list, we know that SYMLINKS and HARDLINK contribute a feature interaction. FIDUS calculates the similarity score between the new feature SCRIPT_WRAPPER and all other features that participate in known interactions in BusyBox version 1.7. Based on the result, as shown in Fig. 3, DONT, HARDLINK, and SYMLINK are the most similar features to SCRIPT_WRAPPER. Therefore, FIDUS places SCRIPT_WRAPPER instead of these similar features in the known unwanted feature interactions and reports them to the developers as possible new, unwanted feature interactions that may need to be resolved.

4   Results

This section presents our results for each of the three research questions. We implement and evaluate our framework, FIDUS, based on three well-performing similarity metrics described in [17]. Table 2 shows the similarity metrics used in our study.

Table 2. Path-based similarity metrics used in this study

Name of similarity metric	Formula	Reference
path	$1/(length shortest path)$	[17]
lch	$-log(length/(2 * D))$	[17]
wup	$2 * depth(LCS)/(depth(F1) + depth(F2))$	[17]

Our goal is to detect pairwise unwanted feature interactions, i.e., those situations in which two features contribute to an interaction and the presence of one causes an unwanted change in the behavior of the second feature [5]. Two-way feature interactions are the majority of interactions in software product lines, so investigating them serves our goal of detecting the majority of feature interactions in the early phases of new product development.

RQ1: *How effectively can we measure the similarity between a new feature and existing features?*
For RQ1, for each of the three similarity metrics, FIDUS evaluates the detection model on BusyBox version 1.17 and reports the performance in terms of Accuracy and Coverage of detection of unwanted feature interactions. *Accuracy* here is defined as the number of correct feature interactions predicted by FIDUS divided by the total predicted. *Coverage* is defined as the number of unique

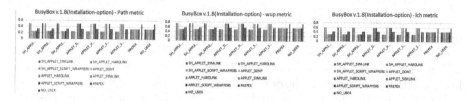

Fig. 3. The Pairwise similarity scores between features in the BusyBox

unwanted feature interactions detected by FIDUS divided by the total unique feature interactions in the software product line. Accuracy and Coverage values are between 0 to 1.

For BusyBox we used a list of 28 XOR constraints in the feature model of BusyBox version 1.17 as the oracle [16]. Since BusyBox is a very large software product line, Fig. 3 shows the similarity scores between features only for the installation-option portion of the BusyBox product line. For all three path-based similarity metrics, the similarity threshold is set here to 0.5. Results presented in Fig. 3 and in Table 3 show that FIDUS effectively measured the similarity between two features using the path similarity measures. High similarity between new features and existing features indicates that they may contribute to similar unwanted feature interactions. The threshold setting and its effect on Accuracy and Coverage are discussed below in RQ3.

RQ2: *To what extent can path similarity between two features detect feature interaction at an early stage of development?*
Table 3 shows the Accuracy and Coverage of FIDUS on BusyBox v.17. As shown in Table 3, with the threshold set to 0.5, the wup metric has the highest Coverage of 100% among the three metrics, with the other two yielding 60% Coverage. This indicates that the wup metric could correctly detect all the known unwanted feature interactions. However, the lch and path metrics have better Accuracy than wup since they are very selective and ignore many new, potential unwanted feature interactions. This reduces the false positives and increases Accuracy while decreasing Coverage. The wup measure allows the similarity formula to consider the locations of both of the two features in the feature model, enabling capture of more context and relaxation of the formula in order to suggest more features as similar features. Considering both Accuracy and Coverage, we see that wup performs better than lch or path measures for our purpose.

Table 3. Accuracy and Coverage of feature interactions detection in BusyBox v1.17

Name of SPL	Similarity measure	Threshold	*Accuracy*	*Coverage*
BusyBox v1.17	wup	0.5	83%	100%
	lch		100%	60%
	path		100%	60 %

RQ3: *To what extent can we effectively predict new feature interactions in a new product in a software product line?*

To answer this question, we consider two different versions of BusyBox, version 1.7 and 1.8. We set BusyBox version 1.7 as the current version of the product line. BusyBox v1.8 serves as the oracle for evaluating the Accuracy and Coverage of FIDUS in predicting new unwanted feature interactions when new features are added. We used a delta comparison between two portions of BusyBox v 1.7 and 1.8 in which unwanted feature interactions occur. A graphical representation of this is shown in Fig. 2, where the green highlighted rectangles are the new features added to BusyBox version 1.7 to constitute BusyBox v1.8. For example, SCRIPT_WRAPPER is added to the APPLET_LINK branch of BusyBox v1.7.

Table 4. Accuracy and Coverage of feature interactions detection in BusyBox v1.8

Name of SPL	Similarity measure	Threshold	*Accuracy*	*Coverage*
BusyBox 1.8	wup	0.5	65%	100%
	lch		100%	20%
	path		100%	20%
	wup	0.3	65%	100%
	lch		65%	100%
	path		30%	20%

As mentioned earlier, we have the list of known unwanted feature interactions for BusyBox version 1.7. We can thus compute the similarity between the new features added to the system in version 1.8, and the existing features known to have contributed to unwanted feature interactions in the past. Finally, we predict the new unwanted feature interactions that involve new features in version 1.8. As shown in Table 4, with the threshold set to 0.5, the wup metric can detect new feature interactions with 65% Accuracy and 100 % Coverage. That is, wup predicted all new feature interactions. The false positive predictions cause it to have a lower Accuracy compared to the Coverage. The lch and path metrics have very weak Coverage, 20%, although they also have 100% Accuracy. These findings indicate that lch and path are overly selective for our purposes at this threshold, identifying as new feature interactions only those with very high potential and thus missing many.

We also investigated a relaxed version of our similarity framework, FIDUS. We defined the relaxation as a decreasing of the threshold from 0.5 to 0.3, and show the FIDUS results in Table 4 on BusyBox version 1.8 with the threshold lowered to 0.3. As shown in Table 4, the lch metric achieved the same results as the wup metric while the path metric still had a weak result. This indicates that, as expected, Coverage will improve when the threshold decreases; however, the Accuracy will decrease. Thus, there is a trade-off between Accuracy and Coverage. However, the wup metric continued to perform acceptably well in both Accuracy and Coverage.

In summary, results from evaluation on the BusyBox software product line indicated that, using the wup measure, the similarity framework, FIDUS, can effectively predict new unwanted feature interactions that are similar to known ones, in the new version of a software product line.

4.1 Threats to Validity

In this section, we describe some threats to the internal and external validity of our model.

Internal Threats. We used a single large software product line case study [16] on which to evaluate our work. However, this case study is a well-regarded benchmark in the software product line literature. Moreover, detection of mutual exclusion feature interactions in BusyBox has continued to be an open problem, so the use of it provides a good challenge to evaluate our technique. Second, we used a limited set of three similarity metrics in our study. However, these three were selected after careful study and achieved good accuracy and reasonable coverage of the targeted feature interactions.

External Threats. Evaluation on additional software product lines in other application areas beyond that presented here is needed. Future work will seek to ascertain whether the results shown on the BusyBox product line hold up and can be generalized to software product lines in other domains.

5 Related Work

In the area of using similarity measures in a software product line, Henard, et al. [13] used similarity measures to prioritize test cases in order to reduce the number of product configurations in software product-line testing. Our study is different in that we used similarity measures to detect feature interaction at the design stage. To our knowledge similarity measures have not been studied in order to detect feature interactions in a new product of a software product line.

Pedersen, et al. [17] evaluated three similarity measures based on path lengths, naming path, lch, and wup in an is-a hierarchy of concepts in order to measure the similarity of two concepts. We use the same similarity measures; however, our study differs in that we use them instead to detect feature interactions.

Nadi, et al. [16] extracted constraints from the code of highly configurable systems and compared them with constraints in the feature model, achieving recovery of 28% of the existing feature model constraints from the code. While we also use the feature model, our work differs from theirs both in that we seek to detect feature interactions early in development and in that we use feature-similarity measures to do so. Atlee, Fahrenberg and Legay [4] used simulation on formal models (featured transition systems) to measure the degree to which a product's behavior differs when a new feature is added. Unlike us, they did not distinguish between intended and unintended feature interactions.

Soares, et al. [19] performed a recent systematic study of works on feature interaction in software product line engineering. They found that 43% of the papers aimed to understand feature interaction at early stages of the software life cycle. Among this 43%, the majority used formal methods, specifically model checking.

Apel, et al. [2,3] proposed feature-aware verification to automate the detection of feature interactions using variability encoding. Our approach to dealing with the feature interaction problem differs from their studies in not requiring formal methods, motivated by the current low uptake of formal methods by industries developing software product lines.

We instead aim in our method to leverage feature similarity to detect unwanted feature interactions in an evolving software product line, since this method is more understandable to the developers and users who work with and maintain the product-line systems.

6   Conclusion

This paper described a similarity-based method to detect feature interactions at the design phase of a new product in an evolving software product line by exploiting knowledge of prior problematic feature interactions and identifying similar new features likely to have the same involvement in those problematic interactions. Results from our evaluation of this approach on a real-world software product line indicate that calculations of path-based similarity between features in a software product line's feature model can help detect unwanted and perhaps risky feature interactions much earlier in the development process for a new product than current testing techniques.

Acknowledgments. We thank Andrei Migunov for feedback on an early draft. The work in this paper was partially funded by National Science Foundation Grant CCF 1513717.

References

1. Apel, S., Atlee, J.M., Baresi, L., Zave, P.: Feature interactions: the next generation (Dagstuhl seminar 14281). Dagstuhl Rep. **4**, 1–24 (2014)
2. Apel, S., Speidel, H., Wendler, P., Von Rhein, A., Beyer, D.: Detection of feature interactions using feature-aware verification. In: ASE, pp. 372–375 (2011)
3. Apel, S., Von Rhein, A., Thüm, T., Kästner, C.: Feature-interaction detection based on feature-based specifications. Comput. Netw. **57**(12), 2399–2409 (2013)
4. Atlee, J.M., Fahrenberg, U., Legay, A.: Measuring behaviour interactions between product-line features. In: 2015 IEEE/ACM 3rd FME Workshop on Formal Methods in Software Engineering (FormaliSE), pp. 20–25. IEEE (2015)
5. Batory, D., Höfner, P., Kim, J.: Feature interactions, products, and composition. In: ACM SIGPLAN Notices, vol. 47, pp. 13–22. ACM (2011)
6. Benavides, D., Segura, S., Ruiz-Cortés, A.: Automated analysis of feature models 20 years later: a literature review. Inf. Syst. **35**(6), 615–636 (2010)

7. Berger, T., et al.: What is a feature?: a qualitative study of features in industrial software product lines. In: SPLC, pp. 16–25. ACM (2015)
8. Berger, T., She, S., Lotufo, R., Czarnecki, K., Wasowski, A.: Feature-to-code mapping in two large product lines. In: SPLC, pp. 498–499. Citeseer (2010)
9. Botterweck, G., Pleuss, A.: Evolution of Software Product Lines. In: Mens, T., Serebrenik, A., Cleve, A. (eds.) Evolving Software Systems, pp. 265–295. Springer, Heidelberg (2014). https://doi.org/10.1007/978-3-642-45398-4_9
10. Bowen, T.F., Dworack, F.S., Chow, C.H., Griffeth, N., Herman, G.E., Lin, Y.: The feature interaction problem in telecommunications systems. In: SETSS 89 (1989)
11. Calder, M., Kolberg, M., Magill, E.H., Reiff-Marganiec, S.: Feature interaction: a critical review and considered forecast. Comput. Netw. 41(1), 115–141 (2003)
12. Chechik, M., Stavropoulou, I., Disenfeld, C., Rubin, J.: FPH: efficient non-commutativity analysis of feature-based systems. In: Russo, A., Schürr, A. (eds.) Fundamental Approaches to Software Engineering, pp. 319–336 (2018)
13. Henard, C., Papadakis, M., Perrouin, G., Klein, J., Heymans, P., Le Traon, Y.: Bypassing the combinatorial explosion: using similarity to generate and prioritize t-wise test configurations for software product lines. TSE 40(7), 650–670 (2014)
14. Jia, Y., Cohen, M.B., Harman, M., Petke, J.: Learning combinatorial interaction test generation strategies using hyperheuristic search. In: ICSE. IEEE (2015)
15. Khoshmanesh, S., Lutz, R.R.: The role of similarity in detecting feature interaction in software product lines. In: ISSREW, pp. 286–292. IEEE (2018)
16. Nadi, S., Berger, T., Kästner, C., Czarnecki, K.: Where do configuration constraints stem from? An extraction approach and an empirical study. IEEE Trans. Softw. Eng. 41(8), 820–841 (2015)
17. Pedersen, T., Patwardhan, S., Michelizzi, J.: WordNet: similarity: measuring the relatedness of concepts. In: HLT-NAACL, pp. 38–41 (2004)
18. Pohl, K., Böckle, G., van Der Linden, F.J.: Software Product Line Engineering: Foundations, Principles and Techniques. Springer, Berlin (2005). https://doi.org/10.1007/3-540-28901-1
19. Soares, L.R., Schobbens, P.Y., do Carmo Machado, I., de Almeida, E.S.: Feature interaction in software product line engineering: a systematic mapping study. Inf. Softw. Technol. 98, 44–58 (2018)
20. Wells, N.: BusyBox: a Swiss army knife for Linux. Linux J. 2000, 10 (2000)

Protein Complex Similarity Based on Weisfeiler-Lehman Labeling

Bianca K. Stöcker[1,2] , Till Schäfer[4] , Petra Mutzel[4] ,
Johannes Köster[1,2,3] , Nils Kriege[4] , and Sven Rahmann[1,4(✉)]

[1] Genome Informatics, Institute of Human Genetics, University Hospital Essen,
University of Duisburg-Essen, 45147 Essen, Germany
Sven.Rahmann@uni-due.de
[2] Algorithms for Reproducible Bioinformatics, Institute of Human Genetics,
University Hospital Essen, University of Duisburg-Essen,
45147 Essen, Germany
[3] Medical Oncology, Dana-Farber Cancer Institute, Harvard Medical School,
Boston, MA 02215, USA
[4] Computer Science XI, TU Dortmund University, 44221 Dortmund, Germany

Abstract. Proteins in living cells rarely act alone, but instead perform
their functions together with other proteins in so-called protein complexes. Being able to quantify the similarity between two protein complexes is essential for numerous applications, e.g. for database searches of
complexes that are similar to a given input complex. While the similarity problem has been extensively studied on single proteins and protein
families, there is very little existing work on modeling and computing the
similarity between protein complexes. Because protein complexes can be
naturally modeled as graphs, in principle general graph similarity measures may be used, but these are often computationally hard to obtain
and do not take typical properties of protein complexes into account.
Here we propose a parametric family of similarity measures based on
Weisfeiler-Lehman labeling. We evaluate it on simulated complexes of the
extended human integrin adhesome network. We show that the defined
family of similarity measures is in good agreement with edit similarity, a similarity measure derived from graph edit distance, but can be
computed more efficiently. It can therefore be used in large-scale studies
and serve as a basis for further refinements of modeling protein complex
similarity.

Keywords: Similarity measure · Protein complexes ·
Weisfeiler-Lehman labeling · Constrained protein interaction networks ·
Jaccard similarity

1 Introduction

Proteins fulfill manifold tasks in living cells, but they rarely act alone. Indeed,
most cellular functions are enabled only when proteins physically interact with

© Springer Nature Switzerland AG 2019
G. Amato et al. (Eds.): SISAP 2019, LNCS 11807, pp. 308–322, 2019.
https://doi.org/10.1007/978-3-030-32047-8_27

other proteins, forming protein complexes. DNA transcription is a typical example, where RNA polymerase II, general transcription factors, cell type specific transcription regulators and mediator proteins interact.

Understanding protein complex formation and function is one of the big challenges of cell biology, approached by both experimental techniques and computational modeling. While the constituent protein sequences can be obtained from the genome, the computational prediction of real protein complexes from protein interaction networks appears to be much more difficult [3,24]. Fortunately, new experimental technologies are about to enhance our understanding of complexes significantly in the near future, e.g. high-resolution protein-protein docking [12,16]. Large scale generation of libraries of cell lines having two or more endogenously tagged fluorescent proteins [4] and recent high-throughput and multiplexed implementations of fluorescence correlation spectroscopy allow us to systematically measure endogenous concentrations, binding constants and high-order complexes in such libraries of cell lines [8,27]. Protein complexes can be made of transient and obligate interactions. The former appear and disappear *in vivo*, whereas the latter are disrupted enzymatically. The approach shown in this work can be applied to both types. However, in our analysis, we focus on the former.

When studying biological entities such as protein sequences or protein complexes, a fundamental task is to define a measure of similarity between two such entities. For protein sequences, there is a well-established theory based on scoring matrices and alignment scores [17]. For protein complexes, it appears that no systematic effort to quantify similarity has been made yet. The purpose of the present article is therefore to discuss the different options to define a similarity measure on protein complexes and to propose a reasonable and computationally tractable definition of protein complex similarity.

Establishing a similarity measure is not only important fundamentally, but there are many immediate applications, such as:

Database search: In the *database search problem* we are given a query complex and a large collection (database) of complexes, and the task is to find the complexes in the database that are most similar to the query.

Comparing predictions: Several complex prediction methods predict putative complexes by locating dense regions in a protein interaction network [6,9,15, 18], and for comparing complexes predicted by different algorithms, it is of interest to compute a maximum-weight matching between the output of two algorithms, where the weighting is given by a similarity function.

Summarizing and clustering: When simulating complex formation based on available knowledge such as possible interactions and interaction constraints, it is helpful to aggregate the simulation output to focus on frequently seen or typical complexes, ignoring small differences. Aggregation or clustering by similarity thereby reduces data size and complexity. Such a task requires a way to quantify the similarity between two protein complexes.

When there are tens of thousands of different complexes subject to pairwise comparison, a similarity measure must be efficiently computable.

Models for Protein Complexes. We first need to discuss models for protein complexes on different levels of detail, namely the *set, multiset,* and *graph* models.

While intuition suggests that protein complexes can be naturally described as graphs with proteins as vertices and physical interactions as edges, there are in fact different ways to formally describe a protein complex. We start with a given set P of all proteins of an organism, the building blocks of the complexes.

Set: In its most simple form, a protein complex can be defined as a set (in the mathematical sense, i.e., without multiplicities) of proteins, i.e., as a subset $\{p_1, p_2, \ldots, p_n\}$ of P. Sets neither capture the multiplicities nor the nature of the physical interactions between the constituent proteins of a complex. Some experimental techniques only give such set-type information, and several existing databases only provide this type of information, e.g. the CORUM database [20].

Multiset: Formally, a multiset is a function $C : P \to \mathbb{N}_0$ that assigns a multiplicity to each protein $p \in P$ with $C(p) = 0$ for proteins p that are not part of the complex. We also use the multiset notation $C = \{\{p_1, p_1, p_2\}\}$ to express that $C(p_1) = 2$, $C(p_2) = 1$ and $C(p) = 0$ for all other $p \in P$. Defining a protein complex as a multiset of proteins gives a more accurate representation of the complex, but still does not consider the interaction topology.

Graph: To add more information, we can define a protein complex as an undirected graph $C = (V, E, \ell)$ with labeled vertices V, such that each vertex $v \in V$ represents a protein and hence has a label $\ell(v) \in P$, each edge $e \in E \subseteq V \times V$ represents a physical interaction between the corresponding proteins, such that E is symmetric and C is connected. The graph description provides the interaction topology. We call this representation a *protein complex graph* and define its *size* as $|C| := |V| + |E|$.

For the set and multiset models, a similarity measure is readily given by the *Jaccard similarity* (see Methods). For graphs, the *graph edit distance* has been proposed for pattern recognition tasks more than 30 years ago [22]. A graph edit distance between graphs C and C' measures the total costs of the edit operations required to transform C into C'. Defining similarity via graph edit operations appears intuitive, but has computational disadvantages, as the graph edit distance generalizes the classical maximum common subgraph problem [5], which is NP-complete [7] and hard to approximate with given guarantees [10]. Recently, a binary linear programming formulation for computing the graph edit distance has been proposed [14], which allows to compare graphs of moderate size using state-of-the art general purpose solvers. However, when we want to compare many complexes, evaluating the edit distance between all pairs becomes infeasible in practice.

In this article, we therefore propose an efficient alternative: We define a family of similarity measures on graphs by resorting to the Jaccard similarity, which is efficiently computed and even more efficiently estimated using established locality-sensitive hashing techniques. Taking the graph structure into account is achieved by so-called Weisfeiler-Lehman labeling of the vertices [28], propagating vertex labels between neighbors. This approach is different from recent work that approximates and bounds the graph edit distance [19] and has the advantage of scaling better to large-scale studies.

The remainder of the article is structured as follows. In the Methods section, we define a parametric family of similarity measures based on Weisfeiler-Lehman labeling and the precise definition of graph edit similarity we compare against. In the Results section, we describe how we obtain pairs of protein complexes, for which we compare Weisfeiler-Lehman similarity and edit similarity. Finally, we discuss limitations and possible extensions of this work.

2 Methods

Our goal is to define a similarity measure between protein complexes that captures not only the (multisets of the) constituent proteins, but also the interaction topology (graph structure). We introduce a parameterized family of similarity measures on protein complexes, which are based on multiset comparisons of vertex labels in the complex graph and take the local neighborhood of each protein into account by using Weisfeiler-Lehman labels.

Jaccard Similarity of Sets and Multisets. To compare sets or multisets, Jaccard similarity coefficients are an established quantity.

Let $M \subseteq U$ and $M' \subseteq U$ be two subsets of a common universe U. Then the *Jaccard similarity* between M and M' is defined as

$$J_{\text{set}}(M, M') := \frac{|M \cap M'|}{|M \cup M'|} \in [0, 1].$$ (1)

This definition is extended to multisets as follows. Recall that multisets M and M' are functions $U \to \mathbb{N}_0$, assigning multiplicities $M(o)$ and $M'(o)$ to each object $o \in U$. (The set definition can be seen as the special case where the value set is only $\{0, 1\}$ instead of $\mathbb{N}_0$.) Then the *Jaccard similarity* between M and M' is defined as

$$J_{\text{multiset}}(M, M') := \frac{\sum_{o \in U} \min\{M(o), M'(o)\}}{\sum_{o \in U} \max\{M(o), M'(o)\}} \in [0, 1].$$ (2)

A Parametric Family of Protein Complex Similarity Measures. Instead of comparing the protein complexes directly by their graph topology and labeling, we extract and compare multisets of features of the protein complexes. Weisfeiler and Lehman developed an iterative label refinement procedure to derive a canonical graph representation for graph isomorphism testing [28]. The same procedure is often used to define graph similarities or graph kernels [23].

Initially, the feature multiset of a graph consists of the union of all vertex labels, i.e., the protein names. After the initialization, the vertex labels are iteratively augmented by the labels of the neighboring vertices from the previous iteration, thereby encoding the (local) graph structure in the vertex labels. Let us now formally define the process.

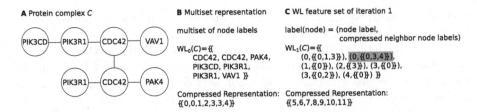

A Protein complex *C*

PIK3CD—PIK3R1—CDC42—VAV1

PIK3R1—CDC42—PAK4

B Multiset representation

multiset of node labels

WL$_0$(C)={{
   CDC42, CDC42, PAK4,
   PIK3CD, PIK3R1,
   PIK3R1, VAV1 }}

Compressed Representation:
{{0,0,1,2,3,3,4}}

C WL feature set of iteration 1

label(node) = (node label,
               compressed neighbor node labels)

WL$_1$(C)={{
   (0,{{0,1,3}}), (0,{{0,3,4}}),
   (1,{{0}}), (2,{{3}}), (3,{{0}}),
   (3,{{0,2}}), (4,{{0}}) }}

Compressed Representation:
{{5,6,7,8,9,10,11}}

Fig. 1. Example of a protein complex and its representations. The colors highlight the labels of an example node in WL$_0$(C) and WL$_1$(C). **A:** Graph representation of protein complex C. **B:** Multiset representation of C which is equal to WL$_0$(C). **C:** Result of the first WL iteration.

Definition 1 (Weisfeiler-Lehman labeling of iteration i for a protein complex graph). Let $C = (V, E, \ell_0)$ be a protein complex graph with label function $\ell_0 : V \to L_0 := P$. Furthermore, let $N(v) := \{u \mid \{v, u\} \in E\}$ denote the neighbors of vertex $v \in V$. Then, the Weisfeiler-Lehman labeling of iteration i is defined as a re-labeling of the protein complex graph: It replaces the labeling function $\ell_0 : V \to L_0$ with a labeling function $\ell_i : V \to L_i$. The value of ℓ_i for a vertex $v \in V$ is recursively defined as

$$\ell_i(v) := (\ell_{i-1}(v), \{\{\ell_{i-1}(u) \mid u \in N(v)\}\}). \tag{3}$$

Note that the second component of the new label is a multiset.

To avoid that the length of labels increases in each iteration, label compression is performed after each step in practice. This is achieved by a one-to-one mapping of the labels $\{\ell_i(v) \mid v \in V\}$ to integer labels. Note that the label compression step must be consistent across multiple graphs in order to construct comparable feature sets.

Given the Weisfeiler-Lehman labeling function of a protein complex graph for some iteration i, we can now define the multiset of Weisfeiler-Lehman features for iteration i.

Definition 2 (Weisfeiler-Lehman feature set of iteration i for a protein complex graph). Let $C = (V, E, \ell_0)$ be a protein complex graph with label function $\ell_0 : V \to L_0 = P$. Then, the Weisfeiler-Lehman features of iteration i are defined as multiset $WL_i(C) = \{\{l_i(v) \mid v \in V\}\}$.

Note that WL$_0$(C) always corresponds to the initial multiset of protein names as described above. Accordingly, WL$_1$(C) integrates the neighborhood labels of each node. Figure 1 shows an example protein complex, together with the associated feature sets WL$_0$(C) and WL$_1$(C). A node and its neighborhood are highlighted in red and blue to demonstrate the relation between WL$_0$(C) and WL$_1$(C).

We use the Jaccard coefficient to obtain a normalized similarity based on multiset intersection. We apply the Jaccard coefficient to the feature sets of each iteration individually and compute a convex combination of the results. Let $w = (w_i)_{i \geq 0}$ be a sequence of non-negative weights with $\sum_{i \geq 0} w_i = 1$. We compare two complexes C and C' by

$$S_w(C, C') := \sum_{i \geq 0} w_i \cdot J_{\text{multiset}}(\text{WL}_i(C), \text{WL}_i(C')), \tag{4}$$

where J_{multiset} is given by Eq. (2). This defines a family of similarity measures between complexes with values in $[0, 1]$, parameterized by the weight vector $w = (w_0, w_1, \dots)$.

It is easy to see that, as long as $w_0 > 0$, we have $S_w(C, C') = 0$ if and only if the protein sets of C and C' are disjoint. If $S_w(C, C') < 1$, the protein complex graphs are not isomorphic. However, $S_w(C, C') = 1$ does not necessarily imply that C and C' are isomorphic even if $w_i > 0$ for all i: There exist examples of non-isomorphic graphs G, G' with $\text{WL}_i(G) = \text{WL}_i(G')$ for all $i \geq 0$. (As a simple example, take G to be a cycle of six vertices, and G' to be two cycles of three vertices, all with the same label.) On the other hand, there exist classes of graphs, such as the so-called CR-graphs, for which the implication "$S_w(C, C') = 1 \Rightarrow C, C'$ are isomorphic" is true if $w_i > 0$ for all i [1]. Moreover, the implication holds with high probability for random graphs (without vertex labels) even when $w_i = 0$ for all $i \geq 3$ [2].

In practice, we may assume that most protein complexes are non-adversarial graphs with sufficiently simple structure and expressive initial labels such that their Weisfeiler-Lehman features are appropriate to characterize their similarity. In fact, we put forward the hypothesis that using a single iteration is frequently sufficient for practical purposes, and we set $w_i := 0$ for $i \geq 2$ in our computational experiments (see Results) and only have a single free parameter $w_0 \in [0, 1]$ that defines $w_1 := 1 - w_0$. In the following, we write ω for w_0. In this case, S_ω is efficiently computable: A proof of the following lemma can be found in the work of [23].

Lemma 1. *For $\omega \in [0, 1]$, each of the one-parameter similarity measures*

$$S_\omega(C, C') := \omega \cdot J_{multiset}(WL_0(C), WL_0(C')) + (1-\omega) \cdot J_{multiset}(WL_1(C), WL_1(C'))$$

can be computed in $O(|C| + |C'|)$ time, where $|C| = |V| + |E|$.

A Similarity Measure Based on Graph Edit Distance. To compare the family of Weisfeiler-Lehman multiset-based similarity measures defined above with graph edit distance, we state a formal definition of the edit-based similarity. We allow the following elementary operations to edit a graph: vertex deletion, vertex insertion, vertex relabeling, edge deletion, and edge insertion. A sequence $(o_1, \dots, o_k)$ of such edit operations that transforms a graph G into another graph H is called an *edit path* from G to H. Each operation o is assigned a cost $c(o)$, which is zero for substituting vertices and edges with the same label. We use

314   B. K. Stöcker et al.

a cost of 1 for all operations except vertex relabeling which has a cost of 2, corresponding to one deletion and one insertion (leaving the edges in place). Note that deleting or inserting a vertex of degree k otherwise has cost $k + 1$ for deleting k edges and the vertex itself. We denote the set of all possible edit paths from G to H by $\Upsilon(G, H)$.

Definition 3. *Let G and H be labeled graphs. The* graph edit distance *from G to H is defined by*

$$d(G, H) = \min \left\{ \sum_{i=1}^{k} c(o_i) \,\middle|\, (o_1, \dots, o_k) \in \Upsilon(G, H) \right\}. \tag{5}$$

Intuitively, the graph edit distance preserves a subgraph G' of G that is also contained in H using zero-cost substitutions, deletes the vertices and edges in G that are not in G' and then inserts vertices and edges to obtain an isomorphic copy of H. Therefore all non-zero costs can be attributed to the elements which are in one of the graphs, but not in their common subgraph. In this sense the graph edit distance is similar to the symmetric difference of two sets. This observation motivates the following normalized similarity measure derived from the graph edit distance. We define the *graph edit similarity* as

$$J_{\text{graph}}(G, H) := \frac{|G| + |H| - d(G, H)}{|G| + |H| + d(G, H)} \in [0, 1], \tag{6}$$

where $|G| := |V(G)| + |E(G)|$. Note that the graph edit distance between G and H is at most $|G| + |H|$, which is achieved by deleting all vertices and edges of G and inserting all vertices and edges of H. In this case the graph edit similarity is zero. Similarly, $J_{\text{graph}}(G, H) = 1$ if and only if $d(G, H) = 0$. In this respect the similarity measure resembles the Jaccard similarity. In fact, if the edges are not taken into account, the graph edit similarity equals the multiset Jaccard similarity.

Lemma 2. *For two protein complexes, let C, D denote their protein multisets and G, H their protein complex graphs. For the edge-free graphs $G' = (V(G), \emptyset)$ and $H' = (V(H), \emptyset)$ it holds that $J_{graph}(G', H') = J_{multiset}(C, D)$.*

Proof. An optimal graph edit path is obtained as follows: We substitute the vertices with common labels free of cost, which are $Z = \sum_{p \in P} \min\{C(p), D(p)\}$ in total. We delete the remaining $|G'| - Z$ vertices in G' and insert $|H'| - Z$ vertices to obtain an isomorphic copy of H' at a total cost of $|G'| + |H'| - 2Z = d(G, H)$. Instead we may also substitute up to $||G'| - |H'||$ vertices, each at cost two, which results in the same total cost. Using the fact that $|G'| = \sum_{p \in P} C(p)$ and $|H'| = \sum_{p \in P} D(p)$, we obtain the result by calculating

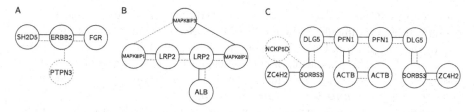

Fig. 2. Three exemplary pairs of protein complexes. Each labeled node is a protein instance, each edge a protein interaction, and solid black vs. dashed red edges distinguish between the two complexes. **A:** Edit similarity 0.714; WL similarity in $[0.4, 0.75]$ depending on weight ω. **B:** Edit similarity 0.838; WL similarity 1.0 (independent of ω). **C:** Edit similarity 0.9; WL similarity in $[0.667, 0.818]$ depending on ω.

$$
\begin{aligned}
J_{\text{graph}}(G', H') &= \frac{|G'| + |H'| - d(G', H')}{|G'| + |H'| + d(G', H')} = \frac{Z}{|G'| + |H'| - Z} \\
&= \frac{Z}{\sum_{p \in P} C(p) + \sum_{p \in P} D(p) - Z} = \frac{\sum_{p \in P} \min\{C(p), D(p)\}}{\sum_{p \in P} C(p) + D(p) - \min\{C(p), D(p)\}} \\
&= \frac{\sum_{p \in P} \min\{C(p), D(p)\}}{\sum_{p \in P} \max\{C(p), D(p)\}} = J_{\text{multiset}}(C, D).
\end{aligned}
$$

Lemma 2 shows that the graph edit similarity can indeed be seen as a natural extension of the multiset Jaccard similarity to graph structured data.

3   Results

Hypothesis. We hypothesize that the Weisfeiler-Lehman based family of similarity measures S_ω defined in Eq. (4) approximates well the edit distance based similarity defined in Eq. (6) for typical protein complexes. The similarity measures S_ω have the advantage that they can be efficiently computed (see Lemma 1).

Experimental Setup. We have implemented the similarity measures based on Weisfeiler-Lehman labeling and the graph edit similarity in Java 8. To compute the exact graph edit distance, we used a recent binary linear programming formulation [14] and solved the instances using Gurobi 7.5.2. All experiments were run on 18-core Intel Xeon E5-2699 CPUs with 2.30 GHz using 64bit Ubuntu Linux 14.04. To ensure the reproducibility of our experiments, the performed data analysis is available as a Snakemake [11] workflow[1].

---

[1] https://doi.org/10.5281/zenodo.1178084.

Data Generation. As mentioned in the Introduction, obtaining real protein complex graphs is difficult at the moment, because experimental techniques that resolve the (graph) topology of the complexes are only being developed. Therefore we resort to the simulation of complexes, based on two types of knowledge: possible physical protein-protein interactions, formalized by a *protein interaction network*, and *constraints between protein interactions*.

Formally, a protein interaction network is an undirected graph $N = (P, I)$, where P is the set of protein types of a cell (or an organism), and $I \subset P \times P$ indicates the pairs of protein types that may potentially physically interact. Since N describes the entirety of possible interactions, any protein complex can be seen as a connected subgraph of N.

It is important to realize that protein interactions are not independent of each other, but interdependent. Those interaction dependencies are generated by two major mechanisms. On the one hand there is allosteric regulation, in which the capability of a protein to bind other proteins is affected by a conformational change upon one interaction [13]. The other key mechanism is steric hindrance that prevents proteins from binding simultaneously to too close or identical protein domains leading to mutual exclusiveness of interactions [21]. The dependencies between interactions constrain the set of possible protein complexes and their assembly paths. One possible model for this are *constrained protein interaction networks*, where the protein interaction network is enhanced by the interaction dependencies (*constraints*) modeled as propositional logic formulas [25]. With constrained protein interaction networks, we can stochastically simulate complex formation based on the available knowledge and obtain a detailed interaction topology (which proteins physically interact) for each complex.

To evaluate the Weisfeiler-Lehman based similarity ("WL similarity") against the edit distance based similarity ("edit similarity"), we computed both similarity measures on pairs of 100 000 protein complexes that have been simulated in a previous study [25]. For the simulation, a *constrained protein interaction network* was generated from the human adhesome protein network and a set of interaction dependencies obtained from protein domain interaction databases and manual curation. Then, protein complex assembly was simulated in a step-wise process, with association and dissociation rates calibrated to fit the complex size distribution of the CORUM database [20], until reaching convergence (for details, see [25]). The obtained complexes mimic the size distribution of known complexes, while also providing information about the actual physical interactions happening inside the complex, an information that is currently not yet available for real data. Since calculating edit similarity is computationally costly, we only considered a subset of all possible pairs for the analysis. To obtain candidate pairs, we considered all pairs of complexes that have at most 20 proteins (larger complexes are so rare that high similarities are unlikely), that have a size difference of protein multisets of at most 10, and that share at least one protein. These were sorted descendingly after the number of shared proteins and then the edit distance based similarity was computed on the first 500 000 candidate pairs. The resulting edit similarity values were classified into bins of width 0.1. Because

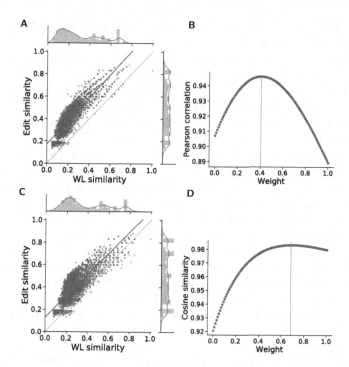

Fig. 3. Comparison of edit similarity and WL similarity. A and **C:** Scatterplot between edit and WL similarity for weights $\omega = 0.41$ (A) and $\omega = 0.69$ (C), including marginal distributions and least-squares regression line. Each point represents a pair of complexes. **B:** Pearson correlation coefficient between edit and WL similarity as function of ω. The maximum correlation (0.946) occurs for $\omega = 0.41$ (A). **D:** Cosine similarity as a function of ω. The maximum cosine similarity (0.983) is at $\omega = 0.69$ (C).

most pairs of complexes share a small number of proteins, we find many pairs with small edit similarity (but none in the range $[0.0, 0.1[$ because we required one common protein) and comparatively few pairs with edit similarity above 0.5. To achieve a uniform distribution among bins for the comparison, we randomly selected 1000 pairs from each bin, excluding the bin $[0.9, 1.0[$ which contained a single pair. This yielded 8000 pairs of complexes from 8 bins.

Similarity Comparison. We first consider three exemplary pairs (Fig. 2 A–C) with edit similarities of approximately 0.7, 0.8 and 0.9, respectively, the latter being the most similar observed pair. Our simulation has been calibrated to yield complexes of a realistic size distribution that additionally reflect all currently known interaction dependencies. However, since this data is likely incomplete and we also did not consider the law of mass action, we do not claim that the particular combination of proteins in these examples is likely to occur in reality. The examples are therefore only meant to illustrate the behavior of the two measures and give an intuition of cases where WL similarity fails to properly approximate edit similarity.

In example A, an additional protein (PTPN3) is added to an existing complex, a linear chain of 3 proteins. The edit similarity is $10/14 = 0.714$, the WL similarity is between 0.75 for $\omega = 1$ and and 0.4 for $\omega = 0$. Because the edit similarity is between the extreme WL similarities, there exists a unique weight $\omega^* \approx 0.898$, for which WL and edit similarities agree for this particular complex pair. Example B is a noteworthy case, because the WL similarity is 1.0, independent of ω, because the vertex labels are identical even after the first Weisfeiler-Lehman iteration. (Further iterations would show a difference.) The edit similarity is $20/24 = 0.83$, which is obtained by attaching ALB to the other LRP2 protein. In example C, one protein is replaced by another one in a fairly large complex. The edit similarity (0.905) is relatively high and outside the WL similarity range between 0.667 for $\omega = 0$ and 0.818 for $\omega = 1$.

Because most protein complexes are small and do not exhibit properties of examples B or C, the overall agreement between WL similarity and edit similarity is high: For each of the selected complex pairs, we computed the exact edit similarity and the WL similarity for each weight $\omega \in W := \{0.0, 0.01, 0.02, \ldots, 1.0\}$. Let e be the vector of edit similarity values and $s(\omega)$ the corresponding vector of WL similarity values using weight ω. To compare the similarity measures, we calculated both the Pearson correlation coefficient and the cosine similarity of e and $s(\omega)$ for all $\omega \in W$. As can be seen from Fig. 3 A and B, the highest values occur for ω between 0.38 and 0.44 and the maximum Pearson correlation coefficient is obtained for $\omega = 0.41$. For the cosine similarity, the maximum value is reached for weight $\omega = 0.69$, but the function is less peaked, and values above 0.4 lead to high agreement (Fig. 3 C and D).

Overall, we find good agreement between edit similarity and WL similarity for sufficiently large values of ω, i.e., if the Jaccard similarity of the constituent protein multiset has sufficiently high weight.

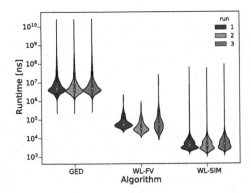

Fig. 4. Running times. Violin plots pf wall-clock times for computing the graph edit similarity (GED), the Weisfeiler-Lehman sets (WL-FV) and their Jaccard coefficients (WL-SIM) in three runs.

Running Time Comparison. To study the practical running time of both similarity measures, a subset of 100 complexes was drawn at random from the data set. We measured the time required to compute all pairwise similarities for this subset. For the Weisfeiler-Lehman based similarities, the computation can be divided into two steps. In the first step, the Weisfeiler-Lehman feature set is computed for each graph. In the second step, these sets are used to compute the quadratic number of similarities between all pairs using the Jaccard coefficient. Thus, the computational costly part is computed only once for each graph, while the quadratic number of comparisons is lightweight. This kind of preprocessing is not possible for the graph edit similarity.

Figure 4 shows the running times obtained experimentally. For the Weisfeiler-Lehman similarity we observe that computing the feature sets indeed dominates the running time in most cases. Most importantly, we observe that the graph edit similarity is about two orders of magnitude slower than the Weisfeiler-Lehman based similarities, on average and in the worst-case. If the overall similarity computation costs dominate the preprocessing costs (e.g., large data sets and a quadratic number of similarity computations) this difference increases to three orders of magnitude. We expect that this tendency is even intensified when considering highly similar pairs only, which are of special interest in many applications. Overall, our experimental results confirm that the proposed similarity measure based on Weisfeiler-Lehman labeling can be computed efficiently.

4   Discussion

Our motivation to consider protein complex similarity was to reduce the complexity of the simulation output of our constrained protein interaction network simulator [25], and we were surprised to see that apparently, no similarity measures have been proposed in the literature. Depending on the underlying representation (set, multiset or graph), different alternatives suggest themselves. However, most graph-based measures are both theoretically and practically hard to compute for larger complexes or for large amounts of complexes. While different tractable graph similarity measures [26], or an approximate graph edit distance [19] have been proposed, none of these appear to be specifically tailored to the properties of protein complexes (often less than ten vertices; sparse). Our proposal to define the similarity as a convex combination of two Jaccard coefficients (protein label multiset and Weisfeiler-Lehman label multiset after one iteration) has two additional advantages. First, using Jaccard coefficients allows to efficiently pre-filter for high similarity using locality-sensitive hashing. In combination with the preprocessing abilities discussed in the experimental running time comparison, this allows for very fast search queries, clusterings, and other applications that rely on intensive distance computations. Second, for weight $\omega = 1$ of the 0-th WL iteration, our measure reduces to the natural similarity measure of the multiset representation. Our framework hence allows for a smooth transition between multiset and graph representation. The comparison to an edit-based similarity seems to indicate that the protein label multiset plays an important role if one wants to approximate the edit similarity.

From a biological point of view, a high similarity between two complexes should indicate a high probability that they share the same function and can substitute each other in a cellular process. If such information were available, we could use it for evaluating similarity measures. At present, when not even the interaction topology of most complexes has been determined, the corresponding data is out of reach, and such an evaluation is not feasible. In this situation, we suggest that edit similarity is a measure that corresponds to intuition about similarity and that any reasonable similarity measure should be close to edit similarity. The measure we propose has this property (for any weight $\omega \in [0,1]$) but offers the advantage that it can be quickly computed and scales to millions of complex pairs. Currently, it is unclear whether the optimal weights presented here will be directly applicable to future data. Instead, we propose to apply the computationally expensive edit similarity based strategy used in this work to obtain optimal weights for WL similarity on a representative subset of the new data, and then use these weights for the efficient calculation of WL similarity on the entire dataset.

Both WL similarity and edit similarity have limitations from a biological point of view because they do not consider similarities between individual proteins: Two proteins are either equal or distinct. However, if two proteins are closely related, should they be treated as equal or distinct? In the former case, we lose resolution. In the latter case, we would benefit from a fine-grained similarity function between proteins (e.g. a modification of p is very similar to p, a protein with some common domains is somewhat similar to p, but a completely disjoint protein in terms of domains has similarity zero). In this sense, the question of how to best measure protein complex similarity is far from settled.

Acknowledgments. This work was supported by Deutsche Forschungsgemeinschaft (DFG) Collaborative Research Center (SFB) 876, projects A6 and C1, and by Mercator Research Center Ruhr (MERCUR), project Pe-2013-0012 (UA Ruhr professorship). The authors thank Eli Zamir for insightful discussions.

References

1. Arvind, V., Köbler, J., Rattan, G., Verbitsky, O.: On the power of color refinement. In: Kosowski, A., Walukiewicz, I. (eds.) FCT 2015. LNCS, vol. 9210, pp. 339–350. Springer, Cham (2015). https://doi.org/10.1007/978-3-319-22177-9_26
2. Babai, L., Kucera, L.: Canonical labelling of graphs in linear average time. In: 20th Annual Symposium on Foundations of Computer Science (SFCS), pp. 39–46. IEEE (1979)
3. Bhowmick, S.S., Seah, B.: Clustering and summarizing protein-protein interaction networks: a survey. IEEE Trans. Knowl. Data Eng. **28**(3), 638–658 (2016)
4. Boutros, M., Heigwer, F., Laufer, C.: Microscopy-based high-content screening. Cell **163**(6), 1314–1325 (2015)
5. Bunke, H.: On a relation between graph edit distance and maximum common subgraph. Pattern Recogn. Lett. **18**(8), 689–694 (1997)

6. Drew, K., et al.: Integration of over 9,000 mass spectrometry experiments builds a global map of human protein complexes. Mol. Syst. Biol. **13**(6), 932 (2017)
7. Garey, M., Johnson, D.: Computers and Intractability: A Guide to the Theory of NP-Completeness. W. H. Freeman, New York (1979)
8. Grecco, H.E., Imtiaz, S., Zamir, E.: Multiplexed imaging of intracellular protein networks. Cytometry A **89**(8), 761–775 (2016)
9. Hernandez, C., Mella, C., Navarro, G., Olivera-Nappa, A., Araya, J.: Protein complex prediction via dense subgraphs and false positive analysis. PLoS ONE **12**(9), e0183460 (2017)
10. Kann, V.: On the approximability of the maximum common subgraph problem. In: Finkel, A., Jantzen, M. (eds.) STACS 1992. LNCS, vol. 577, pp. 375–388. Springer, Heidelberg (1992). https://doi.org/10.1007/3-540-55210-3_198
11. Köster, J., Rahmann, S.: Snakemake - a scalable bioinformatics workflow engine. Bioinformatics **28**(19), 2520–2522 (2012)
12. Kozakov, D., Hall, D.R., Xia, B., Porter, K.A., Padhorny, D., Yueh, C., Beglov, D., Vajda, S.: The ClusPro web server for protein-protein docking. Nat. Protoc. **12**(2), 255–278 (2017)
13. Laskowski, R.A., Gerick, F., Thornton, J.M.: The structural basis of allosteric regulation in proteins. FEBS Lett. **583**(11), 1692–1698 (2009)
14. Lerouge, J., Abu-Aisheh, Z., Raveaux, R., Héroux, P., Adam, S.: New binary linear programming formulation to compute the graph edit distance. Pattern Recogn. **72**, 254–265 (2017)
15. Ma, X., Gao, L.: Discovering protein complexes in protein interaction networks via exploring the weak ties effect. BMC Syst. Biol. **6**(Suppl 1), S6 (2012)
16. Park, H., Lee, H., Seok, C.: High-resolution protein-protein docking by global optimization: recent advances and future challenges. Curr. Opin. Struct. Biol. **35**, 24–31 (2015)
17. Pearson, W.R.: Selecting the right similarity-scoring matrix. Curr. Protoc. Bioinform. **43**, 1–9 (2013)
18. Pellegrini, M., Baglioni, M., Geraci, F.: Protein complex prediction for large protein protein interaction networks with the Core&Peel method. BMC Bioinform. **17**(Suppl 12), 372 (2016)
19. Riesen, K., Ferrer, M., Bunke, H.: Approximate graph edit distance in quadratic time. IEEE/ACM Trans. Comput. Biol. Bioinform. (2015) (epub ahead of print)
20. Ruepp, A., et al.: CORUM: the comprehensive resource of mammalian protein complexes - 2009. Nucleic Acids Res. **38**(suppl 1), D497–D501 (2010)
21. Sánchez Claros, C., Tramontano, A.: Detecting mutually exclusive interactions in protein-protein interaction maps. PLoS ONE **7**(6), e38765 (2012)
22. Sanfeliu, A., Fu, K.S.: A distance measure between attributed relational graphs for pattern recognition. IEEE Trans. Syst. Man Cybern. **13**(3), 353–362 (1983)
23. Shervashidze, N., Schweitzer, P., van Leeuwen, E.J., Mehlhorn, K., Borgwardt, K.M.: Weisfeiler-Lehman graph kernels. J. Mach. Learn. Res. **12**, 2539–2561 (2011)
24. Srihari, S., Yong, C.H., Wong, L.: Computational Prediction of Protein Complexes from Protein Interaction Networks. Association for Computing Machinery and Morgan & Claypool, New York (2017)
25. Stöcker, B.K., Köster, J., Zamir, E., Rahmann, S.: Modeling and simulating networks of interdependent protein interactions. Integr. Biol. **10**, 290–305 (2018)
26. Vishwanathan, S.V.N., Schraudolph, N.N., Kondor, R., Borgwardt, K.M.: Graph kernels. J. Mach. Learn. Res. **11**, 1201–1242 (2010)

27. Wachsmuth, M., Conrad, C., Bulkescher, J., Koch, B., Mahen, R., Isokane, M., Pepperkok, R., Ellenberg, J.: High-throughput fluorescence correlation spectroscopy enables analysis of proteome dynamics in living cells. Nat. Biotechnol. **33**(4), 384–389 (2015)
28. Weisfeiler, B., Lehman, A.A.: A reduction of a graph to a canonical form and an algebra arising during this reduction. Nauchno-Technicheskaya Informatsiya **2**(9), 12–16 (1968). (in Russian)

Multiple Instance Classification in the Image Domain

Ilaria Bartolini, Pietro Pascarella, and Marco Patella$^{(\boxtimes)}$

DISI - Alma Mater Studiorum, Università di Bologna, Bologna, Italy
{ilaria.bartolini,pietro.pascarella,marco.patella}@unibo.it

Abstract. Multiple instance classification (MIC) is a kind of supervised learning, where data are represented as *bags* and each bag contains many *instances*. Training bags are given a label and the system tries to learn how to label unknown bags, without necessarily learning how to label individually each of their instances. In particular, we apply concepts drawn from MIC to the realm of content-based image retrieval, where images are described as bags of visual *local descriptors*. We introduce several classifiers, according to the different MIC paradigms, and evaluate them experimentally on a real-world dataset, comparing their accuracy and efficiency.

1 Introduction

Content-based image retrieval (CBIR) consists in searching for images of interest in large databases, exploiting their visual content, as opposed to concept-based image indexing, which exploits text-based techniques for indexing images, using image captions, surrounding text, keywords, and so on [14]. CBIR can be used per se, e.g., to search for a particular image in a large image dataset (a notable example is the Google Images system, images.google.com), or as a building block for a plethora of other image-related tasks, like browsing [5], annotation [4], classification [1], and so on.

The fundamental concept in CBIR is that of similarity, which is used to compare the image content. Evaluating the similarity between two images involves: (1) automatically extracting relevant *features/descriptors*, summarizing visual content of each image, and (2) compare such features to assess a *similarity score* in $[0, 1]$, with the understanding that higher values indicate high degrees of similarity between images' content.

Approaches to extraction of image features can be broadly classified in global (where descriptors represent visual characteristics of the image as a whole) and local (where features describe visual characteristics of a small set of image pixels), with local features having a major prevalence in recent approaches. Another fundamental ingredient for CBIR is efficient indexing, due to the facts that image databases are usually (very) large and that often a real-time query processing is required. However, as acknowledged in [10], "research on efficient ways to index

© Springer Nature Switzerland AG 2019
G. Amato et al. (Eds.): SISAP 2019, LNCS 11807, pp. 323–331, 2019.
https://doi.org/10.1007/978-3-030-32047-8_28

images by content has been largely overshadowed by research on efficient visual representation and similarity measures".

In this paper, we focus on the task of classifying images, i.e., identifying to which of a set of categories a new image belongs, given a (training) set of images for which the category membership is known. This task is one of the most common ways to measure the effectiveness of any CBIR technique. Indeed, as acknowledged in [11]: "A feature that performs well for the task of classification on a certain data set, it will most probably be a good choice for retrieval of images from that data set, too." Despite the importance of this task, most of the approaches have only focused on establishing the accuracy of image content descriptors (*features*), with a negligent lack of emphasis on classifying techniques and almost no interest to efficiency. The former issue is even more prominent, due to the increasing usage of local features, which opens the way to a plethora of more advanced classification techniques. The latter problem is of utmost importance for those approaches using *lazy learning*, i.e., for which the training data are generalized only (or mostly) when a query is made to the system (when a new image is to be classified). In such cases, the training phase is quite fast, while evaluation is the more costly part of classification (as opposed to *eager learning*, where the opposite happens).

To overcome this deficiency, we propose to combine the realms of multiple instance classification (MIC) and content-based image retrieval, by applying MIC techniques to the task of image classification. To the best of our knowledge, this is the first attempt to combine the world of multiple instance classification to the task of image classification *in a comprehensive way*. Indeed, a few previous attempts [1,9,17,20,21] have used approaches drawn from MIC for classifying images, but without putting them into the proper context (actually, any image classification method based on local features can be thought as exploiting a particular MIC technique). Moreover, among the approaches using lazy learning, we are the first to put an emphasis on efficiency of the classification.

Our goal here is not to propose a novel technique for image classification; rather, we would like to show how the introduction of concepts drawn from Artificial Intelligence could help researchers working in CBIR to evaluate their proposed features and/or indexing techniques in a more structured way, by showing them the existing alternatives. For this, we exemplify the comparison of two well known features, both implemented in the WINDSURF framework [7]. The use of WINDSURF provides us a number of algorithms and indexing data structures for efficient query processing, so that we are able to abstract from the underlying details of feature extraction, data indexing, etc., and to focus on MIC algorithms. A preliminary experimental evaluation, concerning both effectiveness and efficiency, has finally been performed on a real-world dataset, which has been previously used for a similar task [1].

2   Related Work

As already stated in Sect. 1, the task of image classification has received massive attention in the realm of CBIR; it is beyond the scope of this paper to introduce and classify all attempts, and we will limit ourselves to describe the ones most related to the work presented in this paper, in particular those using techniques drawn from MIC.

The first papers [17,18,21] that used MIC techniques for image classification were written by pioneers of multiple instance learning. All papers use the diverse density (DD) algorithm to classify images (the first two in the specific scenario of natural scene images, while the latter paper adopts a more general approach which is applicable to a broader range of images) and differ in the features extracted at the instance level. The DD algorithm is one of the earliest MIC algorithms and follows the SMI assumption (see Sect. 3.1). All the proposed approaches perform a sequential scan in the instance space, so they are clearly not suited for large training sets.

A MIC approach using deep learning is described in [20]. This is clearly an eager learning technique, so it is not comparable to our approach in terms of efficiency, but the accuracy it exhibits is comparable to that obtained here (albeit on a different dataset).

Finally, it is worth noting that the techniques described in [1] are exactly equivalent to some of the MIC approaches we describe here: we will highlight such equivalences when describing our classifiers in Sect. 3. We stress here, however, that in [1] the emphasis wes exclusively on the accuracy of the classification task, while here we advocate the comparison of techniques on both accuracy *and* efficiency.

3   Multiple Instance Classification and Its Application to Images

Multiple instance learning [12] is a branch of supervised learning where, instead of a training set of objects, the learner receives a training set of *bags*, each containing multiple *instances*. Multiple instance classification (MIC) [2] is the name given to the sub-field of MIL focused on classification. MIC includes a number of classification techniques that exploit the fact that the class of each individual bag can be transferred to all (or to some) of its instances.

In image classification, the problem tackled in this paper, bags correspond to *images*, represented by way of n *features* (instances) R_i: $I = \{R_1, R_2, \ldots, R_n\}$. The objective is to estimate a classification function $C(I)$ providing the class of an unknown image I. To this end we are given a *training set* T of M images with corresponding class, $T = \{(I_1, C_1), (I_2, C_2), \ldots, (I_M, C_M)\}$, where C_i is the class of image I_i. From the point of view of MIC, techniques differ in the assumption regarding how the class of each bag is related to the bag instances [13]. The excellent review paper [2] provides the following taxonomy:

Instance Space (IS): it is assumed that the discriminative information lies at the instance level, so that classification is performed on instances and the overall classification is performed by aggregating scores obtained at the instance level.

Bag Space (BS): the main assumption is that discriminative information lies at the bag level, and this cannot be distributed to instances.

Embedded Space (ES): each bag is mapped to a single feature vector, summarizing the relevant information included at the instance level, then a vector-based classifier is exploited.

As we will show in the following sections, it can be easily seen, each of the three alternatives suggests a retrieval model, based on instances (features), bags (images), and vectors, respectively. This also helps researchers proposing novel local features for characterizing image content to define their appropriate model.

3.1   Instance Space Classification

Since it is assumed that information is carried by instances, the retrieval model implies that images retrieved first are those including features that are most similar to features of the query image (this can be made efficient by way of an index built on local features).

Collective Assumption. These methods are based on the premise that all instances in a bag contribute equally to the bag class. In this case the bag class can be estimated by choosing the class maximizing a simple (weighted) average. This was implemented using a two-step confidence-rated IS classifier:

1. First, each image feature R is classified using a feature-level classifier $c(R)$; the classifier also computes a value $\nu(R)$ representing the confidence that c has on its choice $c(R)$.
2. Then, the whole image is classified taking into consideration the class assigned to each of its features.

In particular, for any class C_j a score $s_j(I)$ is computed for image I as the sum of confidences of features classified to each class:

$$s_j(I) = \sum_{R \in I: c(R) = C_j} \nu(R) \tag{1}$$

Then I is classified to the class maximizing the value in Eq. 1:

$$C(I) = \arg\max_j s_j(I) \tag{2}$$

1-NN Classifier - IS: The classifier of feature R and the corresponding score take into account only the nearest neighbor of R only:

$$c(R) = c(NN_1(R)) \qquad \nu(R) = sim(R, NN_1(R))$$

This classifier equals the one called Φ^f in [1]. Efficient retrieval of $NN_1(R)$ is guaranteed by performing a 1-NN query on the feature-based index.

Local Classifier - IS_L: The only difference with respect to the previous classifier is the score which is defined as follows:

$$c(R) = c(NN_1(R)) \qquad \nu(R) = \begin{cases} 1 & \text{if } 1 - \frac{sim(R,NN_{\overline{1}}(R))}{sim(R,NN_1(R))} > c \\ 0 & \text{otherwise} \end{cases}$$

where $NN_{\overline{1}}(R)$ is the nearest neighbor of R of a class different than that of $NN_1(R)$ ([1] calls this classifier Φ^m). The efficient evaluation of this approach is obtained through a sorted access for each query feature R, retrieving instances until a result is obtained whose class is different to that of $NN_1(R)$.

Weighted Local Classifier - IS_{WL}: The classifier of feature R and the corresponding score are defined as follows, taking into account the nearest neighbor of R only:

$$c(R) = c(NN_1(R)) \qquad \nu(R) = 1 - \frac{sim(R, NN_{\overline{1}}(R))}{sim(R, NN_1(R))}$$

With respect to the previous classifier, here the confidence value is not binary, but uses the original "fuzzy" ratio between similarities of the NN of R and of the NN of a different class. In [1], this classifier was denoted as Φ^w.

Weighted k-NN Classifier - IS_W: First, a score for each class is defined according to classes of the nearest k neighbors of feature R:

$$s_j(R) = \sum_{i=1,\dots,k:c(NN_i(R))=C_j} sim(R, NN_i(R))$$

Then, the classifier and the score are defined as:

$$c(R) = \arg\max_j s_j(R) \qquad \nu(R) = 1 - \frac{\overline{\max}s_j(R)}{\max s_j(R)}$$

where $\overline{\max}s_j(R)$ is the score of the second best class for R. Note that this definition of confidence is coherent with the one defined for the weighted local classifier (the two definitions coincide for $k = 1$), but it is different with respect to the one given for the 1-NN classifier. This classifier was named Φ^k in [1]. Efficient evaluation is obtained by way of a k-NN query on the feature index.

Standard Multiple Instance Assumption. Methods in this category suppose that instances of each class are *only* contained in bags of the same class and that every bag contains at least one instance of its class. This is equal to say that, for each bag, one of the instances possesses some "desirable" property making the whole bag of that class, thus we are trying to identify which instance type characterizes each class. The SMI assumption (SMI) was implemented by removing, from the training set, all those instances that would lead to a wrong classification, i.e., whose NN belongs to a different class.[1] This approach has also the advantage of reducing the ground truth size, as also acknowledged in [1], where this approach was called "local features cleaning". Finally, any IS classifier can be exploited on the reduced ground truth to classify each instance.

3.2   Bag Space Classification

Techniques following this paradigm consider each bag as a whole, so that the classification is performed in the space of bags. Typically, methods in this category exploit a distance $d(I_i, I_j)$ obtained by appropriately aggregating distances between correspondent features $\delta(R_h, R_k)$. Examples include the EMD distance (BS_{EMD}) [15,19], the Hausdorff distance (BS_{Haus}), and the Chamfer distance (BS_{Cham}) [8]. It has to be noted that such distances also differ in their time complexity, since the Chamfer and Hausdorff distances are quadratic in the number of instances, while EMD is super-cubic [15] thus it is extremely time consuming, particularly for large bags (we will see that this is the case for SIFT salient point descriptors). Alternatively, one can use a kernel function $K(I_i, I_j) \in [0, 1]$ assessing the similarity between images I_i and I_j. This classifier is called Φ^s in [1].

The retrieval model is the classical one used in CBIR, where images are retrieved for decreasing values of their similarity to the query image and efficient evaluation is obtained exploiting an image-based index. A 1-NN classifier has been used for all implemented alternatives.

3.3   Embedded Space Classification

Methods in this category map each image I to a K-dimensional vector v then exploit a K-dimensional classifier (like SVM or 1-NN) on the so-obtained vector. Mapping from I to v is usually performed by way of a vocabulary V, i.e., a set of K instances (words) $V = \{(w_1, p_1), (w_2, p_2), \ldots, (w_K, p_K)\}$, where each word is characterized by an identifier w and a prototype instance p. Given an image I and a vocabulary V, the mapping function $\mathcal{M}$ produces the vector v, $\mathcal{M}(I, V) = v$.

Histogram-Based Methods. These techniques consider that each component of v is obtained as the average value (for that component) of features in I:

$$v_j = \frac{1}{N} \sum_{R \in I} f_j(R), \quad j \in [1, K] \tag{3}$$

where $f_j(R)$ measures the probability that feature R corresponds to word w_j.

---

[1] Actually, we removed the instance from the training set if its NN in a different bag is in a different class. This was required because it could happen that the NN of an instance belongs to the same bag.

Possible implementations for f_j are:

Bag-of-words with hard assignment: $f_j(R) = 1 \Leftrightarrow j = \arg\min_i \delta(R, p_i)$, otherwise $f_j(R) = 0$. This way, each feature is assigned to one and only one word and the j-th component of v counts how many features of I are assigned to word w_j.

Bag-of-words with soft assignment: $f_j(R) = 1 - \delta(R, p_i)$. $f_j(R)$ represents the similarity between feature R and word w_j and the j-th component of v represents the average similarity of features in I to word w_j.

Distance-Based Methods. These techniques consider that each component of the vector v is obtained as the matching degree between features in I and the corresponding word:

$$v_j = \min_{R \in I} \delta(R, p_j), \quad j \in [1, K] \tag{4}$$

The ES approach makes sense basically whenever the number of instances in a bag is so high to make BS classification (and retrieval) impractical. The retrieval model here consists of comparing image histograms using a vectorial distance (in our experiments, we used the simple Euclidean metric) and indexing histograms using a spatial index.

Finally, we also consider the so-called bag-of-visual-words ($\mathrm{ES_{BOVW}}$) approach, where the size K of the vocabulary is (much) higher than the number of instances in each bag. In this scenario, which is the one commonly used for salient point descriptors, the hard assignment is used, each image is represented as a sparse vector and the Hamming distance is used to compare histograms. Note that this is the de-facto standard for salient point descriptors, for which vector quantization is used to deal with the high number of descriptors in each image (usually, in the order of hundreds) and with the high dimensionality (64–128) of each descriptor.

4 Experiments and Final Discussion

We conducted a preliminary experiment on the Pisa landmark dataset [1], composed of 1227 photos (crawled from Flickr at standard resolution) partitioned in 12 categories, each including a number of images ranging from 46 to 138. The dataset was randomly split into a training set (70%) and a test set (30%). Among local features included in the WINDSURF library (SIFT, SURF, FREAK, ORB, and BRISK), we chose SIFT [16], since it was experimentally proven that such descriptors consistently perform better than the others for image classification [1,6], and the original WINDSURF features [3]. On average, every image contains 4.39 WINDSURF regions and 864.8 SIFT salient points, thus the ground truth has a total of 3186 WINDSURF regions and 627,845 SIFT salient points. All experiments have been performed on a Pentium 4 3.2 GHz machine with 1 GB of RAM under the MS Windows 7 OS. Salient point extraction has been performed by exploiting the OpenCV 2.4.11 library with the Java interface.

Table 1. Accuracy and cost (in seconds) for the implemented approaches: IS (left) and BS/ES (right). Best figures for all classifier types are in **boldface**.

classifier	SIFT accuracy	cost	WINDSURF accuracy	cost
IS	**0.944**	2713	0.804	**3.6**
IS$_L$	0.917	2736	0.810	**3.6**
IS$_{WL}$ = IS$_W$(1)	**0.944**	2758	0.801	**3.6**
IS$_W$(10)	0.806	3682	0.778	3.8
IS$_W$(50)	0.694	3702	0.75	3.8
IS$_W$(SMI,1)	0.8	1836	0.698	**3.6**
IS$_W$(SMI,10)	0.722	2100	0.674	3.8
IS$_W$(SMI,50)	0.639	2274	0.670	3.8

classifier	SIFT accuracy	cost	WINDSURF accuracy	cost
BS$_{Haus}$	0.333	192	0.322	**1.7**
BS$_{EMD}$	0.333	298	**0.402**	2.2
BS$_{Cham}$	0.322	208	0.298	2.8
ES$_H$	**0.778**	**0.62**		
ES$_S$	0.667	0.65		
ES$_D$	**0.778**	0.67		
ES$_{BOVW}$	0.139	3.9		

Results, included in Table 1, show that approaches based on the Instance Space assumption outperform approaches based on the Bag Space assumption on classification accuracy: this somehow confirms results included in [1]. However, the extremely high number of instances per image when using SIFT descriptors makes this alternative extremely inefficient, a fact that was not considered in [1]. On the other hand, the features originally included in WINDSURF allow efficient and accurate classification for the examined dataset. For SIFT, the only viable alternative is to exploit the Embedded Space assumption, with good efficiency at the cost of a slightly reduced precision in classification. It is worth noting that results for BS and ES confirm those obtained in [6] for the similar task of image retrieval. We believe that our results could be helpful to evaluate the effectiveness of any local descriptor to represent the image content, in order to obtain both accurate and efficient classification and/or retrieval.

References

1. Amato, G., Falchi, F.: kNN based image classification relying on local feature similarity. In: SISAP 2010 (2010)
2. Amores, J.: Multiple instance classification: review, taxonomy and comparative study. Artif. Intell. **201**, 81–105 (2013)
3. Ardizzoni, S., Bartolini, I., Patella, M.: Windsurf: region-based image retrieval using wavelets. In: IWOSS 1999 (1999)
4. Bartolini, I., Ciaccia, P.: *Imagination*: exploiting link analysis for accurate image annotation. In: Boujemaa, N., Detyniecki, M., Nürnberger, A. (eds.) AMR 2007. LNCS, vol. 4918, pp. 32–44. Springer, Heidelberg (2008). https://doi.org/10.1007/978-3-540-79860-6_3
5. Bartolini, I., Ciaccia, P., Patella, M.: Adaptively browsing image databases with PIBE. MTAP **31**(3), 269–286 (2006)
6. Bartolini, I., Patella, M.: Windsurf: the best way to SURF (and SIFT/BRISK/ORB/FREAK, too). Multimedia Syst. **24**(4), 459–476 (2018)
7. Bartolini, I., Patella, M., Stromei, G.: The windsurf library for the efficient retrieval of multimedia hierarchical data. In: SIGMAP 2011 (2011)
8. Belongie, S., Malik, J., Puzicha, J.: Shape matching and object recognition using shape contexts. TPAMI **24**(24), 509–522 (2002)

9. Chen, Y., Bi, J., Wang, J.Z.: MILES: multiple-instance learning via embedded instance selection. TPAMI **28**(12), 1931–1947 (2006)
10. Datta, R., Joshi, D., Li, J., Wang, J.Z.: Image retrieval: ideas, influences, and trends of the new age. Comput. Surv. **40**(2), 5 (2008)
11. Deselaers, T., Keysers, D., Ney, H.: Features for image retrieval: an experimental comparison. Inf. Retrieval **11**(2), 77–107 (2008)
12. Dietterich, T.G., Lathrop, R.H., Lozano-Pérez, T.: Solving the multiple instance problem with axis-parallel rectangles. Artif. Intell. **89**(1–2), 31–71 (1997)
13. Foulds, J., Frank, E.: A review of multi-instance learning assumptions. Knowl. Eng. Rev. **25**(1), 1–25 (2010)
14. Lew, M.S., Sebe, N., Djeraba, C., Jain, R.: Content-based multimedia information retrieval: state of the art and challenges. TMCCA **2**(1), 1–19 (2006)
15. Ling, H., Okada, K.: An efficient earth mover's distance algorithm for robust histogram comparison. TPAMI **29**(5), 840–853 (2007)
16. Lowe, D.G.: Distinctive image features from scale-invariant keypoints. IJCV **60**(2), 91–110 (2004)
17. Maron, O., Ratan, A.L.: Multiple-instance learning for natural scene classification. In: ICML 1998 (1998)
18. Ratan, A.L., Maron, O., Grimson, W.E.L., Lozano-Pérez, T.: A framework for learning query concepts in image classification. In: CVPR 1999 (1999)
19. Rubner, Y., Tomasi, C.: Perceptual Metrics for Image Database Navigation. Springer, Boston (2013). https://doi.org/10.1007/978-1-4757-3343-3
20. Wu, J., Yu, Y., Huang, C., Yu, K.: Deep multiple instance learning for image classification and auto-annotation. In: CVPR 2015 (2015)
21. Yang, C., Lozano-Pérez, T.: Image database retrieval with multiple-instance learning techniques. In: ICDE 2000 (2000)

An Image Retrieval System for Video

Paolo Bolettieri, Fabio Carrara, Franca Debole$^{(\boxtimes)}$, Fabrizio Falchi,
Claudio Gennaro, Lucia Vadicamo, and Claudio Vairo

Institute of Information Science and Technologies,
Italian National Research Council (CNR), Via G. Moruzzi 1, Pisa, Italy
{paolo.bolettieri,fabio.carrara,franca.debole,fabrizio.falchi,
claudio.gennaro,lucia.vadicamo,claudio.vairo}@isti.cnr.it

Abstract. Since the 1970's the Content-Based Image Indexing and
Retrieval (CBIR) has been an active area. Nowadays, the rapid increase
of video data has paved the way to the advancement of the technologies
in many different communities for the creation of Content-Based Video
Indexing and Retrieval (CBVIR). However, greater attention needs to be
devoted to the development of effective tools for video search and browse.
In this paper, we present Visione, a system for large-scale video retrieval.
The system integrates several content-based analysis and retrieval mod-
ules, including a keywords search, a spatial object-based search, and a
visual similarity search. From the tests carried out by users when they
needed to find as many correct examples as possible, the similarity search
proved to be the most promising option. Our implementation is based on
state-of-the-art deep learning approaches for content analysis and leverages highly efficient indexing techniques to ensure scalability. Specifically,
we encode all the visual and textual descriptors extracted from the videos
into (surrogate) textual representations that are then efficiently indexed
and searched using an off-the-shelf text search engine using similarity
functions.

Keywords: Content-based image indexing · Neural networks ·
Multimedia retrieval · Similarity search · Object detection

1 Introduction

Video data is the fastest growing data type on the Internet, and because of
the proliferation of high-definition video cameras, the volume of video data is
exploding. Visione [1] is a content-based video retrieval system that participated
for the first time in 2019 to the Video Browser Showdown (VBS) [11], an international video search competition that evaluates the performance of interactive
video retrievals systems. The VBS 2019 uses the V3C1 dataset that consists
of 7,475 video files, amounting for 1000h of video content (1082659 predefined
segments) [15] and encompasses three content search tasks: *visual Known-Item*

Work partially supported by the AI4EU project (EC-H2020 - Contract n. 825619).

G. Amato et al. (Eds.): SISAP 2019, LNCS 11807, pp. 332–339, 2019.
https://doi.org/10.1007/978-3-030-32047-8_29

Search (visual KIS), textual Known-Item Search (textual KIS) and *ad-hoc Video Search (AVS)*. The visual KIS task models the situation in which someone wants to find a particular video clip that he has already seen, assuming that it is contained in a specific collection of data. In the textual KIS, the target video clip is no longer visually presented to the participants of the challenge but it is rather described in details by text. This task simulates situations in which a user wants to find a particular video clip, without having seen it before, but knowing the content of the video exactly. For the AVS task, instead, a textual description is provided (e.g. "A person playing guitar outdoors") and participants need to find as many correct examples as possible, i.e. video shots that fit the given description.

In this paper, we describe the current version of Visione, an image retrieval system used to search for videos, presented for the first time at VBS2019. After the first implementation of the system, as described in [1], we decide to focus our attention on the query phase, by improving the user interaction with the interface. And for that reason, we introduce a set of icons for the object location and, inspired by other system involved in VBS of the previous years (e.g. [10]), we integrate the query-by-color sketch. In the next sections, we describe the main components of the system and the techniques at the bottom of the system.

2   System Components

Visione is based on state-of-the-art deep learning approaches for the visual content analysis and exploits highly efficient indexing techniques to ensure scalability. In Visione, we use the keyframes made available by the VBS organizers (1 million segments and keyframes[1]), focusing our work on the extraction of relevant information on these keyframes and on the design of a clear and simple user interface.

In the following, we give a brief description of the main components of the system: the User Interface and the Search Engine (see Fig. 1).

2.1   User Interface

The user interface, shown in the upper part of Fig. 1, provides a text box to specify the keywords, and a canvas for sketching objects to be found in the target video. Inspired by one of the system on VBS2018, we integrate also the query-by-color sketches, realized with the same interface we used for the objects (canvas and bounding box). The canvas is split into a grid of 7×7 cells, where the user can draw simple bounding boxes to specify the location of the desired objects/colors. The user can move, enlarge or reduce the drawn bounding boxes for refining the search. In the current version of the system, we realize a simple drag & drop on the canvas using icons for the most common objects. Furthermore with the same mechanism we define a color palette available as icons, to facilitate the search

---

[1] https://www-nlpir.nist.gov/projects/tv2019/data.html.

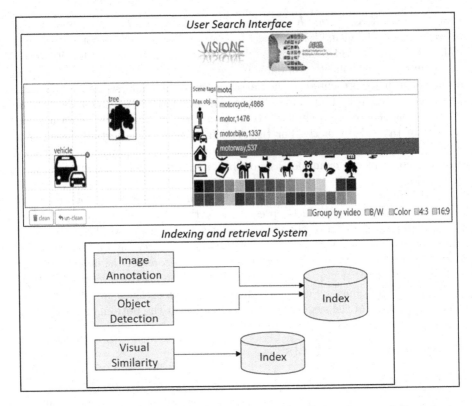

Fig. 1. The main components of Visione: the user search interface, and the indexing and retrieval system

by color: for each cell of the grid (7×7), we calculate the dominant colors using a K-NN approach, largely adopted in color based image segmentation [13].

Moreover, another new functionality added to the old system [1], is the possibility of using some filters, such as the number of occurrences of specific objects, and the type of keyframes to be retrieved (B/W or color, 4:3 or 16:9). At browsing time, the user browsing through the results can use the image similarity to refine the search, or group the keyframes (in the result set) that belong to the same video. Finally, the user interface offers the possibility to show for each keyframe of the result set, all the keyframes of the video of the selected keyframe, and play the video starting from the selected keyframe: this can help to check if the selected keyframe matches the query. A standard search in Visione, for all the tasks, could be done by drawing one or more bounding boxes of objects/colors, or by searching for some keywords, and often by combining them.

2.2 Search Engine

Retrieval and browsing require that the source material is first of all effectively indexed. In our case, we employ state-of-the-art deep learning approaches to extract both low-level and semantic visual features. We encode all the features extracted from the keyframes (visual features, keywords, object locations, and metadata) into textual representations that are then indexed using inverted files. We use a text surrogate representation [6], which was specifically extended to support efficient spatial object queries on large scale data. In this way, it is possible to build queries by placing the objects to be found in the scene and efficiently search for matching images in an interactive way. This choice allows us to exploit efficient and scalable search technologies and platform used nowadays for text retrieval. In particular, Visione relies on the Apache Lucene[2].

In the next section, we describe in detail the techniques employed to obtain useful visual/semantic features.

3   Methodologies

Visione addresses the issues of CBVIR modeling the data using both the simple features (color, texture) and derived features (semantic features). Regarding the derived features, Visione relies heavily on deep learning techniques, trying to bridge the semantic gap between text and image using the following approaches:

- for **keywords search**: we exploit an image annotation system based on different Convolutional Neural Networks to extract scene attributes.
- for **object location search**: we exploit the efficiency of YOLO[3] as a real-time object detection system to retrieve the video shot containing the objects sketched by the user.
- for **visual similarity search**: we perform a similarity search by computing the similarity between the visual features represented using the R-MAC [17] visual descriptor.

Keywords. Convolutional Neural Networks, used to extract the deep features, are able to associate images with categories they are trained from, but quite often, these categories are insufficient to associate relevant keywords/tags to an image. For that reason, then, Visione exploits an automatic annotation system to annotate untagged images. This system, as described in [2], is based on YFCC100M-HNfc6, a set of deep features extracted from the YFCC100M dataset [16], created using the Caffe framework [8]. The image annotation system is based on an unsupervised approach to extract the implicitly existing knowledge in the huge collection of unstructured texts describing the images of YFCC100M dataset, allowing us to label the images without using a training model. The image annotation system also exploits the metadata of the images validated using WordNet [5].

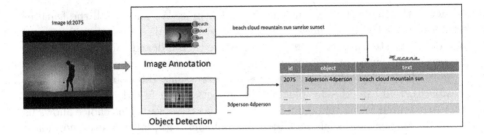

Fig. 2. Search Engine: the index for both object location and keywords.

Object Location. Following the idea that the human eye is able to identify objects in the image very quickly, we decide to take advantage of the new technologies available to search for object instances in order to retrieve the exacted video shot.

For this purpose, we use YOLOv3 [14] as object detector, both because it is extremely fast and because of its accuracy. Our image query interface is subdivided into a 7×7 grid in the same way that YOLO segments images to detect objects. Each object detected in the single image I by YOLO is indexed using a specific encoding *ENC* conceived to put together the location and the class corresponding to the object ($cod_{loc}cod_{class}$). The idea of using YOLO to detect objects within video has already been exploited in VBS, e.g. by [18], but our approach is distinguished by being able to encode the class and the location of the objects in a single textual description of the image, allowing us to search using a standard text search engine. Basically for each image I entry on the index, we have a space-separated concatenation of *ENC*s, one for all the possible cells (cod_{loc}) in the grid that contains the object (cod_{class}) where:

- *loc* is the juxtaposition of *row* and *col* on the grid
- *class* is the name of the object as classified by YOLO.

In practice, through the UI the users can draw the objects they are looking for by specifying the desired location for each of them (e.g., *tree* and *vehicle* in Fig. 1). Meanwhile, for each object, the UI encodes appropriately the request to interrogate the index, marking all the cells in the grid that contain the object. For example, for the query in Fig. 1, we will search for entries I on our index that contain the sequence $p_1 tree \, p_2 tree \, \dots \, p_6 tree$, where p_i is the code of the i-th cell (with $1 \leq i \leq 6$ since the tree icon covers six cells). Note that, a cell of a sketch can contain multiple objects. As showed in Fig. 2, for the image with id 2075, we extract both keywords (*beach, cloud, etcetera*), using the image annotation tool, and object location (*3dperson, etcetera*), exploiting the object detector, and later we index these two features in a single Lucene index.

Visual Similarity. Visione also supports visual content-based search functionalities, which allows users to retrieve scenes containing keyframes visually similar to a query image given by example. To start the search the user can select any keyframe of a video as query. In order to represent and compare the visual content of the images, we use the Regional Maximum Activations of Convolutions (R-MAC) [17]. This descriptor effectively aggregates several local convolutional features (extracted at multiple position and scales) into a dense and compact global image representation. We use the ResNet-101 trained model provided by [7] as feature extractor since it achieved the best performance on standard benchmarks. To efficiently index the R-MAC descriptor, we transform the deep features into a textual encoding suitable for being indexed by a standard full-text search engine, such as Lucene: we first use the Deep Permutation technique [3] to encode the deep features into a permutation vector, which is then transformed into a Surrogate Text Representation (STR) as described in [6]. The advantage of using a textual encoding is that we can efficiently exploit off-the-shelf text search engines for performing image searches on large scale.

4   Results

For the evaluation of our system, we took advantage of the participation to the VBS competition, which was a great opportunity to test the system with both expert and novice users.[4] For each task, a team receives a score based on response time and on the number of correct and incorrect submissions.

KIS Tasks. During the competition, the strategy used for solving both the KIS tasks was mainly based on the use of queries by object locations and keywords. Queries by color-sketch were used sparingly since they resulted to be less stable and sometimes degrades the quality of results obtained with the keywords/object search. As showed in Fig. 3, for our system the textual-KIS task was the hardest, accordingly to the observation done by the organizers of the competition in [12], where they note that textual-KIS task is much harder to solve than visual tasks.

AVS Tasks. In this tasks, keywords/object and the image similarity search functionalities were mainly used. In particular, the image similarity search resulted to be notably useful to retrieve keyframes of different videos with similar visual content.

We experienced how an image retrieval system could be useful for video search, for the (*Textual KIS*) the results were not particularly satisfying, but for the *AVS* task are very promising. A problem on the textual KIS is a too specific categorisation of the object which decreased the recall: sometimes users does not distinguish between car or trunk or vehicle and they may use one of them (as textual query) indistinctly. However, for the YOLO detector the difference is quite significant and this leads to low recall. Globally speaking, one of the main problem was due to a rather simple user interface. In fact, Visione was not supporting functionality like query history, multiple submissions at once, or any form of collaboration between the team members: this leads to redundant submissions and "slow" submission of multiple instances.

---

[4] http://www.videobrowsershowdown.org/example-browsers/infos-and-results-2019/.

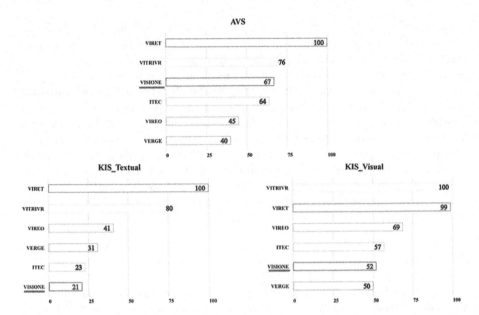

Fig. 3. The VBS2019 competition results for the three tasks AVS, KIS-textual and KIS-visual (score between 0 and 100). The bold line highlights the result of our system.

5   Conclusion

We described Visione, a system presented at the Video Browser Showdown 2019 challenge. The system supports three types of queries: query by keywords, query by object location, and query by visual similarity. Visione exploits state-of-the-art deep learning approaches and ad-hoc surrogate text encodings of the extracted features in order to use efficient technologies for text retrieval. From the experience at the competition, we ascertained a high efficiency regarding the indexing structure, made to support large scale multimedia access but a lack of effectiveness on keywords search. As a result of the system assessment made after the competition, we decide to invest a more effort on the keywords-based search, trying to ameliorate the image annotation part: we plan to integrate dataset of place, concept and categories [4,9,19], and automatic tools for scene understanding. Furthermore, we will improve the user interface to make it more usable and collaborative.

References

1. Amato, G., et al.: VISIONE at VBS2019. In: Kompatsiaris, I., Huet, B., Mezaris, V., Gurrin, C., Cheng, W.-H., Vrochidis, S. (eds.) MMM 2019, Part II. LNCS, vol. 11296, pp. 591–596. Springer, Cham (2019). https://doi.org/10.1007/978-3-030-05716-9_51

2. Amato, G., Falchi, F., Gennaro, C., Rabitti, F.: Searching and annotating 100M images with YFCC100M-HNfc6 and MI-File. In: Proceedings of the 15th International Workshop on Content-Based Multimedia Indexing, CBMI 2017, pp. 26:1–26:4. ACM (2017)
3. Amato, G., Falchi, F., Gennaro, C., Vadicamo, L.: Deep permutations: deep convolutional neural networks and permutation-based indexing. In: Amsaleg, L., Houle, M.E., Schubert, E. (eds.) SISAP 2016. LNCS, vol. 9939, pp. 93–106. Springer, Cham (2016). https://doi.org/10.1007/978-3-319-46759-7_7
4. Awad, G., Snoek, C.G.M., Smeaton, A.F., Quénot, G.: Trecvid semantic indexing of video: a 6-year retrospective. ITE Trans. Media Technol. Appl. 4(3), 187–208 (2016)
5. Fellbaum, C., Miller, G.: WordNet: An Electronic Lexical Database. Language, Speech, and Communication. MIT Press, Cambridge (1998)
6. Gennaro, C., Amato, G., Bolettieri, P., Savino, P.: An approach to content-based image retrieval based on the lucene search engine library. In: Lalmas, M., Jose, J., Rauber, A., Sebastiani, F., Frommholz, I. (eds.) ECDL 2010. LNCS, vol. 6273, pp. 55–66. Springer, Heidelberg (2010). https://doi.org/10.1007/978-3-642-15464-5_8
7. Gordo, A., Almazán, J., Revaud, J., Larlus, D.: End-to-end learning of deep visual representations for image retrieval. Int. J. Comput. Vis. 124(2), 237–254 (2017)
8. Jia, Y., et al.: Caffe: convolutional architecture for fast feature embedding. arXiv preprint arXiv:1408.5093 (2014)
9. Jiang, Y.G., Wu, Z., Wang, J., Xue, X., Chang, S.F.: Exploiting feature and class relationships in video categorization with regularized deep neural networks. IEEE Trans. Pattern Anal. Mach. Intell. 40(2), 352–364 (2018)
10. Lokoč, J., Kovalčík, G., Souček, T.: Revisiting SIRET video retrieval tool. In: Schoeffmann, K., et al. (eds.) MMM 2018. LNCS, vol. 10705, pp. 419–424. Springer, Cham (2018). https://doi.org/10.1007/978-3-319-73600-6_44
11. Lokoč, J., Bailer, W., Schöffmann, K., Münzer, B., Awad, G.: On influential trends in interactive video retrieval: Video browser showdown 2015–2017. IEEE Trans. Multimedia 20(12), 3361–3376 (2018)
12. Lokoč, J., et al.: Interactive search or sequential browsing? A detailed analysis of the video browser showdown 2018. ACM Trans. Multimedia Comput. Commun. Appl. 15(1), 29:1–29:18 (2019)
13. Niraimathi, D.S.: Color based image segmentation using classification of k-NN with contour analysis method. Int. Res. J. Eng. Technol. 3, 1169–1177 (2016)
14. Redmon, J., Farhadi, A.: Yolov3: an incremental improvement. arXiv preprint arXiv:1804.02767 (2018)
15. Rossetto, L., Schuldt, H., Awad, G., Butt, A.A.: V3C – a research video collection. In: Kompatsiaris, I., Huet, B., Mezaris, V., Gurrin, C., Cheng, W.-H., Vrochidis, S. (eds.) MMM 2019. LNCS, vol. 11295, pp. 349–360. Springer, Cham (2019). https://doi.org/10.1007/978-3-030-05710-7_29
16. Thomee, B., et al.: YFCC100M: the new data in multimedia research. Commun. ACM 59(2), 64–73 (2016)
17. Tolias, G., Sicre, R., Jégou, H.: Particular object retrieval with integral max-pooling of CNN activations. arXiv preprint arXiv:1511.05879 (2015)
18. Truong, T.-D., et al.: Video search based on semantic extraction and locally regional object proposal. In: Schoeffmann, K., et al. (eds.) MMM 2018. LNCS, vol. 10705, pp. 451–456. Springer, Cham (2018). https://doi.org/10.1007/978-3-319-73600-6_49
19. Zhou, B., Lapedriza, A., Khosla, A., Oliva, A., Torralba, A.: Places: A 10 million image database for scene recognition. IEEE Trans. Pattern Anal. Mach. Intell. 40(6), 1452–1464 (2017)

Towards Automatic Configuration of Interactive Known-Item Search Systems

Ladislav Peška$^{(\boxtimes)}$, Gregor Kovalčík, and Jakub Lokoč

SIRET Research Group, Department of Software Engineering,
Faculty of Mathematics and Physics, Charles University, Prague, Czech Republic
{peska,lokoc}@ksi.mff.cuni.cz, gregor.kovalcik@gmail.com

Abstract. Searching for one particular scene in a large annotation-free video archive becomes a common task in the multimedia age. Since the task is inherently difficult without knowledge of the scene location, multimedia management systems utilize various notions of similarity and provide both effective retrieval models and interactive interfaces. In this paper, we propose a vision of a simulation framework for automatic configuration of interactive known-item search video retrieval systems. We believe that such framework could help with early, resource-inexpensive evaluations and therefore automatic parameters tuning, detection of effective search strategies and effective configuration of client prototypes.

Keywords: Video retrieval · Known-item search · Simulated users

1 Introduction

With ubiquitous cameras and high-speed connection in personal digital devices, the so-called digital universe has seen rapid increase of multimedia content in recent years. For example, YouTube claims that there are on average 400 hours of video content uploaded to their web servers every minute and every day people generate billions of views.

The current mainstream in multimedia retrieval is highly influenced by significant achievements reached with deep learning. Some examples are automated concepts detection [7], localization of objects [13], semantic segmentation [15], or image/video captioning [17]. However, there are still scenarios, where automated approaches lack effectiveness and users in the loop can make a significant difference. Especially in cases, where an ideal query example is not available and users may provide just clues to find desired content [9]. One important representative of such tasks is known-item search (KIS), where users look for one particular video scene in a given dataset. The task can become difficult once the dataset is large, lacks human annotations and users do not have the information, where the searched scene is located within the dataset structure (e.g., timestamp or folder). With such limitations, users have to resort to various automatic annotation models, content-based query initialization methods and similarity-based retrieval [18] to solve the task.

© Springer Nature Switzerland AG 2019
G. Amato et al. (Eds.): SISAP 2019, LNCS 11807, pp. 340–347, 2019.
https://doi.org/10.1007/978-3-030-32047-8_30

The KIS scenario may appear in many important real-world situations. Fast inspection of videos from police car cameras or wearable LifeLog devices can help to search for memorized moments and speed up the reaction time. In some countries, surgeons are required by law to document their operations, however, the after-inspection of many hours of operations is too time consuming. Video/film productions create hundreds of hours of video, where searching desired scenes in the video editing process is one of the most time consuming tasks. Nonetheless, results of recent Video Browser Showdown (VBS) installments focusing on KIS tasks showed that state-of-the-art interactive video retrieval tools still face problems to solve all KIS tasks within a given time limit [10].

One of the limitations preventing successful KIS task's completion is inability of (especially novice) users to properly select the most suitable search strategies for a given task.[1] Another limitation is a proper parametrization of individual components of the tool, because user-based evaluations are very timely. To face both issues, some model of user behavior describing the query construction process and browsing interactions is desired. Given the pairs of targets and simulated interactions, system designers could obtain first estimates of the performance and promising configurations of the parameter space. Therefore, in order to solve both challenges, we envision a user simulation framework for interactive KIS retrieval tools, describe its key components and concepts, show its possible applications and also discuss open problems to solve.

As a starting point of the envisioned KIS simulations framework, we use a formal framework for the (reproducible) evaluation of interactive IR systems proposed in [19]. We adopt several concepts and methodology defined by the authors and aim on instantiating this generic framework for the KIS scenario.

For some components of interactive KIS retrieval systems such as relevance feedback, there are already devised models of user behavior (e.g., [8]), which can be utilized in the simulation framework. For some other components, e.g., various sketch-based query initialization models, the models of user behavior needs to be proposed. In Sect. 3, we discuss a set of generic user models and its instances for some common components of KIS tools.

2 Models for Interactive Known Item Search

In this section, we would like to describe some models commonly utilized in interactive KIS retrieval tools and their possible extensions. Since most searches start with a set of provided/available "proxy" queries, a pool of suitable initialization models is discussed in Sect. 2.1. Apart from options to sequentially browse results or reformulate the query, the relevance feedback models represents a popular method allowing users to express their feedback and flexibly adapt the underlying ranking models based on it. We review some possible approaches in Sect. 2.2.

---

[1] Note that the interactive systems usually comprise a wide set of available retrieval components and practically unbound list of possible interaction sequences with unknown rewards varying for individual tasks.

2.1 Models for KIS Initialization

In interactive KIS retrieval systems, four categories of query initialization are commonly used – text-based and sketch-based models, query by example image and filtering. Text-based models enable intuitive description of texts obtained from ASR/OCR models or semantic elements detected in searched frames/shots by concept/action detection models [7,21]. The query can be further positioned in models supporting region proposals [13].

For visual KIS tasks, several systems demonstrated that sketch-based models provide an opportunity to effectively express memorized low level features like colors, edges or motion (e.g., the vitrivr system [14]). With the time passing by, however, the users face problems with memorizing and reconstruction of searched scenes. Hence these models have to assume highly noisy inputs and also new types of features should be considered (e.g., from semantic segmentation [15]).

Providing an example image from an external service or the current result set is a popular alternative to the previous two categories once a candidate "proxy" example is available. KIS system may also provide, e.g., self-organized maps of pre-selected representative instances to help users to navigate through the dataset [2]. Usually, features from a deep learning architecture are considered for query by example similarity-based retrieval models [6].

Filtering provides an option to prune the searched collection based on unambiguously detected attributes (e.g., black-and-white or aspect ratio). It can be used also to cut-out a part of the similarity based ordering provided by the first three presented approaches.

All mentioned models as well as their fusion can be instantiated by multiple different retrieval methods, which can be further tuned by their respective hyper-parameters. For instance, a wide range of pre-trained classification networks can be utilized in automated annotations task for keyword-based models. Furthermore, retrieval methods may, e.g., consider implicit or explicit (user-defined) relevance of individual keywords, or assume different (fuzzy) logical operators between them.

2.2 Relevance Feedback Models

Once query initialization attempts fail to target the searched item(s), interactive and systematic browsing [2] represents a viable alternative to the tedious sequential inspection of results. In addition, the browsing can provide a sequence of actions, which could be utilized by some relevance feedback models, i.e., updating preference of items based on the positive and negative examples as expressed by the user. While aiming on the description of user's decision making process, KIS tools may process multiple types of user actions (or inactions [20]) with varying relevance or informativeness including, e.g., detailed observation of items, ignoring their presence or explicit selection as positive or negative candidates.

Relevance feedback models commonly work in two steps: first update relevance scores of individual items based on new feedback and some items' similarity metric and then display a subset of the items to the user. Bayesian relevance feedback models were successfully applied to the first task (e.g., [5]). For the results display, there is a key tradeoff between the focus on current best candidates and capability to cover all suitable candidates. Some inspiration can be found e.g. in the domain of recommender systems, specifically in models aiming on relevance vs. diversity tradeoff [12], exploration vs. exploitation tradeoff [1], or adapted query-based topic modeling approaches [3] combined with some results calibration [16] ensuring the proportional representation of query topics.

Nonetheless, due to the short expected life-span of user's tasks, relevance feedback models have to be highly efficient and capable to update their internal representations online upon the feedback is received.

3 Vision of Automated KIS Configuration

As described in previous sections, current interactive KIS retrieval tools are comprised from multiple components instantiating various search paradigms. For each of the search paradigm, multiple implementations are possible, which can be further tuned by various hyperparameter settings. This situation is challenging not only from the system composition perspective, but also from the user point of view. It is hard or even impossible for users (especially novices) to select, which search strategy leads to the best results. The envisioned simulation framework should contribute towards both of the challenges. With the ability to approximately model user's behavior, some first estimates of performance of individual configurations can be driven. By evaluating performance of different search strategies (e.g., detailed vs. simple query initialization, sequential vs. interactive browsing etc.), we may provide a set of generic suggestions to users, or even design a wizard-style tool.

The core component of the envisioned framework are models of user behavior.[2] We define the user's task as follows: for a given target item t, construct an interaction sequence I, such that the target item is found. At the same time, the cost of interaction sequence $C(I)$ should be as low as possible.[3] As a starting point, we assume that finding the target item could be modelled as a random variable with a probability distribution based on the properties of the current interface card (e.g., volume and size of displayed objects) and the position of the target item among the result set.

We consider the model of user behavior to work in three steps. First, an aggregated user model selects, which component should be interacted with. Then, the behavior model of the respective component is employed and decide, which action(s) are taken. Finally, the aggregated user model process the response of the system, either stating that the target item was found, or continues to the next lap.

---

[2] Note that we adopted several concepts, e.g., user, system, action, interface card, interaction sequence and interaction cost/reward, defined in [19].

[3] $C(I)$ could be defined as a simple sum of all interactions costs, but could also perform some non-linear transformation of them, e.g., impose a threshold on the maximal allowed costs. Nonetheless, $C(I)$ should maintain the non-increasing property for the prefixes of I.

3.1   User Behavior Models for Individual Components

In order to construct a behavior model for individual component c, we need to first define the set of plausible user actions A_c and their costs C_c. The set of actions is component-dependent, but rather straightforward in most cases. For example, in keyword-based query initialization, the set of actions would correspond to insertion/removal of a keyword to the query, or, optionally, modify their weights or logic operators connecting them. In a color-based sketch model (e.g., [11]), the set of actions would correspond to insertion/removal of a colored point to a certain position in the sketch. In relevance feedback models, actions would correspond to marking one of the displayed items as a positive or negative example. Costs C_c should be set according to the considered real-world task, usually based on expected temporal complexity of the action.

Then, the task of the user model is, based on the given target t, assign occurrence probability p_a to each action $a \in A_c$. The key assumption of proposed user models is that, up to some extent, user's perception of data is consistent with component-level data understanding. Although some transformations of system-to-user data understanding are possible, we assume that the lack of data and causality issues would prevent us from learning such models. We consider three types of user models: *posteriori optimal users*, *greedy optimal users* and *noisy optimal users*. Posteriori optimal user model always selects the action that maximizes the probability of finding the target item[4] and therefore provides the theoretical boundaries of the component efficiency.[5] Greedy optimal user model selects the best action according to the data representation featured by the component, e.g., the keyword assigned with the highest probability by an automated annotations in the case of keyword-based query initialization. Finally, noisy optimal user models are based on greedy optimal models, but additionally introduce random noise to the selection process. Note that the introduced noise may be multidimensional, e.g., in color-based sketch, both the color channels and position coordinates may be randomly shifted.

3.2   Aggregated User Behavior Models

While selecting, which component to interact with, multiple strategies can be devised. First, strategies only utilizing a single component (or some fixed components ordering) provide important baseline comparisons and may provide clues to justify utilization of each individual component. Second, similarly as in component-level models, posteriori optimal strategies (or their randomized versions) can be used to provide theoretical boundaries of the tool. Finally, components can be sampled based on their past performance. For instance, multi-armed bandit model with Thompson Sampling, delayed reward and gradually decreasing reward attribution [4] seems suitable for the task. Variations in train set sizes for this model may be used to mimic the differences between novice and

---

[4] Optionally, it could also consider the cost of created interaction sequence $C(I)$.

[5] Note that due to the size of action space in some components, posteriori optimal model cannot be effectively evaluated and some approximations may be necessary.

experienced users and therefore can be utilized in devising individual interfaces for each user group. Apart from long-term component performance, a short-term performance evaluation may be applied as well, preventing a repeated usage of the component, which works badly for the current task.

3.3 Generation of Client Prototypes

Aside from evaluation of the KIS tools and its individual components, an important open challenge is to automatically generate interfaces for (an effective subset of) available components. Preferably, interfaces generation should be personalized w.r.t. target device, user's experience or search style and current status of the task. For example, wizards could be generated to guide the user based on a recommended search strategy (obtained by simulations), presenting a sequence of screens with predefined actions in a given retrieval model. Another option is to tune interface cards according to the simulation results of individual components, or disclose this information to the user. It is also possible to recommend/highlight, which component(s) would be most suitable for the next step, or discourage users from continued usage of ineffective ones. Main challenges of this task are to set a proper level of simplicity vs. effectiveness of the graphical interface (GUI) for individual users and to introduce suitable models transforming simulation results into generated interfaces.

Meta-analysis of the simulations (i.e., which paths commonly lead to the good results) should be a good starting point for wizard-style solutions and component recommendations. While constructing the interface cards, both effectiveness and uniqueness of components (i.e., up to what extent, the query can be mimicked by other components) should be considered. Components' hyperparameter settings should be displayed according to their impact on the results and comprehensibility to the users. In either case, it is desired to propose some intermediate domain specific description language for model-driven development that will allow to encode the GUI composition w.r.t. a standard set of GUI elements provided for the retrieval models. Eventually, sequences of GUI compositions will also be encoded for description of the search strategies.

3.4 Framework Verification and Tuning from User Feedback

Some user behavior models introduce additional hyperparameters (e.g., the magnitudes of noise for noisy optimal models), which may affect simulations' results. Although it is possible to run simulations w.r.t. multiple hyperparameter values and aim to find some globally optimal methods, such results may often be inconclusive and furthermore, it would be extremely challenging to compare results across different components. Nonetheless, the hyperparameter space of behavior models is much less complex than the one of the original components, so only a limited amount of feedback may be sufficient to tune behavior models to mimic the real users well. Alternatively, optimal client prototypes w.r.t. various behavior model settings could be evaluated in standard A/B testing or user studies, which could refer back to the ideal hyperparameter settings.

An open challenge to solve is, up to what extent users behave consistently across different datasets, i.e., whether some transfer learning for behavior models is applicable. Another important challenge is to explore up to what extent, it is necessary to mimic the behavior of real users, i.e., are hereby described models sufficient to provide results consistent with the preference of real users, or more complex models are necessary?

4 Conclusions

In this vision paper, we discussed the importance and challenges of KIS task in large, unannotated video collections. One of the main challenges is the lack of validation options (and their costly acquisition), which leads to insufficiently supported design choices and suboptimal performance of KIS tools. In the envisioned framework, we focus on the possibility to conceptually describe the search process and to propose simulation models suitable to support a range of design choices as well as guide the user through the KIS process. While materializing our vision, we plan to employ a bottom-up process starting at defining and evaluating single-component user models, then advance to aggregated user models and ultimately aim on the automated generation of client prototypes. Finally, a confrontation of the devised approaches in existing interactive known-item search evaluation campaigns (e.g., VBS) represents our ultimate challenge.

Acknowledgements. This paper has been supported by Czech Science Foundation (GAČR) projects Nr. 19-22071Y and 17-22224S.

References

1. Barraza-Urbina, A.: The exploration-exploitation trade-off in interactive recommender systems. In: Eleventh ACM Conference on Recommender Systems, RecSys 2017 (2017)
2. Barthel, K.U., Hezel, N., Jung, K.: Visually browsing millions of images using image graphs. In: 2017 ACM on International Conference on Multimedia Retrieval, ICMR 2017, pp. 475–479 (2017)
3. Blei, D.M., Ng, A.Y., Jordan, M.I.: Latent dirichlet allocation. J. Mach. Learn. Res. **3**, 993–1022 (2003)
4. Brodén, B., Hammar, M., Nilsson, B.J., Paraschakis, D.: Ensemble recommendations via thompson sampling: an experimental study within e-commerce. In: 23rd International Conference on Intelligent User Interfaces, IUI 2018, pp. 19–29. ACM (2018)
5. Ferecatu, M., Geman, D.: A statistical framework for image category search from a mental picture. IEEE Trans. Pattern Anal. Mach. Intell. **31**(6), 1087–1101 (2009)
6. Gordo, A., Almazán, J., Revaud, J., Larlus, D.: Deep image retrieval: learning global representations for image search. In: Leibe, B., Matas, J., Sebe, N., Welling, M. (eds.) ECCV 2016. LNCS, vol. 9910, pp. 241–257. Springer, Cham (2016). https://doi.org/10.1007/978-3-319-46466-4_15

7. He, K., Zhang, X., Ren, S., Sun, J.: Deep residual learning for image recognition. In: 2016 IEEE Conference on Computer Vision and Pattern Recognition (CVPR), pp. 770–778 (2016)
8. Keskustalo, H., Järvelin, K., Pirkola, A.: Evaluating the effectiveness of relevance feedback based on a user simulation model: effects of a user scenario on cumulated gain value. Inf. Retrieval **11**(3), 209–228 (2008)
9. Lokoč, J., Bailer, W., Schoeffmann, K., Muenzer, B., Awad, G.: On influential trends in interactive video retrieval: video browser showdown 2015–2017. IEEE Trans. Multimedia **20**(12), 3361–3376 (2018)
10. Lokoč, J., et al.: Interactive search or sequential browsing? A detailed analysis of the video browser showdown 2018. ACM Trans. Multimedia Comput. Commun. Appl. (TOMM) **15**(1), 29 (2018)
11. Lokoč, J., Phuong, A.N., Vomlelová, M., Ngo, C.-W.: Color-sketch simulator: a guide for color-based visual known-item search. In: Cong, G., Peng, W.-C., Zhang, W.E., Li, C., Sun, A. (eds.) ADMA 2017. LNCS (LNAI), vol. 10604, pp. 754–763. Springer, Cham (2017). https://doi.org/10.1007/978-3-319-69179-4_53
12. Noia, T.D., Rosati, J., Tomeo, P., Sciascio, E.D.: Adaptive multi-attribute diversity for recommender systems. Inf. Sci. **382**, 234–253 (2017)
13. Ren, S., He, K., Girshick, R., Sun, J.: Faster R-CNN: towards real-time object detection with region proposal networks. In: Neural Information Processing Systems (NIPS) (2015)
14. Rossetto, L., Giangreco, I., Gasser, R., Schuldt, H.: Competitive video retrieval with vitrivr. In: Schoeffmann, K., et al. (eds.) MMM 2018. LNCS, vol. 10705, pp. 403–406. Springer, Cham (2018). https://doi.org/10.1007/978-3-319-73600-6_41
15. Shelhamer, E., Long, J., Darrell, T.: Fully convolutional networks for semantic segmentation. IEEE Trans. Pattern Anal. Mach. Intell. **39**(4), 640–651 (2017)
16. Steck, H.: Calibrated recommendations. In: 12th ACM Conference on Recommender Systems, RecSys 2018, pp. 154–162 (2018)
17. Vinyals, O., Toshev, A., Bengio, S., Erhan, D.: Show and tell: lessons learned from the 2015 MSCOCO image captioning challenge. IEEE Trans. Pattern Anal. Mach. Intell. **39**(4), 652–663 (2017)
18. Zezula, P., Amato, G., Dohnal, V., Batko, M.: Similarity Search - The Metric Space Approach. Advances in Database Systems, vol. 32. Kluwer (2006)
19. Zhang, Y., Liu, X., Zhai, C.: Information retrieval evaluation as search simulation: a general formal framework for IR evaluation. In: ACM SIGIR International Conference on Theory of Information Retrieval, ICTIR 2017, pp. 193–200 (2017)
20. Zhao, Q., Willemsen, M.Ç., Adomavicius, G., Harper, F.M., Konstan, J.A.: Interpreting user inaction in recommender systems. In: 12th ACM Conference on Recommender Systems, RecSys 2018, pp. 40–48 (2018)
21. Zoph, B., Vasudevan, V., Shlens, J., Le, Q.V.: Learning transferable architectures for scalable image recognition. CoRR abs/1707.07012 (2017)

Doctoral Symposium Papers

ADAMiSS: Advanced Data Analysis, Mining and Search, System

Jakub Peschel$^{(\boxtimes)}$ and Pavel Zezula

Masaryk University, Brno, Czech Republic
{jpeschel,pzezula}@mail.muni.cz

Abstract. The complexity of contemporary data warrants a need for better analysing tools in investigative areas. Human processing of data is no longer a viable option. We present an architecture of a novel universal system for analysis of graph-structured data, where data-mining and similarity-search operators can be used to discover or search for unknown information. We also present results that were obtained by our prototype implementation on two real-world data collections: the Twitter Higg's boson dataset and the Kosarak dataset.

Keywords: Data analysis · Pattern mining · Similarity search · Advanced data types

1 Introduction

In the investigative fields, such as forensic analysis, there is often a need for analysis of contemporary data gathered from different sources. The main problem that arises is that experts do not know what they are looking for until it is shown to them and then, based on the observation of newly discovered data, there is a need for searching techniques. There exist many techniques for these sub-tasks, such as SNAP [8], SPMF [6], MESSIF [3]. While there are various more specialised tools and approaches that can give even more detailed answer, the forensic analysts rather use a single generic system than multiple ones.

The critical aspect of contemporary data is that it has connections to others. Whether it is a response in the form of a tweet, a picture showing some event or record of purchase, a social network can be created from such information.

Search and mining are two main areas that we are interested in. The main focus on these areas is further narrowed into query by example search in case of searching and into pattern mining for the mining task.

Supervised by P. Zezula.

This work has been supported by the Ministry of the Interior of the Czech Republic under the "Security Research for the Needs of the State Program 2015–2020," through the Project No. VI20172020096, "Complex Analysis and Visualization of Large-scale Heterogeneous Data."

G. Amato et al. (Eds.): SISAP 2019, LNCS 11807, pp. 351–355, 2019.
https://doi.org/10.1007/978-3-030-32047-8_31

In query by example search, one of the main aspects used for searching is similarity between objects [10]. For this task, there exist various types of queries, but we focus on the two widely used ones: k-nearest neighbours query and range query. These two approaches allow for a high range of use-cases.

Pattern mining is a technique of discovery of frequent behaviour in the data. The main emphasis is put on periodicity or frequency. In connection with social networks, the main goals are mining communities and frequent sub-paths. Main techniques are older Apriori algorithm developed by Agrawal *et al.* [1], FP-growth developed by Han *et al.* [7] and GSP developed by Srikant *et al.* [9].

Goal: To address the mentioned problems, we want to propose a system that satisfies these requirements:
- general multipurpose system
- unified data structure
- modularity of system
- efficient processing
- large scalability

We call this system Advanced Data Analysis by Mining and Searching System, shortened as ADAMiSS.

The system for advanced analysis that is presented allows opening research space for the development of new techniques from combined ones.

2   Architecture

In forensic analysis, an expert looks at a high volume of data and tries to uncover new evidence. There is a need to obtain new insight into various data to gain hidden information.

To tackle the problem, we propose an architecture consisting of four components containing data and transformations between them visualised in Fig. 1. Transformations are a series of simple operations that reveal hidden information when combined in the right way. The ordering of operations is dependent on user requirement.

We selected multigraph as the appropriate data structure since a lot of real-world data can be viewed as entities introduced by links. Every element of multigraph is allowed to have attributes, to capture all the information needed for analysis.

ADAMiSS allows working with a broad range of data inputs. Examples of such inputs are interaction networks from Facebook or Twitter, e-mail message network, phone call network, bank transaction network, computer network and many others. The only prerequisite is that transformation into graph must exist.

Because every method needs a specific approach or view on data, ADAMiSS introduces the concept of transaction database. The transaction database is a

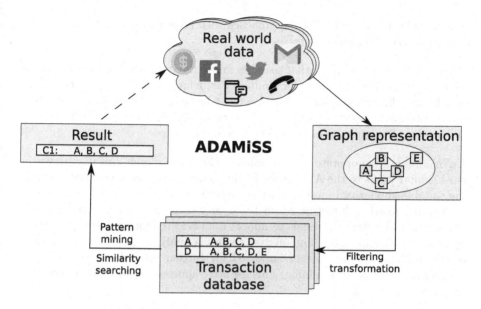

Fig. 1. Overview of an architecture of ADAMiSS

structure used by all the analytic operators. It consists of transactions and is used as a preprocessed data source for analytic operators. By the transformation into this flat structure, lots of the information from graph representation is omitted. Thus, it is necessary to select the right transformation for the desired operation.

Filtering transformation serves as a preprocess of data for analytic operators. During filtering, noisy data is removed, and only important structures are presented. An example of such preprocess is the removal of nodes without connection or taking into account only edges that appeared at least several times. By filtering, transformation can occur and a group node can be introduced by merging several nodes into one. A result of this transformation is a prepared transaction database.

To analyse the data, ADAMiSS contains analytic operators. These operators are basic functions that can be applied to the transaction database to obtain new knowledge. Due to the well-defined structure of transaction database, there is a possibility of implementing a large number of methods. Example operations that are available in the system:

Frequent item-set mining is a method that discovers information by providing frequent patterns of co-occurring items. The frequency of a pattern is determined by a user-defined threshold. In graph analysis, it can be used to find frequent neighbours. With a slight modification, it can be used to mine communities.

Sequence mining serves for searching for frequent paths in a graph. Data in a transaction of transaction database are then taken as a sequence, and ADAMiSS then searches for continuous subsequences. In sequence mining, it is often allowed to have holes in patterns. Both approaches with holes and without holes can be applied to the transaction database.

Range search is an operator that searches for all the items that are similar to the selected query. The threshold for similarity is defined by the user. In the range search, an arbitrary similarity function must be selected. Such a function must satisfy the metric space properties. As an example, similarity can be measured on the sets of neighbours or at selected attribute values.

K-NN search is an operator very similar to the range search, but it retrieves k-most similar items to the ADAMiSS. In the same manner as in the range search, the suitable similarity function must be selected.

The proposed architecture seems to be very useful for an iterative approach to analysis. The main process is to import data selected for analysis, filter them by applying suitable filters and analyse them via analytic operators. The results of the analysis are then again imported for further analysis by searching or another round of pattern mining. This iterative approach is a process of creating an advanced analysis.

3   Prototype Results

Two datasets were used, Twitter Higg's Boson dataset [5] and Kosarak dataset [4], to showcase prototype application.

We applied a frequent item-set mining operator to dataset Twitter Higg's boson. This dataset contains 304691 interactions on Twitter with a topic connected to the discovery of Higg's boson at CERN. A threshold of size 11 was selected, because social networks are quite shallow graphs. TA total of 7 communities of size 12 and 94 communities of size 11 were discovered. Average overlap of all these communities is 80.7%.

For path mining, sequence mining was applied to dataset Kosarak, containing 990000 click-stream data from the logs of an online news portal. The threshold was set to 1024, to obtain quite short frequent paths. There was found 322 paths and only 5 paths containing more than 4 nodes. The longest path contained 16 nodes.

One of the communities of size 12 was picked, and the Jaccard coefficient was computed on its nodes. For efficient processing, ADAMiSS uses MESSIF library developed by Batko et al. [3].

For range queries, the threshold was selected to 0.2. Four nodes (60686, 137247, 137321, 137246) had their most similar items inside the community.

For k-nn queries, 10 was selected as k, and an average amount of most similar nodes from the community for each node was 8.33 node.

4   Conclusion

In this paper we introduced a new system ADAMiSS. This system is used for obtaining new information from large volumes of data that can be transferred into a suitable graph representation. The system is based on unifying graph representation and flat transaction database used for two analytic approaches, pattern mining and similarity searching.

The system is highly modular; new methods can be developed and added. In the near future, we plan to add inference of association rules and to improve pattern mining methods by implementing new method NegFIN [2]. We would also like to improve our definition of community and to develop a suitable method for the discovery of such a structure. We also see a lot of space for research of interconnecting similarity search and pattern mining.

A prototype of the system was used for the analysis of two datasets, Twitter Higg's Boson dataset and Kosarak dataset. The system processed data and provided information about communities, frequent paths through the network and similarities between nodes.

References

1. Agrawal, R., Srikant, R., et al.: Fast algorithms for mining association rules. In: Proceedings of the 20th International Conference on Very Large Data Bases, VLDB, vol. 1215, pp. 487–499 (1994)
2. Aryabarzan, N., Minaei-Bidgoli, B., Teshnehlab, M.: negFIN: an efficient algorithm for fast mining frequent itemsets. Expert Syst. Appl. **105**, 129–143 (2018)
3. Batko, M., Novak, D., Zezula, P.: MESSIF: metric similarity search implementation framework. In: Thanos, C., Borri, F., Candela, L. (eds.) DELOS 2007. LNCS, vol. 4877, pp. 1–10. Springer, Heidelberg (2007). https://doi.org/10.1007/978-3-540-77088-6_1
4. Bodon, F.: A fast apriori implementation. In: FIMI, vol. 3, p. 63 (2003)
5. De Domenico, M., Lima, A., Mougel, P., Musolesi, M.: The anatomy of a scientific rumor. Sci. Rep. **3**, 2980 (2013)
6. Fournier-Viger, P., et al.: The SPMF open-source data mining library version 2. In: Berendt, B., et al. (eds.) ECML PKDD 2016. LNCS (LNAI), vol. 9853, pp. 36–40. Springer, Cham (2016). https://doi.org/10.1007/978-3-319-46131-1_8
7. Han, J., Pei, J., Yin, Y.: Mining frequent patterns without candidate generation. In: ACM Sigmod Record, vol. 29, pp. 1–12. ACM (2000)
8. Leskovec, J., et al.: Stanford network analysis project (2010). http://snap.stanford.edu
9. Srikant, R., Agrawal, R.: Mining sequential patterns: Generalizations and performance improvements. In: Apers, P., Bouzeghoub, M., Gardarin, G. (eds.) EDBT 1996. LNCS, vol. 1057, pp. 1–17. Springer, Heidelberg (1996). https://doi.org/10.1007/BFb0014140
10. Zezula, P., Amato, G., Dohnal, V., Batko, M.: Similarity Search: The Metric Space Approach, vol. 32. Springer, Boston (2006). https://doi.org/10.1007/0-387-29151-2

Feature Similarity: A Method to Detect Unwanted Feature Interactions Earlier in Software Product Lines

Seyedehzahra Khoshmanesh[(✉)] and Robyn R. Lutz

Iowa State University, Ames, IA 50011, USA
{zkh,rlutz}@iastate.edu

Abstract. Software product lines enable the reuse of shared software across a family of products. As new products are built in the product line, new features are added. A feature is a unit of functionality. Unwanted feature interactions, wherein one feature hinders another feature's operation, are a significant problem, especially as large software product lines evolve. Detecting feature interactions is a time-consuming and difficult task for developers. Moreover, feature interactions are often only discovered during testing, at which point costly re-work is needed. The work described here investigates how to discover feature interactions much earlier in the development process. Toward this goal, we propose a similarity-based approach that mines prior feature interactions stored in the software product line's artifacts to predict unwanted interactions between a new feature and existing features. Initial results show that the planned methodology performs well in terms of accuracy and coverage both in experiments on three small software product lines in the literature and in experiments on one large, real-world software product line.

Keywords: Feature interaction · Similarity measures · Software product lines

1 Statement of the Problem

A Software Product Line (SPL) is a family of products that share certain core features (called commonalities) as well as different optional or alternative features (called variabilities). A *feature* is a unit of functionality, such as a service visible to the product's users. SPLs are widely used in industry to reduce the cost and time-to-market of products. It is worthwhile to study all features in a SPL as a set and to reuse the product line's repository of software artifacts to build new products [3,5,9].

Software product lines evolve over time as new features are added [4]. Ensuring that SPL features continue to work as expected in each new product is both

Supervised by: Dr. R. R. Lutz.

essential and difficult [9]. *Feature interaction* in the SPL literature refers to the unwanted situation in which individual features behave as expected in isolation, but when combined in a new product, one or more no longer acts as intended and may even be dangerous. A classic example of unwanted feature interaction is described by Batory et al. Imagine we have a building with two optional features, fire-control, and flood-control. A building with a fire-control feature has sensors that, when they detect fire, activate water sprinklers. A building with a flood-control feature has water sensors that, when they detect standing water, turn off the water main. Either feature operates correctly alone; however, if a new product is built with both, the features interfere with each other and create a hazardous state [2].

Existing approaches to the feature-interaction problem are mostly based on testing. However, this approach can only detect unwanted feature interactions late in the development process after the product has already implemented. Earlier approaches using model checking have been proposed; however, are difficult to apply in practice and are not scalable to actual SPLs [1].

The remainder of the paper is structured as follows. Section 2 describes our similarity-based approach to detecting unwanted feature interactions in a new SPL product. Section 3 describes our contributions to date. Section 4 summarizes the results and discusses our future work.

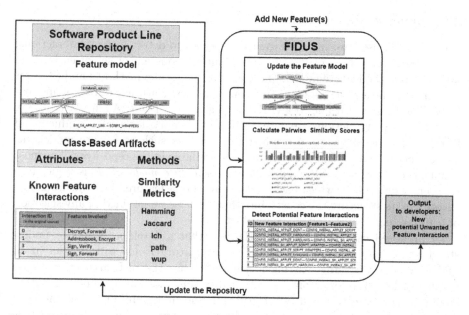

Fig. 1. FIDUS: feature interaction detection using similarity in software ProductLines

2   Outline of the Planned Methodology

Our work aims to discover feature interactions in a new SPL product at an earlier stage of the new product's development. To this end, we have proposed a method that leverages knowledge of prior feature interactions in the SPL together with measures of similarity between existing and new features. This enables design-time discovery of feature interactions in a new product.

Our approach uses the fact that product line repositories typically include documentation of known feature interactions, derived from previous experience and bug reports. We use similarity measures to calculate the similarity between those features known to interact in the software product line and the new features. We then employ this information to build a model that detects similar feature interactions in the new product much earlier in the development process. In a preliminary study [6], we applied our similarity method on a small software product line and obtained results that encourage us to pursue this work. As detecting feature interaction at the code level is partial, costly, and occurs late in development, we concentrate on the early stage artifacts of an SPL consisting of the feature models and the class elements related to each product line feature. Hence, we introduce an efficient framework using similarity measures to detect new feature interactions in evolving software product lines.

2.1   Similarity Framework

Our approach uses two main information sources to detect feature interactions: (1) the SPL repository of known feature interactions and a set of early-stage artifacts, i.e., the feature model, and class attributes and methods; and (2) a set of similarity measures to understand how close new features are to existing features. The proposed framework (FIDUS) is shown in Fig. 1. As displayed there, known feature interactions are derived from bug reports for earlier products and from early-stage elements of product line features such as feature models, as well as class variables and class methods, if available in the repository. The most common artifact in an SPL repository is its *feature model*. This is typically a tree representation in which each node represents a feature. The feature model encodes in its structure the constraints that must be maintained in any selection of features (e.g., XOR) for a new product.

A proposed usage scenario for FIDUS is that a developer wants to understand whether a new product having a new feature *Fnew* will interact in an undesired way with any existing features. We use the SPL artifacts with its capture of known interactions to help answer this question by applying feature similarity measures to *Fnew* and features in known and unwanted interactions. The approach reports to the developer the extent to which the new feature *Fnew* may potentially participate in any known feature interaction. This technique has the potential to reduce feature-interactions inadvertently introduced in a new product, thereby reducing risk as well as saving time and effort in debugging.

2.2 Similarity Measures

The similarity is the pivotal piece of our framework, FIDUS. We use similarity as a heuristic tool to compare two features in the SPL. Table 1 shows the similarity measures applied in our study for computing the similarity score of two features in a small SPL. We selected two set-based similarity measures, Jaccard and Hamming [6], to obtain a similarity score between two features when we have access to the SPL class diagram including class variables and methods. We represent a feature as a set of class methods and class variables used in the design of a feature. Therefore, the feature representation is a class-based artifact since only methods and variables of a coded feature are considered to capture the properties of a feature in an SPL.

Table 1. Class-based similarity metrics used for small SPLs

Name of metric	Category	Case studies
Jaccard	Class	Email, Elevator, MinePump
Hamming	Class	Email, Elevator, MinePump

FIDUS uses different similarity metrics to calculate the similarity between two features depending on the artifact types that are available in an SPL. For very large SPLs, it uses the feature model as the main artifact to locate features and capture similarity between them. We use three path-based metrics on the feature model to obtain similarity measures between features in a very large SPL: path, lch, and wup [8] (Table 2).

Table 2. Path-based similarity metrics used for large SPLs

Name of metric	Formula	Ref.
Path	$1/(lengthshortestpath)$	[8]
lch	$-log(length/(2*D))$	[8]
wup	$2*depth(LCS)/(depth(F1)+depth(F2))$	[8]

3   Experimental Results

In this section, we first introduce the SPL case studies investigated in our work. We then present the results obtained in terms of accuracy and coverage. We applied the similarity-based method proposed in this work on three small to medium size software product lines: Email, Elevator, and Mine Pump, and to different versions of a very large highly configurable software, BusyBox. These SPL are considered as benchmarks in software product line literature [1,7] (Table 3).

Table 3. SPLs used for FIDUS evaluation

SPL name	n(Features)	n(Interactions)	#LOC
Elevator, MinePump, Email [1]	6–9	4–11	580–1233
BusyBox v1.7,8,17 (Configuration, Shells)	40–72	16–28	14.8–19K

Tables 4 and 5 show the results, in terms of accuracy and coverage, of applying the proposed feature-interaction detection framework: (1) on three small SPLs using class-based similarity metrics, and (2) on a very large SPL using path-based similarity metrics. As shown in Table 4, with the threshold set to 0.5, the wup metric gave the highest coverage of 100% among the three metrics, with the other two yielding 60% coverage. Hamming had the highest coverage compared to Jaccard in class-based similarity measures for SPL features.

Table 4. Accuracy,coverage results for detecting feature interactions in three small SPLs

SPL name	Jaccard similarity		Hamming similarity	
	Accuracy	Coverage	Accuracy	Coverage
Email	70%	82%	70%	64%
Elevator	100%	83.3%	100%	83.3%
Mine pump	50%	50 %	37.5 %	100%
Average	73.3%	71.77%	69.17%	82.4%

Table 5. Accuracy and coverage results for detecting feature interactions in a large SPL

SPL name	Similarity measure	Threshold	Accuracy	Coverage
BusyBox v1.17	wup	0.5	83	100
	lch		100	60
	path		100	60

4   Future Work

Planned future work will investigate whether the use of additional product line artifacts such as requirements and design specifications (architectural, state and activity diagrams) can, by identifying similar behaviors among features, improve the performance of our method. Planned future work also will investigate how information in the feature model can be mined to automatically advise minimal repairs to resolve feature interactions flagged by this method.

Acknowledgments. The work in this paper was partially funded by National Science Foundation Grant CCF 1513717.

References

1. Apel, S., Rhein, A.v., Wendler, P., Größlinger, A., Beyer, D.: Strategies for product-line verification: case studies and experiments. In: Proceedings of the 2013 International Conference on Software Engineering, pp. 482–491. IEEE Press (2013)
2. Batory, D., Höfner, P., Kim, J.: Feature interactions, products, and composition. In: ACM SIGPLAN Notices, vol. 47, pp. 13–22. ACM (2011)
3. Berger, T., et al.: What is a feature?: a qualitative study of features in industrial software product lines. In: SPLC, pp. 16–25. ACM (2015)
4. Bosch, J.: Design and Use of Software Architectures: Adopting and Evolving a Product-Line Approach. Pearson Education, London (2000)
5. Kang, K.C., Cohen, S.G., Hess, J.A., Novak, W.E., Peterson, A.S.: Feature-oriented domain analysis (FODA) feasibility study. Technical report, CMU SEI (1990)
6. Khoshmanesh, S., Lutz, R.R.: The role of similarity in detecting feature interaction in software product lines. In: ISSREW, pp. 286–292. IEEE (2018)
7. Nadi, S., Berger, T., Kästner, C., Czarnecki, K.: Where do configuration constraints stem from? An extraction approach and an empirical study. IEEE Trans. Softw. Eng. **41**(8), 820–841 (2015)
8. Pedersen, T., Patwardhan, S., Michelizzi, J.: WordNet: similarity: measuring the relatedness of concepts. In: HLT-NAACL, pp. 38–41 (2004)
9. Soares, L.R., Schobbens, P.Y., do Carmo Machado, I., de Almeida, E.S.: Feature interaction in software product line engineering: a systematic mapping study. Inf. Softw. Technol. **98**, 44–58 (2018)

Author Index

Printed in the United States
By Bookmasters